Microfabrication and Precision Engineering

Related titles

Mechatronics and Manufacturing Engineering
(ISBN 978-0-85709-150-5)

Materials and Surface Engineering
(ISBN 978-0-85709-151-2)

Modern Machining Technology
(ISBN 978-0-85709-099-7)

Materials Forming and Machining
(ISBN 978-0-85709-483-4)

Machining and Machine-Tools
(ISBN 978-0-85709-154-3)

Woodhead Publishing
in Mechanical Engineering

Microfabrication and Precision Engineering

Research and Development

Edited by

J. Paulo Davim

ELSEVIER

AMSTERDAM • BOSTON • HEIDELBERG • LONDON
NEW YORK • OXFORD • PARIS • SAN DIEGO
SAN FRANCISCO • SINGAPORE • SYDNEY • TOKYO
Woodhead Publishing is an imprint of Elsevier

WP
WOODHEAD PUBLISHING

Woodhead Publishing is an imprint of Elsevier
The Officers' Mess Business Centre, Royston Road, Duxford, CB22 4QH, United Kingdom
50 Hampshire Street, 5th Floor, Cambridge, MA 02139, United States
The Boulevard, Langford Lane, Kidlington, OX5 1GB, United Kingdom

Notice
Knowledge and best practice in this field are constantly changing. As new research and experience broaden our understanding, changes in research methods, professional practices, or medical treatment may become necessary.

Practitioners and researchers must always rely on their own experience and knowledge in evaluating and using any information, methods, compounds, or experiments described herein. In using such information or methods they should be mindful of their own safety and the safety of others, including parties for whom they have a professional responsibility.

To the fullest extent of the law, neither the Publisher nor the authors, contributors, or editors, assume any liability for any injury and/or damage to persons or property as a matter of products liability, negligence or otherwise, or from any use or operation of any methods, products, instructions, or ideas contained in the material herein.

British Library Cataloguing-in-Publication Data
A catalogue record for this book is available from the British Library

Library of Congress Cataloging-in-Publication Data
A catalog record for this book is available from the Library of Congress

ISBN: 978-0-85709-485-8 (print)
ISBN: 978-0-85709-486-5 (online)

For information on all Woodhead Publishing visit our website at https://www.elsevier.com

Publisher: Matthew Deans
Acquisition Editor: Glyn Jones
Editorial Project Manager: Harriet Clayton
Production Project Manager: Omer Mukthar
Cover Designer: Greg Harris

Typeset by MPS Limited, Chennai, India

Contents

List of contributors

B. Bhattacharyya Jadavpur University, Kolkata, West Bengal, India

C. Gao Nanjing University of Aeronautics and Astronautics, Nanjing, Jiangsu, China

R.G. Jasinevicius University of São Paulo, São Carlos, Brazil

N.E. Karkalos National Technical University of Athens, Athens, Greece

G. Kibria Aliah University, Kolkata, West Bengal, India

A.S. Kuar Jadavpur University, Kolkata, West Bengal, India

S. Kumar Indian Institute of Technology Patna, Bihar, India

Z. Liu Nanjing University of Aeronautics and Astronautics, Nanjing, Jiangsu, China

A.P. Markopoulos National Technical University of Athens, Athens, Greece

S. Mitra Jadavpur University, Kolkata, West Bengal, India

S. Prakash Indian Institute of Technology Patna, Bihar, India

R. Rentsch Bremen University, Bremen, Germany

A.R. Rodrigues University of São Paulo, São Carlos, Brazil

N. Roy Jadavpur University, Kolkata, West Bengal, India

B. Xue Harbin Institute of Technology, Harbin, Heilongjiang, China

J. Yan Keio University, Yokohama, Japan

Y.D. Yan Harbin Institute of Technology, Harbin, Heilongjiang, China

J. Zhao Nanjing University of Aeronautics and Astronautics, Nanjing, Jiangsu, China

X.S. Zhao Harbin Institute of Technology, Harbin, Heilongjiang, China

T. Zhou Beijing Institute of Technology, Beijing, P.R. China

About the editor

J. Paulo Davim received his PhD in Mechanical Engineering from the University of Porto in 1997, the Aggregate title from the University of Coimbra in 2005 and a DSc from London Metropolitan University in 2013. Currently, he is Professor at the Department of Mechanical Engineering of the University of Aveiro. He has about 30 years of teaching and research experience in manufacturing, materials and mechanical engineering with special emphasis in Machining & Tribology. Recently, he has also interest in Management/Industrial Engineering and Higher Education for Sustainability. He has worked as an evaluator of projects for international research agencies as well as an examiner of PhD thesis for many universities. He is the Editor in Chief of eight international journals, guest editor of journals, book editor, book series editor and Scientific Advisory for many international journals and conferences. Presently, he is an editorial board member of 30 international journals and acts as reviewer for more than 80 prestigious Web of Science journals. In addition, he has also published as author and co-author more than 10 books, 60 book chapters and 350 articles in journals and conferences (more than 200 articles in journals indexed in Web of Science/h-index 35+ and SCOPUS/h-index 42+).

Preface

In recent years, microfabrication and precision engineering is an area of technology that has been gaining in importance. The term micro-electro-mechanical-systems (MEMS) emphasize the miniaturization of both electronic and mechanical components. Microsystems products can be classified by application area, (e.g., medical, automotive, aerospace, optical, military, alternative energy, microelectronics packaging, etc.) Microsystems technology is a general term that refers to the products as well as the fabrication technologies to produce them. Microfabrication and precision engineering can be applied to metallic, polymers, ceramics, composites, biomaterials and complex materials.

This research book aims to provide information on microfabrication and precision engineering for modern industry. The initial chapter of the book provides modelling of micro and nanoscale cutting. Chapter two is dedicated to machining scale (workpiece gran size and surface integrity in micro end milling). Chapter three presents micromachining technique based on the orbital motion of diamond tip. Chapter four covers implementation of innovative machining strategies in microelectrical discharge machining of Ti-6Al-4V. Chapter five is dedicated to the principle and capabilities of microelectrochemical machining. Chapter six contains information on microchannel fabrication via direct Laser writing. Chapter seven is dedicated to underwater pulsed Laser beam cutting with a case study. Finally, the last chapter of the book is dedicated to glass molding process for microstructures.

The current book can be used as research book for final undergraduate engineering course or as a topic on fabrication and precision engineering at the postgraduate level. Also, this book can serve as a useful reference for: academics, researchers, mechanical, manufacturing, industrial and materials engineers as well as professionals in microfabrication and precision engineering. This book presents scientific interest for the industry, centers of the research, laboratories and universities throughout the world.

The Editor acknowledges WoodHead/Elsevier for this opportunity and for their professional support. Finally, I would like to thank all the chapter authors for their availability for this work.

August 2016

J. Paulo Davim
University of Aveiro, Portugal

Modeling of micro- and nano-scale cutting

1

R. Rentsch[1], *A.P. Markopoulos*[2] *and N.E. Karkalos*[2]
[1]Bremen University, Bremen, Germany, [2]National Technical University of Athens, Athens, Greece

1.1 Introduction

Technologies for processing various materials and manufacturing components that possess features from a few nm to a few hundreds of μm are in use in contemporary industry. These micro- and nano-machining processes shape parts by removing unwanted material, carried away from the workpiece, usually in the form of chips. Evaporation or ablation may take place in some machining operations. The specific term *cutting* describes chip formation by the interaction of a wedge-shaped tool with the workpiece surface; the chip forms as a result of their relative movement. These machining operations include processes such as turning, milling, and drilling, usually described by the accompanying prefix micro- or nano-, depending on the scale of reference. In contemporary industry, abrasive processes, such as grinding, have great importance in cutting. Micro- and nano-cutting are more advantageous than other processes, since it is possible to machine a variety of materials in complex shapes with excellent surface finish and tight tolerances.

Until 2016, cutting and grinding at micro- and nano-scale have been studied theoretically and experimentally (Alting, Kimura, Hansen, & Bissacco, 2003; Brinksmeier et al., 2006; Byrne, Dornfeld, & Denkena, 2002; Corbett, McKeown, Peggs, & Whatmore, 2000; Dornfeld, Min, & Takeuchi, 2006; Madou, 2002; Mamalis, Markopoulos, & Manolakos, 2005; Masuzawa, 2000; Rentsch, 2009). The small dimensions of workpieces, cutting tools, and cut depths bring up a number of issues that may play no significant role in traditional machining but are of significance in micro- and nano-cutting. For example, in microcutting, features known as minimum chip thickness and size effect influence the underlying mechanisms of chip formation. It is not always feasible to carry out experimental work in order to overcome micro- and nano-scale manufacturing component problems. Moreover, increased demand, innovation, reliability, and cost reduction requirements need to be satisfied. Modeling and simulation techniques exist to aid engineers and scientists who use them in a variety of ways. These include: reducing experimental time and testing, giving insight into complex phenomena, exploring possibilities, reducing complexity and learning cycles of a process, increasing accuracy, and optimizing processes and products. However, in the related processes, the actual material removal can be limited to the surface of the workpiece (i.e., only a few atoms or layers of atoms).

Microfabrication and Precision Engineering. DOI: http://dx.doi.org/10.1016/B978-0-85709-485-8.00001-2

Inherent measurement problems and the lack of more detailed experimental data limit the possibility to develop analytical and empirical models as more assumptions need to be made. Therefore, modeling and simulation with advanced and specialized methods are employed. The following paragraphs focus on modeling with the finite elements method (FEM) for microscale cutting and molecular dynamics (MD) for nanoscale cutting. The main principles of the aforementioned techniques, the fields of application, limitations, considerations, and an up-to-date bibliography are provided within this chapter.

1.2 Modeling of microscale cutting

1.2.1 Minimum chip thickness and size effect

The set-up used in the modeling of microscale cutting is similar to the one used in the macroscale traditional cutting processes (i.e., a wedge-like tool is removing material from a surface). All the geometrical features and kinematic characteristics of the tool and workpiece are identifiable. However, downscaling all phenomena in order to apply the same theories in both the micro and macro regime proves to be inadequate. There are features of machining and phenomena that are considerably different in micromachining and do not allow for such a simplification. These differences arise when considering the chip formation process, the resulting cutting forces, the surface integrity, and tool life.

In Fig. 1.1, the orthogonal cutting model corresponding to microscale cutting may be observed. At this level, the depth of cut may be well below 10 μm with an anticipated surface roughness of only a few nm. The cutting edge, no longer be considered sharp, has a cutting edge radius comparable in size to the uncut

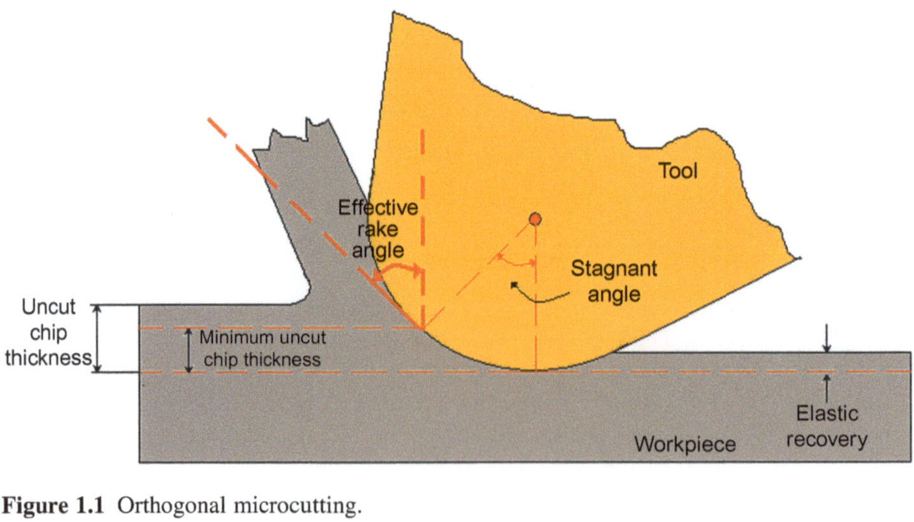

Figure 1.1 Orthogonal microcutting.

chip thickness. Although the rake angle of the tool indicated in Fig. 1.1 is positive, the actual the effective rake angle participating in the processes is negative. In this case, the elastic-plastic deformation of the workpiece material and ploughing need to be taken into account, as well as the elastic recovery at the clearance face.

As explained above, one may deduct the existence of a removable minimum chip thickness from the workpiece surface in a mechanical micromachining operation. A stagnation point above which a chip is formed and below only elastic-plastic deformation takes place is assumed. The stagnation point is connected to a stagnant angle θ_m, which with the tool edge radius determines the value of the minimum uncut chip thickness, h_m (Malekian, Mostofa, Park, & Jun, 2012):

$$h_m = r_e(1 - \cos\theta_m) \tag{1.1}$$

The minimum chip thickness determines whether a chip is formed or not, because if the depth of cut for a microcutting operation is set below this minimum, then the cutting edge is expected to plastically deform the workpiece material without producing a chip. This is known as the ploughing mechanism. In addition to the obvious effect on the surface integrity and quality of the finished workpiece, it significantly alters the cutting forces, and thus the process stability, in microscale machining and makes the force prediction methods ineffective. The minimum chip thickness was identified in grinding by Finnie (1963) and also in microturning and micromilling by several other researchers. Ikawa, Shimada, Tanaka, and Ohmori (1991), using a turning diamond tool with an edge radius of about 10 nm, found the uncut chip thickness is approximately 1/10 the cutting edge radius. Weule, Huntrup, and Tritschle (2001) studied the minimum chip thickness for micromilling. In this case, during the same pass of a cutter's single tooth from the workpiece on a chip of varying thickness, the material removal mechanism may change from shearing to ploughing or vice versa, resulting in a saw tooth-like surface profile and a deteriorating surface finish. The minimum chip thickness was also reported in micromilling by Kim, Bono, and Ni (2002) by comparing the process with nominal chip volume and the workpiece surface feed marks with the feed per tooth. It was concluded that a chip was not formed with every pass of a cutting tooth, and this was attributed to the minimum chip thickness.

Although ploughing may exist in machining, the study of its effect on the overall process may be neglected. The effect caused by the cutting edge radius is important in microscale cutting. Many researchers consider this to be the primary cause of the size effect, the nonlinear increase in the specific energy, and the specific cutting force with decreasing depth of cut, all of which are observed in microcutting. Albrecht (1960, 1961) argues that there are two areas where ploughing occurs, one on the rake face and another around the tool edge. Masuko (1956) introduces a new effect that acts independently from ploughing and calls it indenting. According to this theory an indenting force causes the cutting edge to penetrate the workpiece; this force is held responsible for the size effect according to this theory.

The size effect was identified in metal cutting operations as early as 1952 by Backer et al. The researchers processed specimens made of SAE 1112 steel with grinding, micromilling, turning, and tension tests. These data correlate chip thickness with resisting shear stress, and clearly show the size effect, which is attributed to the significantly reduced amount of imperfections—namely crystallographic defects such as grain boundaries, missing and impurity atoms, and inhomogeneities present in all commercial metals—encountered when deformation takes place in a small volume. With smaller uncut chip thickness, the material strength is expected to reach its theoretical value of strength. Many more investigators have acknowledged the size effect experimentally and theoretically (Kim & Kim, 1995).

Although size effect, such as minimum chip thickness, is present in metal cutting, it is of special importance in microscale cutting. Aside from the explanations already mentioned, there are several other discussions regarding the size effect appearance. It is also attributed to material strengthening due to increasing the strain rate in the primary shear zone (Larsen-Basse & Oxley, 1973) or to decreasing temperature in the tool-chip interface (Kopalinsky & Oxley, 1984) with decreased chip thickness. Atkins (2003) proposes that the size effect is due to the energy required for new surface creation via ductile fracture. Another explanation is based on the size effect appearing in micro-nano-indentation and its extension to machining (Dinesh, Swaminathan, Chandrasekar, & Farris, 2001). The increased hardness of a material with reduced indentation depth is a result of the dependence of material flow stress on the strain gradient in the deformation zone; strain gradient plasticity can be the reason of size effect in machining because of the intense strain gradients observed. The criticism of this theory stems from the fact that the size effect's impact on hardness in indentation is related to size effect in cutting when the von Mises criterion is applicable. However, this assumption is not compatible with the experiments of Merchant (Shaw & Jackson, 2006). However, some works using analytical and FEM models and experimental validation based on strain gradient plasticity have been published (Joshi & Melkote, 2004; Liu & Melkote, 2007).

From the literature review, it is evident that many reasons for the size effect in machining and microscale cutting have been reported. It is not clear which of the above mechanisms is dominant, or whether there could be more than one mechanism acting simultaneously. Even when multiple mechanisms act together, there may be additional influences that alter the contribution of each factor in each case. For further reading on the size effect in machining, and especially in micromachining, the works of Liu, DeVor, Kapoor, & Ehmann (2004), Shaw (2003), and Shaw & Jackson (2006) are suggested.

1.2.2 FEM modeling of microscale cutting

In this section, the aspects of modeling microscale cutting with FEM are explored. FEM modeling of machining presents similarities in its macro- and micro-scale forms. General examples of FEM applications in manufacturing technology, and in particular, cutting, can be found in the works of Dixit & Dixit, 2008; Klocke et al., 2002; Mamalis,

Manolakos, Ioannidis, Markopoulos, and Vottea, 2003. The following paragraphs provide a detailed description of the modeling of cutting, with focus on the modeling of microscale cutting.

1.2.3 FEM basics

In FEM, the basic principle is discretization (i.e., the replacement of a continuum by finite elements forming a mesh). Each finite element is simpler in geometry and therefore easier to analyze than the actual structure. Every finite element possesses nodes where the initial problem and boundary conditions are applied, and the degrees of freedom are calculated. Furthermore, the finite elements are connected to one another in nodes. The problem variables as well as properties applied on the nodes of each element are assembled, and global relations are formatted.

Two different time integration strategies address nonlinear and dynamic models. These refer to implicit and explicit schemes. The former approach solves the set of finite element equations by using a central difference rule to integrate the equations of motion through time. While the latter is realized by solving the set of finite element equations and performing iterations, until a convergence criterion is satisfied for each increment. Another topic pertains to the use of a certain numerical formulation. Currently, the ones used in metal cutting FEM models are of three types: Eulerian, Lagrangian, and the newer arbitrary Lagrangian-Eulerian (ALE) analysis. In the Eulerian approach, the finite element mesh is spatially fixed and covers a control volume where the material flows through it in order to simulate the chip formation. In the Lagrangian approach, the elements are attached to body-centered meshes (e.g., that of the workpiece). The workpiece is deformed due to the action of the cutting tool, and consequently so is the mesh, providing a more realistic simulation. Disadvantages of the Lagrangian formulation is connected to the large mesh deformation observed during the simulation and the use of chip separation criteria. The updated Lagrangian analysis has overcome the disadvantage of a chip separation criterion by applying continuous re-meshing and adaptive meshing, dealing at the same time with the mesh distortion. Finally, the ALE formulation has also been proposed with the aim to combine the advantages of the two aforementioned methods. More details on the subjects discussed in this paragraph can be found in the work of Markopoulos (2013).

1.2.4 FEM cutting models

As expected, the geometrical characteristics of the tool and the workpiece greatly influence the outcome of the microscale cutting models. Specifically, the tool edge radius is connected to the size effect, minimum chip thickness, effective rake angle, stagnation point, and ploughing mechanism, as discussed in the previous section. In microcutting, the simulation of the process with a sharp tool is of no interest because the size effect would not be accounted for; the size of the cutting forces and the chip formation would be unrealistic. Some researchers have investigated the influence of the tool edge radius on the size effect. Weber et al. (2007) performed an analysis using similarity mechanics and various values for the tool edge radius. Woon, Rahman, Neo, and Liu (2008) also investigated the tool edge radius and its influence on the material deformation and the contact length between the tool and workpiece and then validated their numerical results with the aid of a small field-of-view photography technique. Liu and Melkote (2007) concluded from their analysis that the tool edge radius accounts for only part of the size effect in microcutting, and the material strengthening is associated with a temperature drop in the secondary deformation zone for higher cutting speeds.

Another topic of interest pertains to the boundary conditions applied in the initial mesh, and specifically, the manner thermo-mechanical coupling is considered. In cutting processes, heat generation originates from the two deformation zones (i.e., the primary and the secondary) due to inelastic and frictional work. Addressing this problem involves a nonlinear analysis of the associated strain hardening and thermal softening in these zones. This feature of the model is important for microscale cutting, since the size effect is attributed to material strengthening due to a temperature decrease in the tool–chip interface with a decreased chip thickness (Kopalinsky & Oxley, 1984).

Moriwaki, Sugimura, and Luan (1993) developed a thermo-mechanical model of copper micromachining and calculated the stress, strain flow of the cutting heat, and temperature of the tool and the workpiece. Effective strain hardening and thermal softening are integrated by employing tribological and material models that are functions of mechanical and thermal behavior with strain, strain-rate, and temperature.

1.2.5 Friction modeling

Microscale cutting uses the same assumptions as macroscale cutting, regarding the secondary deformation zone in friction modeling at the interface of the chip and rake face of the tool. Many researchers utilize Coulomb's law (i.e., the frictional sliding force is proportional to the applied normal load, and the ratio of these two is the coefficient of friction μ which is constant in all the contact length between chip and tool). The relation between frictional stresses τ and normal stresses may be expressed as:

$$\tau = \mu\sigma \tag{1.2}$$

However, as the normal stresses increase and surpass a critical value, this equation fails to give accurate predictions. Experimental analysis verifies that two contact regions may be distinguished in dry machining, namely the sticking and sliding regions. Zorev's (1963) stick-slip temperature independent friction model is the one commonly used. In this model, there is a transitional zone with distance ℓ_c from the tool tip that signifies the transition from sticking to sliding regions. Near the tool cutting edge and up to ℓ_c (i.e., the sticking region), the shear stress is equal to the shear strength of the workpiece material, k; while in the sliding region, the frictional stress increases according to Coulomb's law.

$$\tau = \begin{cases} k, 0 \leq \ell \leq \ell_c \\ \mu\sigma, \ell > \ell_c \end{cases} \tag{1.3}$$

In machining based on Zorev's model, other approaches have been reported that include the defining of an average friction coefficient on the rake face or different coefficients for the sliding and sticking regions (Childs & Maekawa, 1990; Iwata, Osakada, & Terasaka, 1984; Sekhon & Chenot, 1993; Usui, Maekawa, & Shirakashi, 1981; Yang & Liu, 2002). For more information on friction models in cutting, the reader is encouraged to read the work of Markopoulos (2013).

1.2.6 Material modeling

Material modeling in microscale machining is of particular interest, especially the extensively studied flow properties of the workpiece material and the corresponding equations included in FEM. These constitutive equations describe the flow stress or instantaneous yield strength at which work material starts to plastically deform or flow. Some of the many constitutive equations employed in metal cutting are discussed here. The first is the relation by Usui et al. (1981):

$$\sigma = B\left[\frac{\dot{\varepsilon}}{1000}\right]^{M} e^{-kT}\left[\frac{\dot{\varepsilon}}{1000}\right]^{m}\left\{\int_{\text{Path}} e^{kT/N}\left[\frac{\dot{\varepsilon}}{1000}\right]^{-m/N} d\varepsilon\right\}^{N} \tag{1.4}$$

In this equation, B is the strength factor, M is the strain-rate sensitivity, N the strain hardening index, all functions of temperature are T, and k and m are constants. The integral term accounts for the historical effects of strain and temperature in relation to strain rate. In the absence of these effects, the last equation is reduced to (Childs, Otieno, & Maekawa, 1994):

$$\sigma = B\left[\frac{\dot{\varepsilon}}{1000}\right]^{M} \varepsilon^{N} \tag{1.5}$$

Among the most used material models is the Johnson–Cook model (1983). The equation consists of three terms: the elasto-plastic term represents strain hardening, viscosity demonstrates that material flow stress increases for high-strain rates, and the temperature softening term for the softening of the material due to temperature increase. It is a thermo-elasto-visco-plastic material constitutive model described as:

$$\sigma = (A + B\varepsilon^{n})\left[1 + C\ln\left(\frac{\dot{\varepsilon}}{\dot{\varepsilon}_{o}}\right)\right]\left[1 - \left(\frac{T - T_{a}}{T_{m} - T_{a}}\right)^{m}\right] \tag{1.6}$$

where T_a is the ambient temperature; T_m the melting temperature; and A, B, C, n, and m are constants that depend on the material and are determined by material tests (Jaspers & Dautzenberg, 2002; Lee & Lin, 1998) or predicted (Özel & Karpat, 2007). Umbrello, M'Saoubi, and Outeiro (2007) investigated the influence of the Johnson-Cook constants on the outcome of machining modeling and found that FEM results are sensitive to these inputs, which in turn are strongly related to the test method used to derive the constants. On the other hand, the results from a test method can be plugged into different constitutive equations, and the selection of the material model can influence the predicted results (Liang & Khan, 1999; Shi & Liu, 2004). A review of material models used in manufacturing processes, including microcutting, can be found in the work of Dixit, Joshi, and Davim (2011).

All the above mentioned FEM material models refer to isotropic materials. No crystallographic effects are considered in the modeling process. Zerilli and Armstrong (1987) developed a constitutive model based on dislocation-mechanics theory and consideration of materials' crystal structures. They suggested two different models, one for body-centered cubic(BCC) and the other for face-centered cubic (FCC) lattice structure, respectively:

$$\sigma = C_o + C_1 \exp[-C_3T + C_4T \ln(\dot{\varepsilon})] + C_5\varepsilon^n \tag{1.7}$$

$$\sigma = C_o + C_2\varepsilon^n \exp[-C_3T + C_4T \ln(\dot{\varepsilon})] \tag{1.8}$$

where C_i, $i = 0-5$, and n are material constants determined experimentally (e.g., by the SHPB method) (Meyer & Kleponis, 2001).

In microscale cutting, the cutting tool radius is comparable to the size of the grains of the material being processed. Furthermore, in materials with surface defects or multiphase materials, such as cast iron, the microcutting mechanism is quite different in comparison to nonheterogeneous materials due to the encounter of the cutting tip with these features of the material during the course of the process. The model prepared by Chuzhoy, DeVor, and Kapoor (2003) simulates cutting at a microstructure level. Ductile iron and two of its constituents, namely pearlite and ferrite, are modeled in the same continuum, taking into account the microstructural composition, the grain size, and the distribution of each material. The model's position of predict stresses, strains, and temperatures. Simoneau, Ng, and Elbestawi (2006a, 2006b, 2007a, 2007b) studies involve the machining of 1045 steel and consider its microstructure. In their models, the initial mesh of the material is divided into bands of material A, with a pearlite-like behavior, and bands of material *B*, with a ferrite-like behavior; the material bands are of appropriate size. Material plasticity is formulated with a strain dependent Johnson-Cook model. The results of the analysis show good correlation between experimental and numerical results regarding the morphology of the chip and the behavior of each phase of the material. The predicted strain result was larger in a heterogeneous FEM model in comparison to a homogeneous one due to strain localization at each phase.

However, when nanoscale cutting is considered, material modeling cannot be adequately represented by continuum mechanics; a more detailed view on the materials microstructure and the interaction between the workpiece, the tool, and the chip is required. For this reason, the MD method, presented in detail in the next part, is applied for cutting process simulation at this scale.

1.3 Modeling of nanoscale cutting

Nanoscale cutting material removal involves a few atom layers of the workpiece, thus an atomistic modeling is required for the simulation of the process. MD, which can simulate the behavior of materials in an atomic scale, is used for simulating nanometric cutting. FEM based on the principles of continuum mechanics and at nanometric level is considered a drawback. MD is a modeling method in which atoms interact for a period of time, by means of a computer simulation. Particle interactions are described by potential functions and very large particle numbers are needed to simulate a molecular systems; therefore, a vast number of equations need to be solved repeatedly to describe the properties of these systems as they develop along the time domain. Laws and theories from mathematics, physics, and chemistry constitute the backbone of this multidisciplinary method. In order to deal with these problems, numerical methods, rather than analytical ones, are used, and algorithms from computer science and information theory are employed. Although the method was originally intended to be exploited in theoretical physics, today it is mostly applied in materials science, the production of biomolecules, and in nanomanufacturing.

The MD method was introduced in the simulation of micro- and nano-processing in the early 1990s (Ikawa et al., 1991; Rentsch & Inasaki, 1995; Stowers et al., 1991). Komanduri and Raff (2001) presented a review paper on MD simulation of machining at the atomic level. The results of the related works indicated that MD is a possible modeling tool for the nanocutting process. Since in atomistic modeling, the atomic interactions and microstructure are explicitly included in the simulation, it can provide better representation of micro- and nano-level material characteristics than other modeling techniques. Therefore, MD can stimulate phenomena that cannot be investigated with continuous mechanics. MD models are used for the investigation of the chip removal mechanisms, tool geometry optimization, cutting force estimations, subsurface damage identification, burr formation, and surface roughness and integrity predictions. The following sections provide a thorough description of modeling techniques used in the MD simulation of nanoscale cutting. Specifically, the basic aspects of MD models graphically presented in Fig. 1.2, such as geometry, material properties and potential functions, boundary conditions, and integration schemes are described in the following subsections.

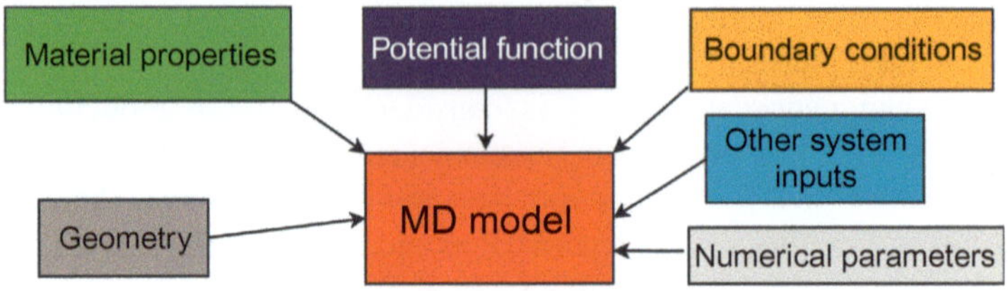

Figure 1.2 Elements for the setting up of an MD model.

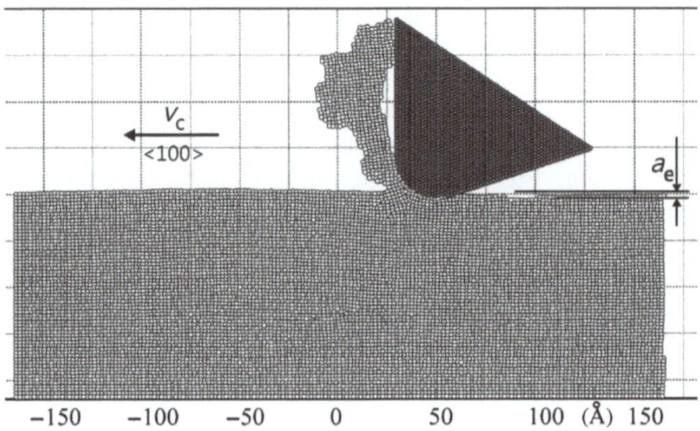

Figure 1.3 MD model set-up of nanoscale cutting (Rentsch, 2001) (workpiece: Cu, surface {001}, width 1 unit cell, $a_e = 0.3$ nm, cutting length = 60 nm, 15 million time steps).

1.3.1 Model geometry and material microstructure

One of the most fundamental subjects in simulations is the accurate description of the geometric entities that constitute the domain on which the solution of governing equations is required. Realistic geometric modeling can produce more reliable results, if combined with the appropriate material properties, boundary conditions, and other numerical details, and is required in order to conduct comparisons with experimental studies. Fig. 1.3 depicts a general model set-up, resembling orthogonal cutting, which is often used in MD simulations.

The majority of workpiece geometry cases are simplistic and have the form of a rectangular box. Usual sizes extend up to several tenths of nanometers, depending on the material type, and the number of atoms is in the order of 10^5. It is noteworthy to mention that due to the direct correlation of model size to computational requirements, model sizes are restricted by the available computational power. However, cases exist in the literature where models with almost a million atoms are reported (Ren, Zhao, Dong, & Liu, 2015; Tong, Liang, Jiang, & Luo, 2014). Also, multimillion atom models were created in order to study, among other topics, the effect of workpiece size on the results of the simulation (Pei, Lu, & Lee, 2007; Xiao, To, & Zhang, 2015). Large-scale MD simulations with model sizes up to ten million atoms can also be found in the literature (Pei, Lu, Lee, & Zhang, 2009). The arrangement of the atoms in workpiece strictly follows the atomic structure of each material. On rare occasions (Eder, Bianchi, Cihak-Bayr, Vernes, & Betz, 2014; Li, Fang, Zhang, & Liu, 2015; Rashid, Goel, Luo, & Ritchie, 2013), a modified type of the surface is used as the effect of surface roughness on the outcome of the machining process is desired. Moreover, the material can be modeled as a multigrain bulk or various crystal orientations may exist in the workpiece (Goel, Luo, Reuben, & Pen, 2012; Junge & Molinari, 2014; Solhjoo & Vakis, 2015). Furthermore, there are some rare cases where lubricants or cutting fluids are taken into consideration in the simulation (Chen, Cian, Yu, & Huang, 2014; Ren et al., 2015; Rentsch & Brinksmeier, 2005; Rentsch & Inasaki, 2006). Obviously, knowledge of crystal structures is fundamental when modeling a workpiece in MD simulations. Crystal structures are considered as unique arrangements of atoms, molecules, or atoms within a crystalline material. The most common lattice systems employed in nanocutting atomistic simulation are BCC, FCC, and body-centered tetragonal, which correspond to the most common engineering materials.

Tool geometry is generally somewhat more complex and diverse than the workpiece geometries. The majority of nanomanufacturing simulations involve the use of a cutting tool as they represent general nanocutting processes. Usually, the tool shape is process-dependent; the tool is portrayed as close as possible to a real cutting tool. In many cases, the cutting tool is considered rigid, and the tool atoms retain their relative positions while travelling at constant speed relative to the workpiece. The simplest form of cutting tool has a round or atomically sharp edge (Gao et al., 2009; Lin, Lin, & Hsu, 2014; Rentsch & Brinksmeier, 2005), and its geometric representation includes the definition of rake and clearance angles. Studies with various rake angles are also conducted (Markopoulos, Karkalos, Kalteremidou, Balafoutis, & Manolakos, 2015a; Pei, Lu, Fang, & Wu, 2006), even with negative ones (Komanduri, Chandrasekaran, & Raff, 1999). Furthermore, in nanogrinding simulations, the cutting tool is represented as having single (Markopoulos, Savvopoulos, Karkalos, & Manolakos, 2015b; Rentsch & Inasaki, 1994; Rentsch & Inasaki, 1995) or multiple grains with a cubic (Brinksmeier et al., 2006), ellipsoid (Eder et al., 2014), or spherical shape (Li, Fang, Liu, & Zhang, 2014).

1.3.2 Potential function

Potential functions are employed in MD simulations as means of representing material behavior at the atomistic scale. A variety of potential functions has been developed during the past decades with a view to simulate interaction between atoms of the same or different type. Historically, Lennard-Jones and Morse potentials were the first to be employed. However, over time, more complex potentials, such as embedded atom method (EAM) potential or bond-order potentials, are developed with significantly better accuracy. All of these potentials involve the determination of a set of parameters, calculated either from experimental findings or from simulations with more fundamental methods. When the potential function is selected, force can be directly calculated, as the first derivative of potential energy with respect to position.

The Lennard-Jones potential is a simple simulation model describing interaction between a pair of uncharged atoms or molecules as a function of the distance between their centers. Its most common expression is the so-called 12-6 Lennard-Jones potential:

$$V = 4\varepsilon\left[\left(\frac{\sigma}{r}\right)^{12} - \left(\frac{\sigma}{r}\right)^{6}\right] = \varepsilon\left[\left(\frac{r_{\rm m}}{r}\right)^{12} - 2\left(\frac{r_{\rm m}}{r}\right)^{6}\right] \tag{1.9}$$

where ε is the energy constant (or well depth, depicting interaction strength), σ is effectively the diameter of one of the atoms (or the distance at which the two particles are at equilibrium), and $r_{\rm m}$ represents the distance when potential energy is minimum and is equal to $2^{1/6}\sigma$. The Lennard-Jones potential curve is depicted in Fig. 1.4.

The Lennard-Jones potential is considered a good approximation for a preliminary study of an atomistic system, and is particularly sufficient in modelling noble gases. Lennard-Jones potential assumes that interactions are valid for an infinite range. Although the interatomic forces fade over a large distance, the definition of a cut-off distance is necessary to reduce the unnecessary computational cost for calculations between atoms separated by a large distance. The cut-off distance is often set at 3σ. In terms of nanoscale cutting, Lennard-Jones potential is rarely employed (Eder et al., 2014; Oluwajobi & Chen, 2011a), especially in cases where potential functions do not exist for a material or to model a more complex material, as in Solar et al. (2011).

Morse potential is another two-body potential function for the calculation of interatomic interactions. This potential function provides a good approximation of molecular vibration (vibrational excitations of a chemical bond). It is considered the simplest potential function that can simulate dissociation between two atoms, which would be impossible to be simulated with a harmonic oscillator model.

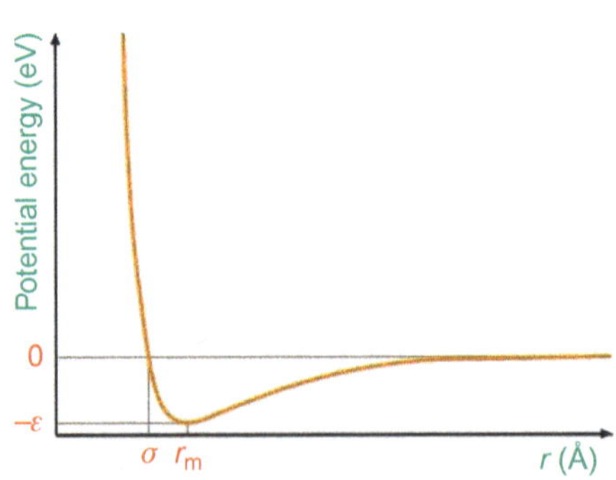

Figure 1.4 Lennard-Jones potential curve.

Morse potential has a simple formulation and requires the definition of only three parameters to provide the potential energy of a diatomic molecule:

$$V = D_e\left(e^{-2a(r-r_e)} - 2e^{-a(r-r_e)}\right) \tag{1.10}$$

where r is the distance between the atoms (or internuclear distance), r_e is the equilibrium bond distance (bond length), D_e is the well depth (relative to the dissociated atoms, dissociation energy measured from the equilibrium position, from the minimum of the curve), and parameter α controls the "width" of the potential (the smaller α is, the larger the well); and thus is considered as an adjustable shape parameter. The potential curve of Morse potential function is depicted in Fig. 1.5. In terms of nanoscale cutting, Morse potential was initially employed to model a variety of materials. Afterward, it was used to model cubic metals, and after the introduction of more complex potential functions, it is used to model interactions between different materials as long as potential parameters for other potential functions are not available.

In MD simulations, bond order potentials are a class of empirical potential functions capable to model complex material behavior. Bond order is defined as the number of chemical bonds between a pair of atoms. Bond length is related to bond order as more electrons participating in bond formation lead to a shorter bond. Potential functions such as Tersoff Potential, Brenner Potential, ReaxFF, reactive empirical bond order, analytic bond order potential, and adaptive intermolecular reactive empirical bond order belong to this class. These potentials can simulate different bonding states of atoms and even describe chemical reactions. The general concept of this class of potential functions is that the strength of the chemical bond is not constant but depends on the local bonding condition, focusing on parameters

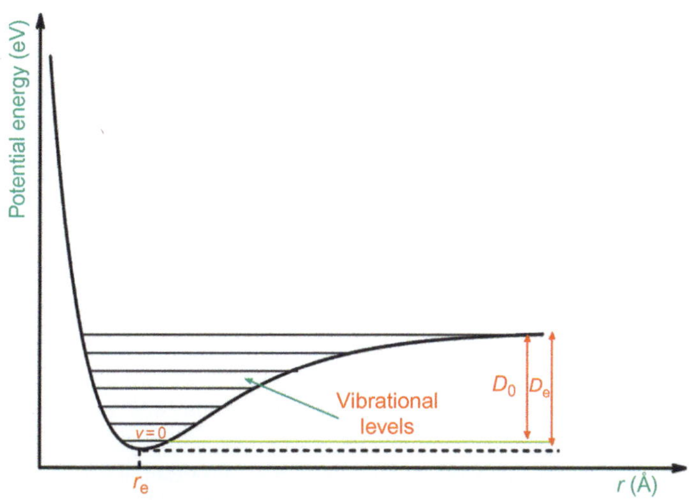

Figure 1.5 Morse potential curve.

such as the number of bonds, bond angles, and bond length rather than on atomic distance only. The general form of bond potentials is:

$$V_{ij}(r_{ij}) = V_{\text{repulsive}}(r_{ij}) + b_{ijk} V_{\text{attractive}}(r_{ij}) \tag{1.11}$$

This formula demonstrates that potential energy is dependent upon the interatomic distance r_{ij}, as in simple potential functions (i.e., the Lennard-Jones function). However, the bond strength is also affected by the environment of the atom through the b_{ijk} term.

One of the most popular bond-order models used to model nonmetallic and ceramic materials is the Tersoff potential function, which is often used to simulate carbon, silicon, or SiC. It is formulated in the following way:

$$E = \sum_i E_i = \frac{1}{2} \sum_{i \neq j} V_{ij} \tag{1.12}$$

$$V_{ij} = f_C(r_{ij})[f_R(r_{ij}) + b_{ijk} f_A(r_{ij})] \tag{1.13}$$

$$f_R(r_{ij}) = A_{ij} e^{-\lambda_{ij} r_{ij}} \tag{1.14}$$

$$f_A(r_{ij}) = -B_{ij} e^{-\mu_{ij} r_{ij}} \tag{1.15}$$

$$f_c(r_{ij}) = \begin{cases} 1, r_{ij} < R_{ij} \\ \frac{1}{2} + \frac{1}{2} \cos\left[\pi \frac{r_{ij} - R_{ij}}{S_{ij} - R_{ij}}\right], S_{ij} > r_{ij} > R_{ij} \\ 0, r_{ij} > S_{ij} \end{cases} \tag{1.16}$$

$$b_{ij} = \chi_{ij} \left(1 + \beta_i^{n_i} \zeta_{ij}^{n_i}\right)^{-\frac{1}{2n_i}} \tag{1.17}$$

$$\zeta_{ij} = \sum_{k \neq i,j} f_c(r_{ik}) \omega_{ik} g(\theta_{ijk}) \tag{1.18}$$

$$g(\theta_{ijk}) = 1 + \frac{c_i^2}{d_i^2} - \frac{c_i^2}{(d_i^2 + h_i - \cos(\theta_{ijk}))} \tag{1.19}$$

The EAM potential is a many-body potential function. The necessity of employing such a potential in nanoscale cutting simulations emerges from the fact that previous, simpler potential functions fail to account for the physics of metallic bonding. In fact, EAM-type potentials are now the most frequently used in simulations of metallic or alloy systems. This potential calculates the potential energy of the system as a combination of pair terms and an embedding function that is employed to determine the local energy density. This formulation considers the energy of the system as a function of electron density, where the density is assumed to be a superposition of the local atomic densities. Usually, an EAM potential is constructed by means of three functions: the aforementioned embedding function, an electronic density function, and a function that models potential due to pair exchange. The equation:

$$E_i = F_\alpha\left(\sum_{i\neq j}\rho_\beta(r_{ij})\right) + \frac{1}{2}\sum_{i\neq j}\varphi_{\alpha\beta}(r_{ij}) \tag{1.20}$$

where $\varphi_{\alpha\beta}$ is a suitable pairwise potential function, ρ_β the contribution to the electron charge density from atom j of type β at the location of atom i, and F is a function representing the energy required to place atom i of type α into the electron cloud.

1.3.3 Boundary conditions and input parameters

As in the case of most types of simulation, boundary conditions are essential for the completion of the definition of a problem after the governing equations are properly derived. In the case of atomistic simulations, three types of boundary conditions exist in most cases: fixed boundary, thermal boundary, and periodic boundary conditions.

Fixed boundary conditions, one of the most common boundary condition in MD simulations of nanoscale cutting, resembles the fixed support in continuum mechanics problems or the behavior of a perfect rigid body (Shimizu, Eda, Zhou, & Okabe, 2008). In fact, these boundary conditions are implemented by keeping several layers at the sides of the workpiece fixed in their initial lattice positions throughout the simulation, as if these atoms were not affected by the process (Cheng, Luo, Ward, & Holt, 2003). This boundary condition provides a simple way to reduce some edge or more generally, boundary effects, and it helps to maintain the symmetry of the lattice (Goel, Luo, & Reuben, 2013; Yan, Sun, Dong, Luo, & Liang, 2006); although the workpiece dimensions also need to be large enough to reduce the impact of the fixed boundaries on the simulation results. These considerations can become essential to prevent an undesired rigid body motion caused by the interaction with other bodies, such as the cutting tool (Liang, Wang, Chen, Chen, & Tong, 2011; Rentsch & Brinksmeier, 2005; Zhang & Tanaka, 1997).

Thermal boundary conditions are also essential to almost every MD simulation of nanocutting. As the majority of cases is assumed to be conducted at vacuum conditions (Shimizu et al., 2008), there is no heat exchange between the workpiece and its environment. However, thermal effects within the workpiece are considerable. Specifically, during the nanocutting process, a significant amount of energy is converted into heat. To avoid artificial over-heating and temperature rise due to the finite volume modelled, the heat needs to be properly dissipated from the system to prevent affecting simulation results. Thus, it is important to define a suitable boundary condition that can act as a heat sink and also control the temperature of the workpiece.

In conventional machining, it is considered that a significant portion of cutting heat produced as a result of shear and friction energy can be carried away by lubricant or chip (Cheng et al., 2003; Goel, Luo, & Reuben, 2011). It is also shown

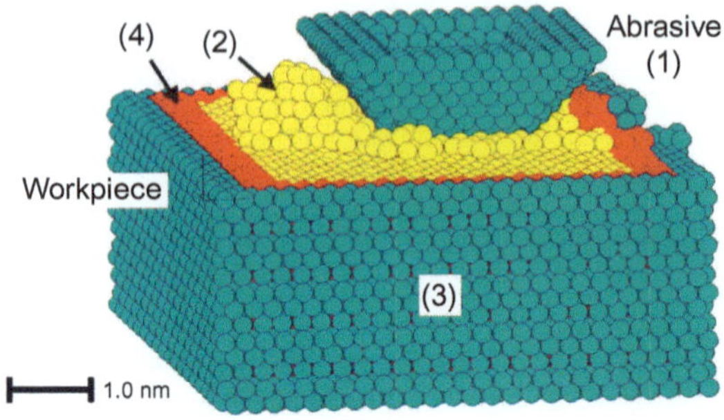

Figure 1.6 Types of atoms used in MD simulations: (1) cutting tool or abrasive; (2) workpiece – Newtonian atoms; (3) fixed boundary atoms; (4) thermostat atoms (Rentsch & Inasaki, 1995).

(Narulkar, Bukkapatnam, Raff, & Komanduri, 2009) that from the heat attributed to plastic deformation of the workpiece, 10% is dissipated to the tool, 10% is dissipated to the workpiece, and almost 80% is carried away by the chip. Since most nanometric models are smaller than the systems they simulate, for reasons of computational cost, they are also not capable of sufficiently dissipating the produced heat within the modelled body (Goel et al., 2013; Narulkar et al., 2009). Therefore, layers of thermostat atoms are defined within the workpiece as a temperature-controlling layer (Cai, Li, & Rahman, 2007; Huang, 2013), usually between the fixed boundary layers of atoms and the surrounding, freely moving so-called Newtonian atoms, as seen in Fig. 1.6.

Newtonian atoms constitute the main, deformable region of the workpiece that interacts actively with the cutting tool or an energy source. This region forms a free boundary or is surrounded by layers of thermostat and fixed boundary atoms (Tong et al., 2014). The motion of the atoms within this region is computed from the direct numerical integration of Newton's second law, according to the interatomic forces calculated from the potential function using a suitable and efficient algorithm to determine the list of atoms that interact with each other (e.g., neighbor-list) (Cai et al., 2007; Cheng et al., 2003; Goel et al., 2013; Oluwajobi & Chen, 2011b; Rentsch, 1996).

Another important type of boundary conditions are periodic boundary conditions. These involve taking a boundary particle—that has left the side where it belonged—and "returning" it to the opposite side of workpiece, in an effort to approximate a very large system, as it can be seen in Fig. 1.7. Relatively large systems in MD simulations serve to reduce boundary effects on simulation results (Cai et al., 2007; Rentsch & Brinksmeier, 2005). For these reasons, some researchers avoid using fixed boundary conditions, and periodic boundary conditions are often applied to the sides of workpiece along a specific direction (Goel et al., 2012; Hosseini, Vahdati, & Shokuhfar, 2011; Tanaka, Shimada, & Anthony, 2007; Yan et al., 2006; Ye, Biswas, Morris, Bastawros, & Chandra, 2003), or rarely, along two directions.

Figure 1.7 2D schematic of periodic boundary conditions.

Although cutting speed can be considered a simple input parameter in most macroscopic machining simulations, it requires specific attention in nanoscale cutting simulations. As clearly seen in almost every work reported in the literature, most researchers consider the cutting speed employed, usually 20–200 m/s, to be significantly higher than the ones employed in real experiments; nanoscratching experiments are often used, usually in the range of 1–10 m/s (Gao et al., 2009; Ye et al., 2003). The latter are considered too slow to resemble realistic cutting conditions but can be employed for verification or calibration, with precaution and consideration of the experimental approach particularities. The primary reason for choosing a high cutting speed in modeling is that nanocutting simulations would either need extremely long time periods to complete or a considerable investment in computational infrastructure. Some investigations show clearly that there is an effect of speed on the results (Rentsch & Brinksmeier, 2005; Zhu, Qiu, Fang, Yuan, & Shen, 2014). However, in cases of high-speed cutting, in the range of 200–500 m/s, these large tool velocity values are acceptable and can be employed to simulate the processes in a realistic way.

1.3.4 Numerical integration and equilibration

The initial steps of modeling involves the determination of the workpiece and cutting tool geometry. Furthermore, process parameters, materials, and potential functions are chosen. Then starts the preliminary part of the calculation. The initialization of atom positions is first conducted during the geometry set-up, where atoms, assigned with an initial random velocity, are chosen relative to the temperature and other characteristics of the case and follow a Maxwell-Boltzmann distribution (Cheng et al., 2003; Goel et al., 2011; Maekawa & Itoh, 1995). After each atom's initial velocities are assigned in the system and just before the simulation begins, it is also essential to relax the system for a specific period of time in order to adjust atom positions to close to equilibrium for a preset temperature. At that initial equilibration stage, the total energy is not conserved and potential energy can be transformed into kinetic energy and vice-versa by altering velocities and positions of atoms that move around their initial lattice positions until temperature fluctuations have ceased (Guo et al., 2010; Goel et al., 2011). It is strongly advised that atom trajectories at that stage should not be included in any kind of calculations.

When the equilibration process has been completed, the total energy is preserved by means of the thermostat layer (Cai et al., 2007; Goel et al., 2011; Luo, Tong, & Liang, 2014; Tong et al., 2014), where a suitable velocity reset method is applied after a certain number of steps (Cai et al., 2007; Narulkar et al., 2009; Oluwajobi & Chen, 2011b; Yan et al., 2006), or when the temperature exceeds the predefined bulk temperature by five to ten K (Guo et al., 2010; Hosseini et al., 2011; Ye et al., 2003).

The numerical integration scheme is an important aspect of an MD simulation. Numerical solution of Newtonian equations requires a time-dependent integration method. For this reason, a variety of methods can be used, as in nonatomistic simulations. However, it is crucial to consider that not all integration schemes are effective in atomistic simulations, as a method that would require calculations of interatomic forces multiple times per timestep—like the Runge-Kutta family of methods, often suitable for many time-dependent simulations—contributes to a significant increase of computational time. Consequently, the most frequently employed integration methods are Verlet-type schemes and in some cases, predictor–corrector schemes.

Verlet method is the most widely used integration algorithm, often directly associated with MD simulations and specifically with the integration of Newtonian equations of motion. This algorithm is shown to exhibit good numerical stability, while having no significant cost than simpler methods such as Euler method. Verlet method is very popular in atomistic simulations, as it requires only a single calculation of forces per time-step, which is something already mentioned as very significant for this type of simulation. One formulation of Verlet algorithm (Störmer-Verlet formulation) is:

$$\vec{x}_1 = \vec{x}_0 + \vec{v}_0 \Delta t + \frac{1}{2}\vec{a}_0 \Delta t^2 \tag{1.21}$$

$$\vec{x}_{n+1} = 2\vec{x}_n - \vec{x}_{n-1} + \vec{a}_n \Delta t^2, \; n = 1, 2..N \tag{1.22}$$

where $\vec{x}_i$ represents the position vectors of system atoms at the time-step indicated by subscript i (ranging from 0 to N), $\vec{v}_i$ represents the velocity vectors of system atoms, $\vec{a}_i$ represents the acceleration vector at time-step n (equal to interatomic force vector $\vec{F}_i$ if all quantities are divided by atomic mass), Δt is the integration time-step, and N is the number of total time-steps in the simulation.

Furthermore, the Leapfrog scheme is a similar second order integrating scheme involving the calculations per time-step:

$$\vec{x}_n = \vec{x}_{n-1} + \vec{v}_{i-1/2} \Delta t \tag{1.23}$$

$$\vec{a}_n = F(\vec{x}_n) \tag{1.24}$$

$$\vec{v}_{n+1/2} = \vec{v}_{n-1/2} + \vec{a}_n \Delta t \tag{1.25}$$

Predictor–corrector integration schemes are numerical integration schemes with a more complex formulation and consist of two major steps: a prediction step where a simpler integration method is used to provide a first approximation of the result and then a more robust method is applied to "correct" this result. This is usually accomplished by using a suitable combination of an explicit and an implicit method, for prediction and correction purposes, respectively, so as the process converges effectively.

Finally, it should be noted that various limitations and particularities render MD simulation computationally demanding, even for relatively moderate atomistic systems. It is worth noting that multiscale models, involving FEM and MD combination analysis can be found in the literature (Pen, Bai, Liang, & Chen, 2009). This approach exploits the capabilities of continuum and micromechanics models and simultaneously reduces the computation cost of a full atomistic simulation. More effective algorithms and state-of-the-art computational systems capabilities need to be taken seriously when novel and demanding studies with greater and more realistic models and fewer assumptions are employed.

Conclusions

This book chapter presents an overview of the micro- and nanoscale cutting modeling methods and techniques. The main body of the book chapter is divided into two parts. The first section describes the modeling of microscale cutting when employing the FEM. The second pertains to the modeling of nanoscale cutting, while using MD.

Finite elements is a powerful modeling and simulation tool, already used for the simulation of cutting for many years. The microcutting modeling presents similarities with cutting. However, downsizing also involves the consideration of phenomena such as minimum chip thickness and size effect; these phenomena are described and analyzed in detail.

The steps and requirements for the construction of a FEM model for microcutting are provided. The model building procedure involves the selection of the model formulation and integration strategy, the insertion of geometrical characteristics of the tool and workpiece in the model, and the application of boundary conditions to the model. Furthermore, two important aspects for a successful simulation are discussed, namely friction and material modeling.

In nanoscale, continuous mechanics do not represent the phenomena taking place at this scale adequately; MD is proposed as an alternative. For this method, the model geometry, including the tool and workpiece, and the size of the model are discussed. Furthermore, of special interest in nanoscale modeling is the material microstructure representation and the potential function used in MD. Additionally, the boundary conditions, input parameters (e.g., cutting speed), and equilibration for a realistic model of nanoscale cutting are presented. Finally, the topic of numerical integration is discussed.

Both techniques presented can provide important results and insights into the related physics that would be difficult or impossible to obtain experimentally; such results pertain to cutting forces, stresses, stains and strain rates, and temperatures. The models are able to predict chip formation accurately. Although the computational costs for modelling are significant, especially for 3D simulations, the advances in computer technology make it possible to have a large number of research groups involved in the presented modeling techniques today. The increasing number of publications is indicative of the usefulness and interest in modeling for the simulation of micro- and nano-scale cutting.

References

Albrecht, P. (1960). New developments in the theory of metal cutting processes: Part I-the ploughing process in metal cutting. *ASME Journal of Engineering for Industry*, *82*, 348–357.

Albrecht, P. (1961). New developments in the theory of metal cutting processes: Part II-the theory of chip formation. *ASME Journal of Engineering for Industry*, *83*, 557–568.

Alting, L., Kimura, F., Hansen, H. N., & Bissacco, G. (2003). Micro engineering. *Annals of the CIRP*, *52*(2), 635–657.

Atkins, A. G. (2003). Modeling metal cutting using modern ductile fracture mechanics: Quantitative explanations for some longstanding problems. *International Journal of Mechanical Sciences*, *45*, 373–396.

Backer, W. R., Marshall, E. R., & Shaw, M. C. (1952). The size effect in metal cutting. *Transactions of ASME*, 74-61.

Brinksmeier, E., Aurich, J. C., Govekar, E., Heinzel, C., Hoffmeister, H.-W., Klocke, F., ... Wittmann, M. (2006). Advances in modeling and simulation of grinding processes. *Annals of the CIRP*, *55*(2), 667–696.

Byrne, G., Dornfeld, D., & Denkena, B. (2002). Advancing cutting technology. *Annals of the CIRP*, *52*(2), 483–507.

Cai, M. B., Li, X. P., & Rahman, M. (2007). Characteristics of "dynamic hard particles" in nanoscale ductile mode cutting of monocrystalline silicon with diamond tools in relation to tool groove wear. *Wear*, *263*, 1459–1466.

Chen, C. S., Cian, H. J., Yu, C. H., & Huang, C. W. (2014). Friction coefficient calculation and mechanism analysis for MoS2 nanoparticle from molecular dynamics simulation. *Procedia Engineering*, *79*, 617–621.

Cheng, K., Luo, X., Ward, R., & Holt, R. (2003). Modeling and simulation of the tool wear in nanometric cutting. *Wear*, *255*, 1427–1432.

Childs, T. H. C., & Maekawa, K. (1990). Computer-aided simulation and experimental studies of chip flow and tool wear in the turning of low alloy steels by cemented carbide tools. *Wear*, *139*, 235–250.

Childs, T. H. C., Otieno, A. M. W., & Maekawa, K. (1994). The influence of material flow properties on the machining of steels. In *Proceedings of the third international conference on the behaviour of materials in machining* (pp. 104–19).

Chuzhoy, L., DeVor, R. E., & Kapoor, S. G. (2003). Machining simulation of ductile iron and its constituents, part 2: Numerical simulation and experimental validation of machining. *Transactions of the ASME–Journal of Manufacturing Science and Engineering*, *125*, 192–201.

Corbett, J., McKeown, P. A., Peggs, G. N., & Whatmore, R. (2000). Nanotechnology: International developments and emerging products. *Annals of the CIRP*, *49*(2), 523–545.

Dinesh, D., Swaminathan, S., Chandrasekar, S., & Farris, T.N. (2001). An intrinsic size-effect in machining due to the strain gradient. In *Proceedings of the 2001 ASME IMECE* (pp. 197–204). Nov. 11–16, NY.

Dixit, P. M., & Dixit, U. S. (2008). *Modeling of metal forming and machining processes by finite element and soft computing methods*. Springer-Verlag.

Dixit, U. S., Joshi, S. N., & Davim, J. P. (2011). Incorporating of material behavior modeling of metal forming and machining processes: A review. *Materials and Design*, *32*, 3655–3670.

Dornfeld, D., Min, S., & Takeuchi, Y. (2006). Recent advances in mechanical micromachining. *Annals of the CIRP*, *55*(2), 745–768.

Eder, S. J., Bianchi, D., Cihak-Bayr, U., Vernes, A., & Betz, G. (2014). An analysis method for atomistic abrasion simulations featuring rough surfaces and multiple abrasive particles. *Computer Physics Communications*, *185*, 2456–2466.

Finnie, I. (1963). A comparison of stress strain behavior in cutting with that in other materials tests. *International Research in Production Engineering ASME*, 76–82.

Gao, Y., Lu, C., Huynh, N. N., Michal, G., Zhu, H. T., & Tieu, A. K. (2009). Molecular dynamics simulation of effect of indenter shape on nanoscratch of Ni. *Wear*, *267*, 1998–2002.

Goel, S., Luo, X., & Reuben, R. L. (2011). Molecular dynamics simulation model for the quantitative assessment of tool wear during single point diamond turning of cubic silicon carbide. *Computational Materials Science*, *51*, 402–408.

Goel, S., Luo, X., & Reuben, R. L. (2013). Wear mechanism of diamond tools against single crystal silicon in single point diamond turning process. *Tribology International*, *57*, 272–281.

Goel, S., Luo, X., Reuben, R. L., & Pen, H. (2012). Influence of temperature and crystal orientation on tool wear during single point diamond turning of silicon. *Wear*, *284–285*, 65–72.

Hosseini, S. V., Vahdati, M., & Shokuhfar, A. (2011). Effect of tool nose radius on nano-machining process by molecular dynamics simulation. *Defect and Diffusion Forum*, *312–315*, 977–982.

Huang, J. C. (2013). The study on the nanocutting by rigid body tool and elastic body tool using molecular dynamics simulations. In *Proceedings of the world congress on engineering 2013 Vol II*, WCE 2013, UK.

Ikawa, N., Shimada, S., Tanaka, H., & Ohmori, G. (1991). Atomistic analysis of nanometric chip removal as affected by tool-work interaction in diamond turning. *Annals of the CIRP*, *40*(1), 551–554.

Iwata, K., Osakada, K., & Terasaka, Y. (1984). Process modeling of orthogonal cutting by the rigid plastic finite element method. *Journal of Engineering Materials and Technology*, *106*(2), 132–138.

Jaspers, S. P. F. C., & Dautzenberg, J. H. (2002). Material behaviour in conditions similar to metal cutting: Flow stress in the primary shear zone. *Journal of Materials Processing Technology*, *122*, 322–330.

Johnson, G.R., & Cook, W.H. (1983). A constitutive model and data for metals subjected to large strains, high strain rates and high temperatures. In *Proceedings of the 7th international symposium on ballistics* (pp. 541–547).

Joshi, S. S., & Melkote, S. N. (2004). An explanation for the size-effect in machining based on strain gradient plasticity. *Journal of Manufacturing Science and Engineering, Transactions of the ASME, 126*, 679–684.

Junge, T., & Molinari, J. F. (2014). Plastic activity in nanoscratch molecular dynamics simulations of pure aluminium. *International Journal of Plasticity, 53*, 90–106.

Kim, C. J., Bono, M., & Ni, J. (2002). Experimental analysis of chip formation in micro-milling. *Transactions NAMRI/SME, 30*, 1–8.

Kim, J. D., & Kim, D. S. (1995). Theoretical analysis of micro-cutting characteristics in ultra-precision machining. *Journal of Materials Processing Technology, 49*, 387–398.

Klocke, F., Beck, T., Hoppe, S., Krieg, T., Müller, N., Nöthe, T., ... Sweeney, K. (2002). Examples of FEM application in manufacturing technology. *Journal of Materials Processing Technology, 120*, 450–457.

Komanduri, R., Chandrasekaran, N., & Raff, L. M. (1999). Some aspects of machining with negative-rake tools simulating grinding: A molecular dynamics simulation approach. *Philosophical Magazine Part B, 79*(7), 955–968.

Komanduri, R., & Raff, L. M. (2001). A review on the molecular dynamics simulation of machining at the atomic scale. *Proceedings of the Institution of Mechanical Engineers, Part B: Journal of Engineering Manufacture, 215*, 1639–1672.

Kopalinsky, E. M., & Oxley, P. L. B. (1984). Size effects in metal removal processes. *Institute of Physics Conference Series, 70*, 389–396.

Larsen-Basse, J., & Oxley, P. L. B. (1973). *Effect of strain rate sensitivity on scale phenomena in chip formation. Proceedings of the 13th international machine tool design and research conference* (pp. 209–216). University of Birmingham.

Lee, W. S., & Lin, C. F. (1998). High-temperature deformation behavior of Ti6Al4V alloy evaluated by high strain-rate compression tests. *Journal of Materials Processing Technology, 75*, 127–136.

Li, J., Fang, Q., Liu, Y., & Zhang, L. (2014). A molecular dynamics investigation into the mechanicms of subsurface damage and material removal of monocrystalline copper subjected to nanoscale high speed grinding. *Applied Surface Science, 308*, 331–343.

Li, J., Fang, Q., Zhang, L., & Liu, Y. (2015). The effect of rough surface on nanoscale high speed grinding by a molecular dynamics simulation. *Computational Materials Science, 98*, 252–262.

Liang, R., & Khan, A. S. (1999). A critical review of experimental results and constitutive models for BCC and FCC metals over a wide range of strain rates and temperatures. *International Journal of Plasticity, 15*, 963–980.

Liang, Y., Wang, Z., Chen, M., Chen, J., & Tong, Z. (2011). Potential analysis in nanoturning of single crystal silicon using molecular dynamics. *Advanced Materials Research, 239–242*, 3236–3239.

Lin, Z. C., Lin, M. H., & Hsu, Y. C. (2014). Simulation of temperature field during nanoscale orthogonal cutting of single-crystal silicon by molecular statics method. *Computational Materials Science, 81*, 58–67.

Liu, K., & Melkote, S. N. (2007). Finite element analysis of the influence of tool edge radius on size effect in orthogonal micro-cutting process. *International Journal of Mechanical Sciences, 49*, 650–660.

Liu, X., DeVor, R. E., Kapoor, S. G., & Ehmann, K. F. (2004). The mechanics of machining at the microscale: Assessment of the current state of the science. *Journal of Manufacturing Science and Engineering-Transactions of the ASME, 126*, 666–678.

Luo, X., Tong, Z., & Liang, Y. (2014). Investigation of the shape transferability of nanoscale multi-tip diamond tools in the diamond turning of nanostructures. *Applied Surface Science, 321*, 495–502.

Madou, M. J. (2002). *Fundamentals of microfabrication: The science of miniaturization*. CRC Press.

Malekian, M., Mostofa, M. G., Park, S. S., & Jun, M. B. G. (2012). Modeling of minimum uncut chip thickness in micro machining of aluminum. *Journal of Materials Processing Technology*, *212*, 553–559.

Mamalis, A. G., Manolakos, D. E., Ioannidis, M. B., Markopoulos, A., & Vottea, I. N. (2003). Simulation of advanced manufacturing of solids and porous materials. *The International Journal for Manufacturing Science & Production*, *5*(3), 111–130.

Mamalis, A. G., Markopoulos, A., & Manolakos, D. E. (2005). Micro and nanoprocessing techniques and applications. *Nanotechnology Perceptions*, *1*, 31–52.

Markopoulos, A. P. (2013). *Finite element method in machining processes*. Springer.

Markopoulos, A. P., Karkalos, N. E., Kalteremidou, K. A. L., Balafoutis, A., & Manolakos, D. E. (2015a). The influence of cutting conditions and cutting tool geometry on the atomistic modeling of precision cutting. *Engineering Solid Mechanics*, *3*(3), 195–206.

Markopoulos, A. P., Savvopoulos, I. K., Karkalos, N. E., & Manolakos, D. E. (2015b). Molecular dynamics modeling of a single diamond abrasive grain in grinding. *Frontiers of Mechanical Engineering*, *10*(2), 168–175.

Masuko, M. (1956). Fundamental research on the metal cutting: Second report. *Bulletin of Japan Society of Mechanical Engineers*, *22*, 371–377.

Masuzawa, T. (2000). State of the art of micromachining. *Annals of the CIRP*, *49*(2), 473–488.

Meyer, H. W., Jr., & Kleponis, D. S. (2001). Modeling the high strain rate behavior of titanium undergoing ballistic impact and penetration. *International Journal of Impact Engineering*, *26*, 509–521.

Moriwaki, T., Sugimura, N., & Luan, S. (1993). Combined stress, material flow and heat analysis of orthogonal micromachining of copper. *Annals of the CIRP*, *42*(1), 75–78.

Narulkar, R., Bukkapatnam, S., Raff, L. M., & Komanduri, R. (2009). Graphitization as a precursor to wear of diamond in machining pure iron: A molecular dynamics investigation. *Computational Materials Science*, *45*(2), 358–366.

Oluwajobi, A., & Chen, X. (2011a). The effect of interatomic potentials on the molecular dynamics simulation of nanometric machining. *International Journal of Automation and Computing*, *8*(3), 326–332.

Oluwajobi, A. O., & Chen, X. (2011b). The effect of depth of cut on the molecular dynamics (MD) simulation of multi-pass nanometric machining. In *Proceedings of the 17th international conference on automation & computing*, UK.

Özel, T., & Karpat, Y. (2007). Identification of constitutive material model parameters for high-strain rate metal cutting conditions using evolutionary computational algorithms. *Materials and Manufacturing Processes*, *22*, 659–667.

Pei, Q. X., Lu, C., Fang, F. Z., & Wu, H. (2006). Nanometric cutting of copper: A molecular dynamics study. *Computational Materials Science*, *37*, 434–441.

Pei, Q. X., Lu, C., & Lee, H. P. (2007). Large scale molecular dynamics study of nanometric machining of copper. *Computational Materials Science*, *41*(2), 177–185.

Pei, Q. X., Lu, C., Lee, H. P., & Zhang, Y. W. (2009). Study of materials deformation in nanometric cutting by large-scale molecular dynamics simulations. *Nanoscale Research Letters*, *4*(5), 444–451.

Pen, H., Bai, Q., Liang, Y., & Chen, M. (2009). Multiscale simulation of nanometric cutting of single crystal copper-effect of different cutting speeds. *Acta Metallurgica Sinica (English Letters)*, *22*(6), 440–446.

Rashid, W. B., Goel, S., Luo, X., & Ritchie, J. M. (2013). The development of a surface defect machining method for hard turning processes. *Wear*, *302*, 1124–1135.

Ren, J., Zhao, J., Dong, Z., & Liu, P. (2015). Molecular dynamics study on the mechanism of AFM-based nanoscratching process with water-layer lubrication. *Applied Surface Science*, *346*, 84–98.

Rentsch, R. (1996). Process modeling by means of molecular dynamics (MD). *VDI-Berichte*, *1276*, 175–195.

Rentsch, R. (2001). On the on-set of chip formation and the process stability in cutting at atomic level. In I. Inasaki (Ed.), *Initiatives of precision engineering at the beginning of a millennium* (pp. 274–278). Yokohama, Japan: Kluwer Academic Publ.

Rentsch, R. (2009). Nanoscale cutting. In J. P. Davim (Ed.), *Nano and micromachining* (pp. 1–26). London, UK: ISTE Ltd.

Rentsch, R., & Brinksmeier, E. (2005). Tribology aspects in state-of-the-art MD cutting simulations. In *Proceedings of the 8th CIRP international workshop on modeling of machining operations* (pp. 401–408). Chemnitz, Germany.

Rentsch, R., & Inasaki, I. (1994). Molecular dynamics simulation for abrasive processes. *Annals of the CIRP*, *43*(1), 327–330.

Rentsch, R., & Inasaki, I. (1995). 'Investigation of surface integrity by molecular dynamics simulation. *Annals of the CIRP*, *44*(1), 295–298.

Rentsch, R., & Inasaki, I. (2006). Effects of fluids on the surface generation in material removal processes-molecular dynamics simulation-. *Annals of the CIRP*, *55*(1), 601–604.

Sekhon, G. S., & Chenot, J.-L. (1993). Numerical simulation of continuous chip formation during non-steady orthogonal cutting. *Engineering Computations*, *10*(1), 31–48.

Shaw, M. C. (2003). The size effect in metal cutting. *Sadhana-Academy Proceedings in Engineering Sciences*, *28*(5), 875–896.

Shaw, M. C., & Jackson, M. J. (2006). The size effect in micromachining. In M. J. Jackson (Ed.), *Microfabrication and nanomanufacturing*. CRC Press.

Shi, J., & Liu, C. R. (2004). The influence of material models on finite element simulation of machining. *Transactions of the ASME – Journal of Manufacturing Science and Engineering*, *126*, 849–857.

Shimizu, J., Eda, H., Zhou, L., & Okabe, H. (2008). Molecular dynamics simulation of adhesion effect on material removal and tool wear in diamond grinding of silicon wafer. *Tribology Online*, *3*(5), 248–253.

Simoneau, A., Ng, E., & Elbestawi, M. A. (2006a). Chip formation during microscale cutting of a medium carbon steel. *International Journal of Machine Tools & Manufacture*, *46*, 467–481.

Simoneau, A., Ng, E., & Elbestawi, M. A. (2006b). Surface defects during microcutting. *International Journal of Machine Tools & Manufacture*, *46*, 1378–1387.

Simoneau, A., Ng, E., & Elbestawi, M. A. (2007a). Modeling the effects of microstructure in metal cutting. *International Journal of Machine Tools & Manufacture*, *47*, 368–375.

Simoneau, A., Ng, E., & Elbestawi, M. A. (2007b). Grain size and orientation effects when microcutting AISI 1045 steel. *Annals of the CIRP*, *56*(1), 57–60.

Solar, M., Meyer, H., Gauthier, C., Benzerara, O., Schirrer, R., & Baschnagel, J. (2011). Molecular dynamics simulations of the scratch test on linear amorphous polymer surfaces: A study of the local friction coefficient. *Wear*, *271*, 2751–2758.

Solhjoo, S., & Vakis, A. I. (2015). Single asperity nanocontacts: Comparison between molecular dynamics simulations and continuum mechanics models. *Computational Materials Science*, *99*, 209–220.

Stowers, I. F., Belak, J. F., Lucca, D. A., Komanduri, R., Moriwaki, T., Okuda, K., ... Drescher, J. D. (1991). Molecular- dynamics simulation of the chip forming process in single crystal copper and comparison with experimental data. In *Proceedings ASPE Annual Meeting* (pp. 100–104).

Tanaka, H., Shimada, S., & Anthony, L. (2007). Requirements for ductile-mode machining based on deformation analysis of mono-crystalline silicon by molecular dynamics simulation. *Annals of the CIRP*, *56*(1), 53–56.

Tong, Z., Liang, Y., Jiang, X., & Luo, X. (2014). An atomistic investigation on the mechanism of machining nanostructures when using single tip and multi-tip diamond tools. *Applied Surface Science*, *290*, 458–465.

Umbrello, D., M'Saoubi, R., & Outeiro, J. C. (2007). The influence of Johnson-Cook material constants on finite element simulation of machining of AISI 316L steel. *International Journal of Machine Tools and Manufacture*, *47*, 462–470.

Usui, E., Maekawa, K., & Shirakashi, T. (1981). Simulation analysis of built-up edge formation in machining low carbon steels. *Bulletin of the Japan Society of Precision Engineering*, *15*, 237–242.

Weber, M., Hochrainer, T., Gumbsch, P., Autenrieth, H., Delonnoy, L., Schulze, V., ... Fleischer, J. (2007). Investigation of size-effects in machining with geometrically defined cutting edges. *Machining Science and Technology*, *11*, 447–473.

Weule, H., Huntrup, V., & Tritschle, H. (2001). Micro-cutting of steel to meet new requirements in miniaturization. *Annals of the CIRP*, *50*(1), 61–64.

Woon, K. S., Rahman, M., Neo, K. S., & Liu, K. (2008). The effect of tool edge radius on the contact phenomenon of tool-based micromachining. *International Journal of Machine Tools and Manufacture*, *48*, 1395–1407.

Xiao, G., To, S., & Zhang, G. (2015). Molecular dynamics modelling of brittle–ductile cutting mode transition: Case study on silicon carbide. *International Journal of Machine Tools and Manufacture*, *88*, 214–222.

Yan, Y. D., Sun, T., Dong, S., Luo, X. C., & Liang, Y. C. (2006). Molecular dynamics simulation of processing using AFM pin tool. *Applied Surface Science*, *252*, 7523–7531.

Yang, X., & Liu, C. R. (2002). A new stress-based model of friction behaviour in machining and its significant impact on the residual stresses computed by finite element method. *International Journal of Mechanical Sciences*, *44*, 703–723.

Ye, Y. Y., Biswas, R., Morris, J. R., Bastawros, A., & Chandra, A. (2003). Molecular dynamics simulation of nanoscale machining of copper. *Nanotechnology*, *14*, 390–396.

Zerilli, F. J., & Armstrong, R. W. (1987). Dislocation-mechanics-based constitutive relations for material dynamics calculations. *Journal of Applied Physics*, *61*, 1816–1825.

Zhang, L., & Tanaka, H. (1997). Towards a deeper understanding of wear and friction on the atomic scale- a molecular dynamics analysis. *Wear*, *211*, 44–53.

Zhu, P. Z., Qiu, C., Fang, F. Z., Yuan, D. D., & Shen, X. C. (2014). Molecular dynamics simulations of nanometric cutting mechanisms of amorphous alloy. *Applied Surface Science*, *317*, 432–442.

Zorev, N. N. (1963). Interrelationship between shear processes occurring along tool face and on shear plane in metal cutting. In *Proceedings of the international research in production engineering conference, ASME* (pp. 42–49).

Machining scale: workpiece grain size and surface integrity in micro end milling

2

A.R. Rodrigues and R.G. Jasinevicius
University of São Paulo, São Carlos, Brazil

2.1 Introduction

Manufactured microcomponents were first presented in the 1960s by the watch industry. Considered as a strategic field in the 1970s, precision mechanics required miniaturization of equipment and devices. The 1980s was marked by the world growing of research on microtechnology, known as *microelectromechanical systems* (MEMS) in United States and as *microsystems technology* in Europe. In the 1990s, the technologies based upon chemical machining of silicon were broadly applied in the electronic industry (Arai, 2008; Ferreira & Fernandes, 2003). From the year 2000 to the present day, it is noticeable that research and applications are primarily focused on mechanical micromachining, with experimental and numerical studies based upon chip formation mechanism, scale or size effect, elimination or burrs attenuation, and machining of complex tridimensional microparts.

Today, the demand for miniaturized parts and products is rapidly increasing in several industrial segments, such as aeroespacial, biomedics, optics, electronics, communications, and automotive (Chen, Ni, Wang, & Jiang, 2012; Klocke, Gerschwiler, & Abouridouane, 2009). Among everyday products, inkjet printer heads, electronic auto injection sensors, magnetic disks heads, portable devices for cholesterol blood dosage, fiber optic connectors, and filters, as well as optical lens used in cameras for mobile phones, are just few examples of components or products that use micromachining technology to manufacture them (Ferreira & Fernandes, 2003; Lekkala et al., 2011).

Large-scale production of microcomponents in the coming years has been identified as a challenge. Chae, Park, and Freihet (2006) noticed that most manufacturing processed applied to microcomponents are based on semiconductor processing techniques, where silicon based materials are photoengraved by chemical etching or dry processing. However, these methods are considered slow, limited to some silicon based materials, restricted to planar geometries, and have higher production costs.

Micromachining processes most employed in the production of microcomponents are electrical discharge (micromolds), laser (microholes), ultrasound (micropenetration in brittle materials), photolithography (planar microwalls) and electrochemical (smooth microsurfaces on metals). These machining techniques are known as nonconventional machining processes, even though presenting higher costs and some limitations, they represent the most employed microcomponent fabrication processes (Masuzawa, 2000).

Microfabrication and Precision Engineering. DOI: http://dx.doi.org/10.1016/B978-0-85709-485-8.00002-4

Mechanical micromachining (microcutting) may be considered a good alternative when creating miniaturized components with dimensions ranging from tens of micrometres to a few millimitres (Cardoso & Davim, 2010; Chae et al., 2006). According to Masuzawa (2000) and Weule, Hüntrup, and Tritschler (2001), several cutting operations may be employed in mechanical micromachining, such as drilling of microholes, milling of microchannels and molds (complex 3D geometries), and micropin turning.

The most attractive advantage of micromilling is the possibility to fabricate 3D microstructures made from different materials, characterized by high aspect ratio and geometrical complexity (Afazov, Ratchev, & Segal, 2012; Li et al., 2010). Furthermore, it can produce more complex geometry than that obtained by lithography, and it is potentially useful in combination with MEMS lithography based processing techniques. Micromilling is also important for the production of parts in mesoscale (parts in the order of 1 mm to 1 cm), which are too large for lithographic techniques but too small for many other traditional manufacturing techniques (Lee & Dornfeld, 2004). When machining micromolds, micromilling overcomes other machining processes such as electrical discharge and laser machining, since it presents a higher material removal rate, good compliance with dimensional tolerances, and better surface quality (Schmidt & Tritschler, 2004).

Accordingly, micromilling, often using cutting tools with diameters smaller than 1 mm, has become an important method for the manufacturing of complex microparts that require a variety of materials, shapes, and functional interfaces (Aramcharoen & Mativenga, 2009). Fig. 2.1 illustrates some microstructures milled at microscale for engineering applications.

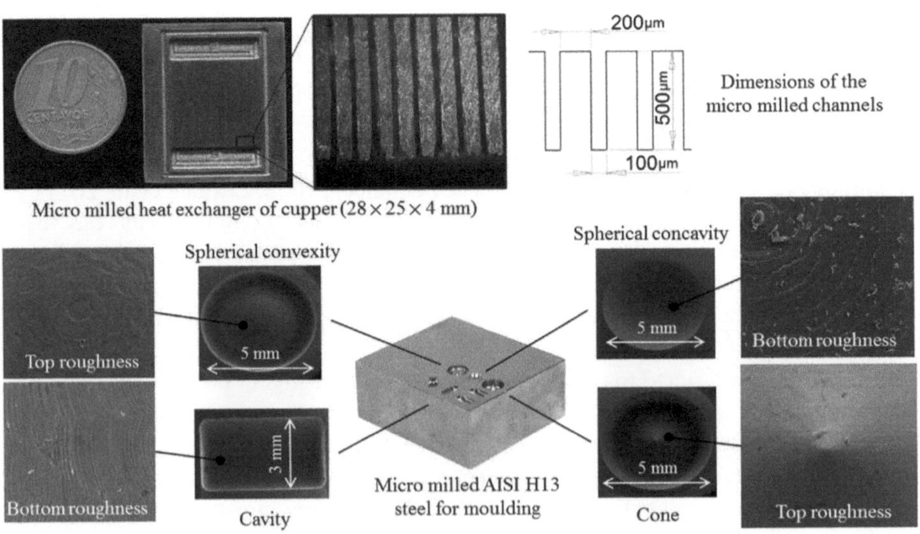

Figure 2.1 Micromilled structures for engineering applications (heat exchanger and mold for plastic injection).

Alting, Kimura, Hansen, and Bissaco (2003) defined micromachining as a process dealing with the development of products with functional characteristics, or that at least have a dimension in the micrometre scale. According to Masuzawa (2000), micro in micromachining indicates 'micrometre' and represents the range of 1 to 999 μm. Aramcharoen and Mativenga (2009) described machining in microscale, where the nominal cutting thickness ranges from namometres to a few micrometres. Dornfeld, Min, and Takeuchi (2006) and Câmara, Campos Rubio, Abrão, and Davim (2012) stated that mechanical micromachining uses cutting tools smaller than 1 mm, presenting a defined geometry cutting edge.

The definitions of micromachining reported in literature define the term as material removal processes in which the dimensions of the tool, component and cutting parameters are in the micrometre range (Simoneau, Ng, & Elbestawi, 2007).

However, the above proposed definition is somewhat generic or incomplete. Consequently, questions may arise such as: (1) How are micrometric ranges effectively set? (2) Which tool dimensions or products' functional characteristics, smaller than 1 mm, can be classified as micrometric?

Thus, the definiton of microcutting should not consider only workpiece and/or tool dimensions (even lesser than 1 mm), but the cutting parameter ranges which affect effectively the chip removal mechanism. This could be considered a more coherent way of defining microachining based on the interaction scale from material removal mechanism.

The cutting parameters effectively involved in microcutting such as tool feedrate f (or, equivalently, the nominal cutting thickness h) and the depth of cut a_p are in the range of about 5 to 20 μm and 1 to 100 μm, respectively. Consequently, machining in the micrometre scale considers the typical area bounded by the chip cross section (A_c) ranging from 100 to 999 μm. These values are much larger than the tool edge radius which is typically smaller than 5 μm (Fig. 2.2). Thus, this order of magnitude implies

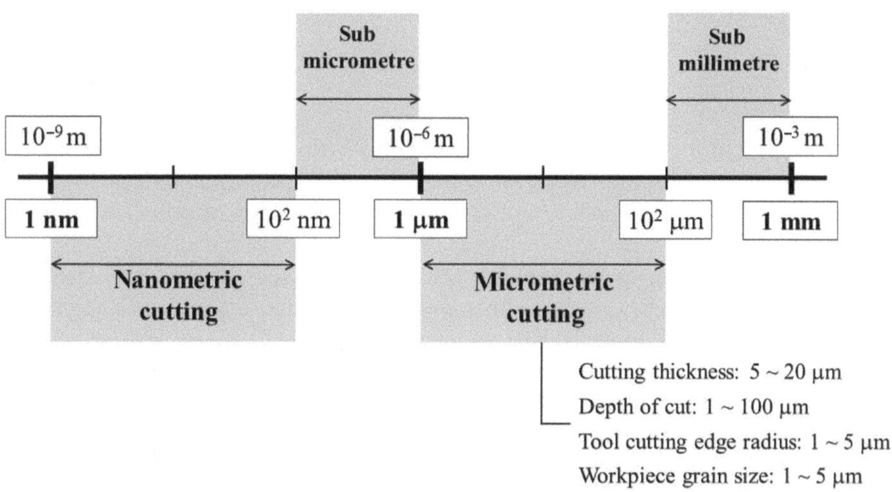

Figure 2.2 Microcutting scales.

that material removal process be governed by shear mechanism, reducing to a minimum the metal pressed into the machined surface due to ploughing mechanism and mitigating surface residual stresses. The concepts of minimum cutting thickness, size effect, and surface integrity will be discussed in the next sections.

Moreover, by assuming that the cutting parameters magnitude as the main factor to define micromachining, material microstructure should have a direct influence at the dimensions found in micromachining as well. In this case, the material is not considered continuous, homogeneous, and/or isotropic as in macromachining. These characteristics can only be achieved when the average grain size (GS) is reduced to match similar A_c/GS ratio found in conventional machining. During micromachining of conventional GS (some tens of square micrometres), the cutting tool portion interacting with material microstructure is sometimes smaller than the nominal GS. This means that the length of the tool cutting edge will cut within a crystal grain. In this case the tool–material interaction will be affected by factors such as grain boundaries, crystallographic orientation, and flaws and impurities, such as hard inclusions. Therefore, the microstructural scale of the workpiece material should be suitable to chip cross section dimensions. This will be better detailed in Section 2.4.

2.2 Specific cutting energy

During the machining process, the cutting power can be determined by directly multiplying the cutting force F_c by cutting speed v_c. However, as many cutting parameters directly influence the amount of energy consumed, power is typically normalized by dividing it by the rate at which material is removed from the part, or the product of the uncut chip thickness h and width of cut w and cutting speed v_c, as shown in Eq. (2.1).

$$u = \frac{F_c v_c}{h w v_c} = \frac{F_c}{A_c} \tag{2.1}$$

By keeping cutting speed v_c in the expression [W/(m^3/s)], u can be understood as the energy per unit volume (King, 1986). This term is the power required from the machine tool spindle in *Watt* to remove 1 m^3 of material per second (Kalpakjian & Schmid, 2000).

From the perspective of workpiece mechanical behavior, Shaw (1995) states that the specific cutting energy u is an intensive quantity that characterizes the cutting resistance offered by a material just as tensile stress and hardness characterize the strength and plastic deformation resistance of a material, respectively.

The specific cutting energy is a very useful concept not only for machining with tools that present defined and undefined geometry but also for forming processes (Shaw, 1995). In machining, cutting forces can be estimated that, in terms of machine tool design, determine the structural needs of machine, and in terms of fabrication process, influence the dimensional and geometric qualities of the finished product.

The specific cutting energy can also be used to estimate the power required in a machining operation of a given material (Ersoy & Atici, 2004). In this case, the engine efficiency should be considered in the calculations in order to take into account the inertia and frictional effects. High specific cutting energy values generate high heat transfer rates and high residual stresses in the part, which can lead to a poor surface integrity (Salmon, 1992). Low specific cutting energy levels generated during the process cause less damage to the part and indicate the quality and cutting tool efficiency (Ersoy & Atici, 2004). It is worth mentioning that, in the latter case, the chip cross section is larger than the cutting edge radius.

Thus, the goal should be to mitigate the specific cutting energy during machining. The specific cutting energy is also considered a good indicator of machinability index of materials and allows us to understand the entire cutting process (Polini & Turchetta, 2004).

The specific cutting pressure is defined as the ratio between the cutting force and the area of the chip cross section, which is analogous to the definition of specific cutting power, found in Eq. (2.1).

The terminology found in the literature does not provide standardization for the term specific cutting power. Power, energy, pressure, and specific cutting force are terms commonly found. In general, force or specific cutting pressure are used more in machining with defined geometry tools, while energy or specific cutting power are applied to processes with undefined geometry tools (typically abrasive). The most usual dimensional relationship between the various forms of specific cutting power, using the international system, are shown in Fig. 2.3.

Relying on the specific cutting energy concept, the plastic deformation and failure mechanisms inherent in machining processes can be better comprehended. Around ¾ of the total energy is associated with shear in the primary zone (in the direction of shear plane), and ¼ of the remainder stems from secondary shear zone (the interface between the chip and the rake surface of the cutting tool). According to Melo et al. (2003), the energy consumed in cutting is essentially converted into heat. Portions of this heat is transferred to the chip (90%), to workpiece (5%), and to the tool (5%), respectively.

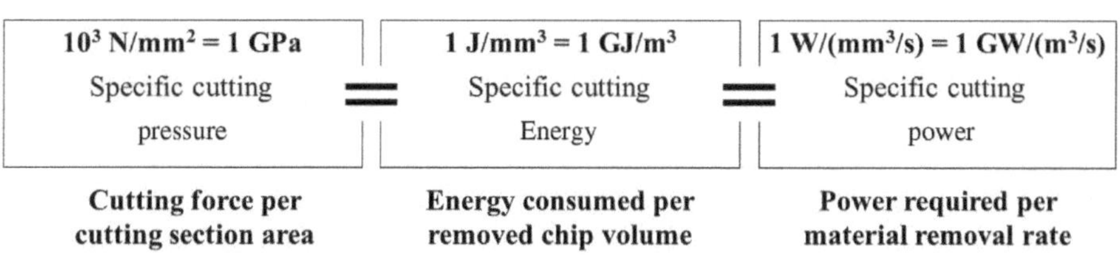

Figure 2.3 Usual units of the specific cutting pressure, energy, and power.

The specific cutting energy can be divided into four parts: shear energy per unit volume, u_s; friction energy per unit volume, u_f; kinetic energy per unit volume, u_m; and surface energy per unit volume, u_a (Black, 1989; Shaw, 1984).

The energy per unit volume resulting from the shear process (u_s) can be estimated by replacing the cutting force and the cutting speed in Eq. (2.1) by the shear force and shear rate in the primary shear zone.

The energy per unit volume, due to the effect of friction (u_f), is consumed by the friction of the chip on the cutting tool rake surface. The energy component may be determined by considering the chip flow rate on the tool rake surface and the frictional force at the chip-tool interface.

The kinetic energy per unit volume (u_m) required to accelerate the chip is often overlooked, but it becomes especially important when machining at high cutting speeds. This component is also responsible for changing the momentum that the chip suffers when passing through the shear plane. It can be obtained by the momentum force and the shear rate of the chip.

Finally, the energy per unit volume (u_a) demanded to produce two new surfaces not machined is given by the ratio between the surface energy of the material being machined and the uncut chip thickness, h. The specific energy (u_a) is required, because it is associated to the ionic bonds that support the formation of solid material, but this component is also generally neglected. In conclusion, for applications in macromachining with defined geometry tool in speeds below 900 m/min, the total energy per unit volume required for cutting material can be well approximated by the sum of the shear and frictional portions ($u_s + u_f$). Fig. 2.4 shows the shear zones inherent to any cutting process.

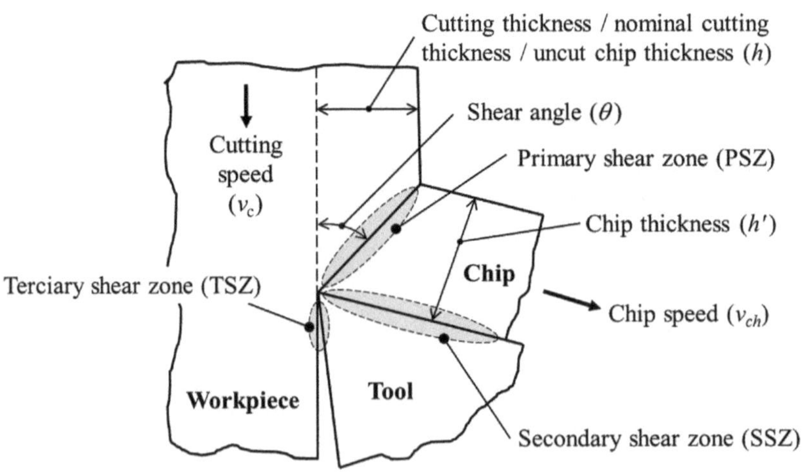

Figure 2.4 Shear zones in mechanical machining for nonabrasive processes.

According to Rodrigues (2005), several factors influence the behavior of specific cutting energy, including those arising from the workpiece material (hardness, shear stress, type of material, chemical composition, and microstructure), tool (substrate and coating, geometry, and wear), cutting conditions (machining process, operation, cutting parameters, and cutting fluid).

In general, the higher the material hardness, the greater its specific cutting energy, but similar hard materials may have different specific cutting energies due to differences in the microstructure and alloy chemical composition. Sulfur, boron, and lead tend to decrease specific cutting energy because they form low resistance coating during chip formation and act as a solid lubricant, reducing friction.

Materials' tensile strength is directly related to the specific cutting energy, since the chip formation involves shear (i.e., shear deformation in the primary shear zone and friction at the chip-tool interface (secondary shear zone)). However, when machining a material with high shear strength, the specific cutting energy will not necessarily be higher, since its plasticity decreases.

The use of coated carbide tools along with the application of cutting fluid decrease the specific cutting energy, since the friction component is reduced. The specific cutting energy may also be affected by tool geometry. The use of more positive rake and back rake angles decrease the cutting force since they reduce the chip-tool contact length by increasing the shear angle.

Increasing depth of cut, feedrate, and cutting speed tend to reduce the specific cutting energy due to a higher cutting ratio ($r = h/h'$) and greater thermal softening of the workpiece, respectively. Among the cutting parameters used in machining, the most influential in specific cutting energy is the uncut chip thickness. Finally, finishing operations tend to raise the specific cutting energy due to removal of the smaller chip cross section and the geometry of the tool edge (in general, more rounded, as in abrasive processes). Further details can be found in Degarmo, Black, and Kohser (1997), Gorczyca (1987), Shaw (1984), Daves (1989), Chou and Song (2004), Cohen (1989), Boothroyd and Knight (1989), Kalpakjian and Schmid (2000), and Groover (2002).

Several mathematical models, derived from empirical tests carried out to determine the specific cutting energy with defined geometry cutting tools, have been proposed. Therefore, it is possible to find in the literature a compendium of specific cutting energy values for various materials, machining processes, and cutting conditions. However, the literature is still relatively scarce of tabulated specific cutting energy data in the microscale mechanical machining process. Table 2.1 presents data measured for metallic alloys, including ultrafine grain materials.

All these models are based basically on the materials' specific constants studied and cutting parameters, such as feedrate and depth of cut. Among the most known models proposed for conventional machining, Taylor, ASME, AWF, Hucks, Kronenberg (1966), Shaw (1995), and Kienzle (1952) are the most commonly found. The Kienzle model in Eq. (2.2) is one of the most widely used today, where the constants k_{s1} and z are, respectively, the specific material constant for chip cross section of 1 mm^2, z is angular coefficient of the slope of the line, and both are obtained from di-logarithmic linearized plot of the specific

Table 2.1 Specific cutting energy data for some metallic alloys obtained by micro- and macro end milling (dry cutting)

Workpiece material	Hardness	Main chemical composition and GS	Main cutting parameters (microscale)	u (J/mm^3)		Main cutting parameters (macroscale)
				Microscale	Macroscale	
Aluminum RSA 6061 micrograin Rapid solidification aluminum Mirrors fabrication	117 HV	98.2% Al, 0.3% Si, 0.5% Mg GS ≈ 1.0 μm	$v_c = 60$ m/min $f_z = 3$ μm/tooth $a_p = 0.16$ mm $a_e = 1.0$ mm $r_e = 2.7$ μm ∅ tool = 1.0 mm	5.6	0.9	$v_c = 800$ m/min $f_z = 300$ μm/tooth $a_p = 0.5$ mm $a_e = 2.0$ mm $r_e = 25$ μm ∅ tool = 16 mm
COSAR 60 steel ultrafine grain size (UFGS) High strength Thermomechanically modified for grain refinement Microparts applications	322 HV	0.15%C, 0.28%Cr, 0.008%Ni, 0.048%Nb, 0.044%V, 0.16%Ti GS = 0.7 μm (ferrite)	$v_c = 60$ m/min $f_z = 3$ μm/tooth $a_p = 0.16$ mm $a_e = 1.0$ mm $r_e = 2.7$ μm ∅ tool = 1.0 mm	13.6	4.9	$v_c = 100$ m/min $f_z = 200$ μm/tooth $a_p = 3.0$ mm $a_e = 2.0$ mm $r_e \approx 30$ μm ∅ tool = 25 mm
COSAR 60 steel conventional grain size (CS) High strength As received material Structural applications	198 HV	0.15%C, 0.28%Cr, 0.008%Ni, 0.048%Nb, 0.044%V, 0.16%Ti GS = 11 μm (ferrite)	$v_c = 135$ m/min $f_z = 5$ μm/tooth $a_p = 0.032$ mm $a_e = 0.13$ mm $r_e = 5$ μm ∅ tool = 0.8 mm	19.9	3.8	$v_c = 100$ m/min $f_z = 200$ μm/tooth $a_p = 3.0$ mm $a_e = 2.0$ mm $r_e \approx 30$ μm ∅ tool = 25 mm
VP100 steel Uniform hardening by quenching Mold and die fabrication	329 HV	0.22%C, 0.25%Mo, 0.4%Si, 0.58%Cr, 0.4%Ni GS ≈ 100 μm (bainite)	$v_c = 60$ m/min $f_z = 3$ μm/tooth $a_p = 0.16$ mm $a_e = 1.0$ mm $r_e = 2.7$ μm ∅ tool = 1.0 mm	14.4	3.0	$v_c = 200$ m/min $f_z = 100$ μm/tooth $a_p = 0.5$ mm $a_e = 17.5$ mm $r_e \approx 30$ μm ∅ tool = 25 mm
AISI 1045 steel Mild carbon steel Structural applications	98 HV	0.46%C, 0.72%Mn, 0.24%Si GS = 12 μm (ferrite) and 43 μm (pearlite)	$v_c = 60$ m/min $f_z = 3$ μm/tooth $a_p = 0.16$ mm $a_e = 1.0$ mm $r_e = 2.7$ μm ∅ tool = 1.0 mm	14.9	3.3	$v_c = 60$ m/min $f_z = 10$ μm/tooth $a_p = 0.16$ mm $a_e = 2.0$ mm $r_e = 2.7$ μm ∅ tool = 2.0 mm

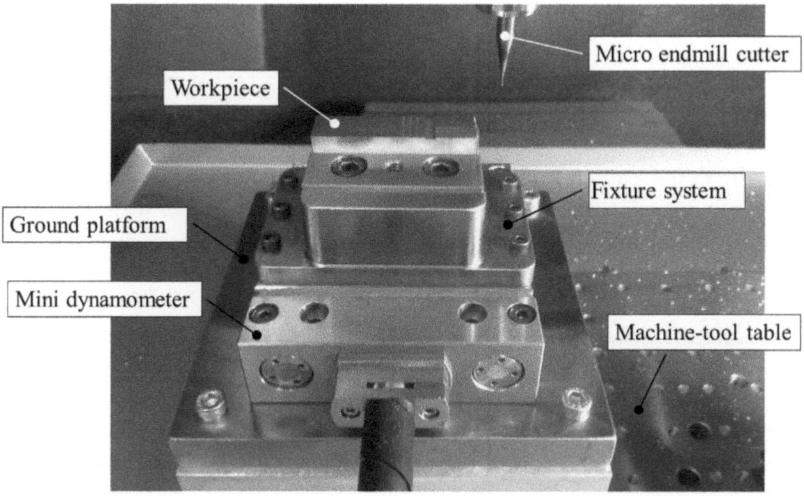

Figure 2.5 Typical 3-component piezoelectric dynamometer applied to measure machining forces at microscale.

cutting energy (u) versus uncut chip thickness (h). Corrections of the constants can be implemented to adapt the models to mechanical micromachining (Dib, 2013; Oliveira et al., 2015).

$$k_s = \frac{k_{s1}}{h^z} \tag{2.2}$$

For mechanical micromachining processes, it is appropriate to employ specific piezoelectric dynamometers for small cutting force amplitudes, typically ranging between negative and positive 300 N, and higher sensitivity. A typical mini dynamometer commonly used for micromilling is shown in Fig. 2.5.

When the machining process generates a constant chip cross section, such as microturning and microdrilling, the cutting force signal is constant; and it is possible to determine the specific cutting pressure simply by the ratio of the cutting force and the area of the chip cross section ($A_c = w \cdot h = a_p \cdot f$). However, in micromilling, the chip section is variable, so the cutting force signal presents a cyclical response. In this case, a good approach for the force or specific cutting pressure is to divide the average cutting force (RMS signal) by the average chip thickness (h_m).

The magnitudes of cutting force during micromachining operation are in the range of some tenths of a Newton to levels lower than 100 N, because it depends upon thermo mechanical properties of workpiece material and cutting parameters. For example, the specific cutting energy value is similar to that achieved by abrasive grains, generally larger than 10 J/mm^3 up to about 100 J/mm^3, as will be seen in Section 2.3.

2.3 Size effect

The concept of size effect in machining operations is directly influenced by trade-off between material properties and cutting tool microgeometry. According to Saxena (2000), the size effect may be defined by the effect that nominal dimensions of the workpiece material cause to mechanical properties and manufacturing variables. In general, the mechanical properties affect material response at a different dimension scale. For example, in the field of mechanical testing, particularly the Charpy impact test, the average impact strength decreases with the specimen size, however the energy per unit volume increases. (Saxena, 2000).

In machining processes, the size effect theory has been formulated for the paradoxical result and inversely proportional between the uncut chip thickness (h) and the specific cutting energy (u). In other words, its proposal has emerged to explain the apparently antagonistic consequence, where reduced cutting thickness causes an increase in the cutting zone's shear stress levels. This theory is based on the increase of yield stress due to the smaller probability of shear take place within very small volumes (Malkin, 1989).

According to Shaw (1996), an increase in u, resulting from the size effect, reflects the decreasing probability of finding microstructural defects in the shear plane. Schroeter (1999) explains the increase the specific cutting energy by the decrease in the number of dislocations with sliding capacity found with decreasing the cutting thickness.

Boothroyd and Knight (1989) formulated that the size effect arises due to the geometrical relationship between the tool cutting edge radius and cutting thickness or the uncut chip thickness. During machining, the resultant force of the cutting process is distributed over the areas where the tool is in contact with the workpiece (i.e., the rake surface and clearance face).

No cutting tool edge is perfectly sharp, as shown in Fig. 2.6. Measurements via *scanning electron microscopy* (SEM) show that microtools edge radii range from tenths of micrometres to about 5 μm. In the case of nano-machining (e.g., a diamond turning tool) the edge radius reaches a few tens of nanometres. Therefore, there is a tool fillet radius in this region that joins the rake and flank surfaces. In these areas, the action of a constant force is only responsible for deforming the workpiece material and does not contribute to chip removal. This force is called plowing or plowing force. Please, see tool cutting edge radius in an end mill cutter in Fig. 2.6.

For large magnitudes of uncut chip thickness (h), the resultant plowing deformation force acting on the tool edge rounding region is proportionally small compared to the total shear force. However, for small values of h, the deforming force is proportionately larger and cannot be neglected. With an increase in specific cutting energy, represented by the size effect, the deforming force is translated by the large ratio between the tool edge radius and cutting thickness. In this case, the removed chip volume is small when compared to the amount of deformation experienced by the workpiece. A typical size effect curve is shown in Fig. 2.7.

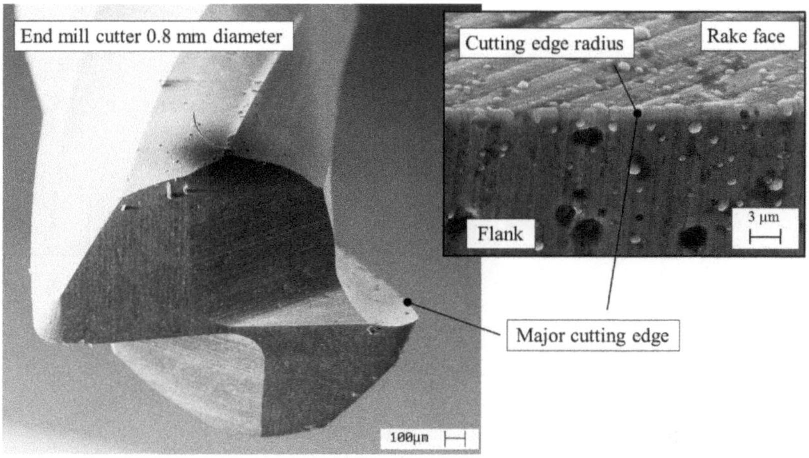

Figure 2.6 Example of end mill cutter with cutting edge radius.

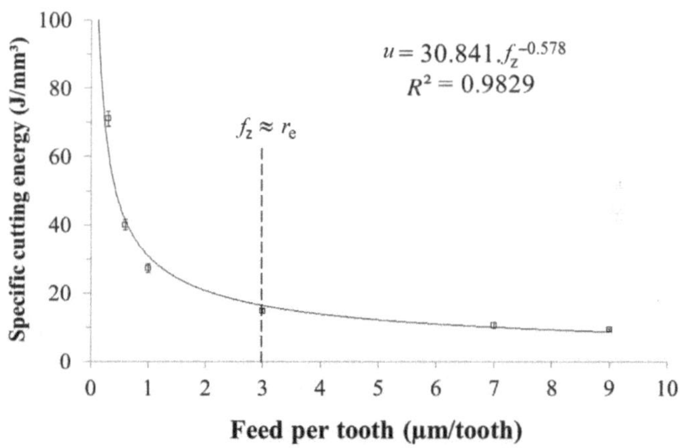

Figure 2.7 Example of typical size effect curve in micromilling processes (Oliveira et al., 2015).

It is important mentioning that most of the tool manufacturers and suppliers for microscale mechanical machining do not provide any information on the values of edge radii. It is evident that the feedrate recommended for microtools (per revolution or per cutting edge) are likely large enough to be larger than the minimum cutting thickness, and consequently, larger than the tool edge radius. To measure the edge radius, it is important to employ a suitable optical system that has computational measurement and image processing tools. The most suitable equipment are the confocal microscopes and optical profilers, as they have great focal field,

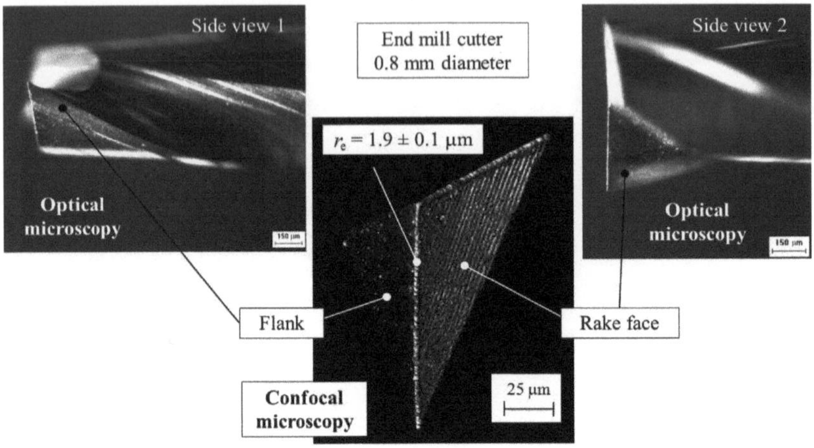

Figure 2.8 Measurement of tool cutting edge radius by using confocal microscopy.

Table 2.2 **Specific cutting energy for mild steel (Shaw, 1995)**

Machining process	***h*** **(μm)**	***u*** **(J/mm^3)**
Defined geometry tool	250	2.1
Rough grinding	25	13.8
Finishing grinding	1.25	68.9

and allow to set accurately measurement plans. Fig. 2.8 shows how the cutting edge radius of end mill is measured by confocal optical profiler.

Shaw (1995) adds more accuracy to the size effect theory, considering metallurgical aspects in grinding processes. These data indicate that the resulting force between the abrasive particle and the workpiece material determines the size and orientation of the elastoplastic deformation zone on the workpiece, in the same direction of the resultant force. This mechanism can be better realized considering a single abrasive particle of finishing grinding, illustrating the size effect on the specific cutting energy, which is much higher in grinding than in turning operation.

Thus, in finishing grinding, a very small amount of material must be deformed under high specific energy to remove an effective chip volume. Consequently, specific cutting energy will be much higher than what would be achieved during machining with defined geometry tools. Table 2.2 brings a comparative of specific cutting energy values to defined and undefined geometry tools.

By assuming grinding process a reference for finishing operations, the cutting thickness in relation to the tool cutting edge radius is so small that the chip formation model proposed by Merchant for orthogonal cutting becomes inappropriate.

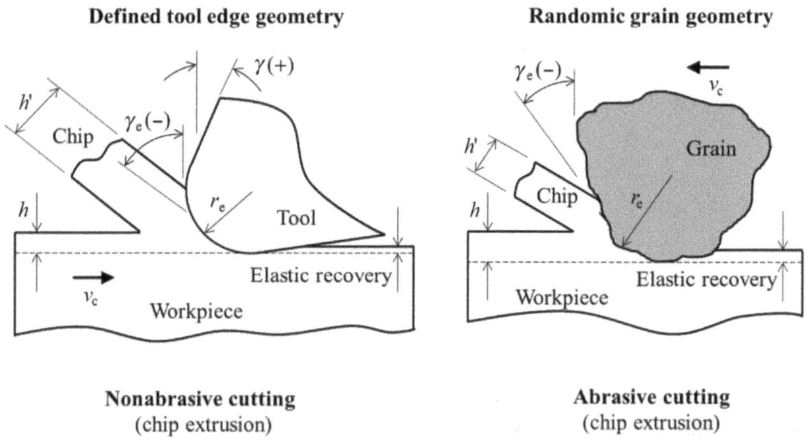

Figure 2.9 Similarity of the chip formation mechanism between microcutting and abrasive processes.

When the effective rake angle becomes extremely negative the chip formation mechanism may be similar to the extrusion process.

As mentioned in Section 2.2, the specific cutting energy levels in the microcutting with defined geometry tools are very similar to abrasive processes. The reason is that, even for nonabrasive machining, the edge radius of the tool is of the same order of magnitude as the cutting thickness, making the effective rake angle negative, similar to the grinding, and leading to chip formation in an extrusion mechanism. Fig. 2.9 illustrates this similar mechanism of chip formation.

When the volume of material removed decreases in micromachining, some issues take place such as the tool edge radius microgeometry, workpiece material GS, and crystallographic orientation effects; these are considered less influential in macrocutting and become more influential factors on the microcomponents' dimensional precision and surface integrity (Brinksmeier, Riemer, & Twardy, 2010; Komatsu et al., 2012). Furthermore, from the perspective of the cutting tool, especially in micromilling, the low cutting speed and tool rigidity hinder proper chip formation (Kim, Bono, & Ni, 2002; Robinson & Jackson, 2005).

When cutting scale decreases, the cutting thickness may be compared to the radius of the tool cutting edge, as shown in Fig. 2.9. As a result, cutting can occur with a highly negative rake angle, and a small variation in cutting thickness can influence the cutting process and the chip formation (Malekian et al., 2012; Sooraj & Mathew, 2011). The relationship between the cutting thickness and the cutting edge radius defines the active mechanism of chip removal, which must occur by shearing and cutting of the material. The interaction between cutting thickness and tool cutting edge radius in micromachining can only promote deformation of the material, with no shear and material removal, which directly influences the final surface roughness (Yun et al., 2011). This process of elastoplastic deformation, to which the material is subjected, causes a substantial increase in the value of specific cutting energy.

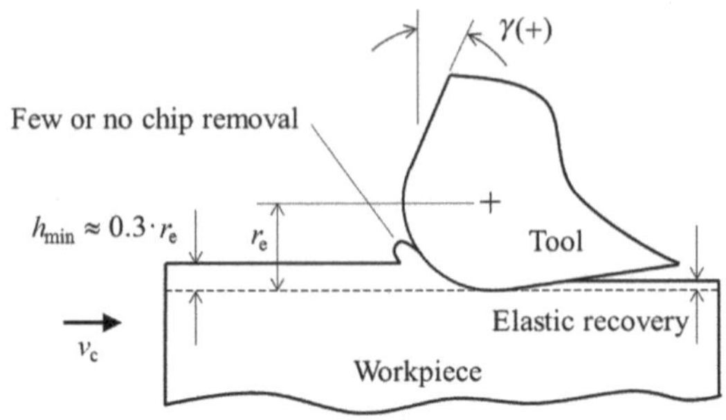

Figure 2.10 Minimum cutting thickness (h_{min}) for chip removal.

Chae et al. (2006) and Aramcharoen and Mativenga (2009) presented the concept of minimum cutting thickness (h_{min}), whereby the chip will not form unless the uncut chip thickness (h) is greater than h_{min}. When the uncut chip thickness becomes smaller than h_{min}, as seen in Fig. 2.10, the material undergoes an elastoplastic sliding ocurring at the flank face and plowing due to the tool edge radius. As the uncut chip thickness becomes larger than h_{min} value, a compression of the part material ahead of the tool promotes shear in the primary shear zone (PSZ). and the whole uncut chip thickness is removed, forming the chip completely.

According to Jackson (2008) and Mian, Driver, & Mativenga (2011), there is an exponential increase in the specific cutting energy with decreasing chip thickness during micromachining. As previously mentioned, it is believed that this increase derives from the lower probability of detecting defects, such as grain boundaries and impurities, which act as stress concentrators in smaller volumes of material to be removed. Thus, the stress required to promote shearing of the material should be higher, causing an increase in specific cutting energy.

This phenomenon has a major impact on cutting forces, process stability, and surface finishing during machining in microscale (Liu, Devor, & Kapoor, 2004).

2.4 Workpiece microstructure scale

Policrystalline solids are composed of small crystals termed grains, which in turn are formed by repetition of the unit cell (i.e., the smallest tridimensional grouping of atoms representative of a given crystalline structure). During the material solidification nuclei are originated in aleatory positions and orientations within material volume which grow up to reach each other at adjacent ends named grain boundaries. These regions are characterized by significant crystallographic disarrangement (Callister, 2007). A pearlite microstructure with evidence of the grain boundaries is demonstrated in Fig. 2.11.

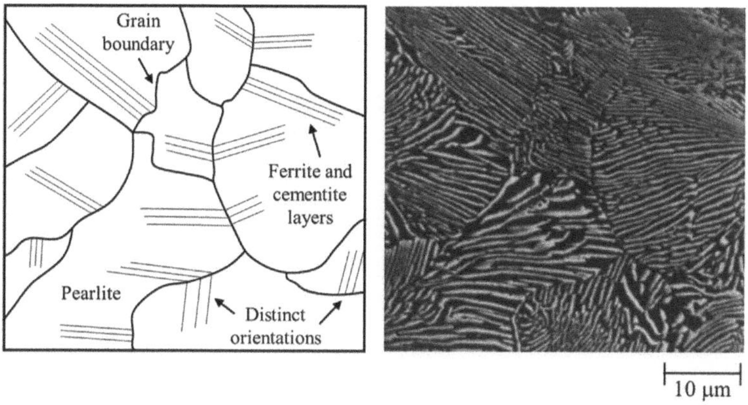

Figure 2.11 Pearlite steel photo (Wayman & Juleff, 1999) and schematic drawing to define grains and grain boundaries.

Materials' mechanical property, such as Young Modulus, present different values because of its dependence on the grain crystallographic orientation. Thus, anisotropy can be defined as the directionality of a given material's properties. Conversely, isotropic solids present properties which do not depend on a measurement direction (Dieter, 1988).

During the macroscale mechanical machining, the tool cutting edge cuts or removes a great amount of workpiece grains by means of intra- and intergranular fracture since the magnitude of the cutting parameters (e.g., tool feed and depth of cut) are much bigger than GS, even when compared with coarser grains (hundreds of micrometres). In this particular condition, grain boundaries, crystallographic orientations, or crystalline defects are minimized or absent, given that the amount of grains intercepted by tool cutting section area is large. In other words, the workpiece material behaves as continuous, homogeneous, and isotropic during machining because the interferences on dynamic of chip removal and on workpiece surface integrity do not take place.

When machining at microscale, micrometric dimensions of tool feed (f) and depth of cut (a_p) can also produce a micrometric chip cross section area (A_c) depending on parameter amplitudes. This dimensional scale, resulting from cutting parameters, is strongly susceptible to the effects of GS, crystallographic orientations, grain boundaries, inclusions, and voids, which cause the material response to be heterogeneous, anisotropic, and discontinuous. Accordingly, undesirable effects can be minimized or eliminated by using material with a finer microstructure. Fig. 2.12 shows an inappropriate relation between the tool cutting edge radius and part GS.

On the other hand, microcutting applied to workpiece materials with coarser GS makes the chip formation and its removal significantly anisotropic and dependent on crystallographic orientations and grain boundaries, whereas finer or ultrafine grains restore isotropic behavior in the microcutting that is typically found in macrocutting. Fig. 2.13 presents a real case where microcutting has grain density (GD) (grains per cutting section area) equivalent to macrocutting.

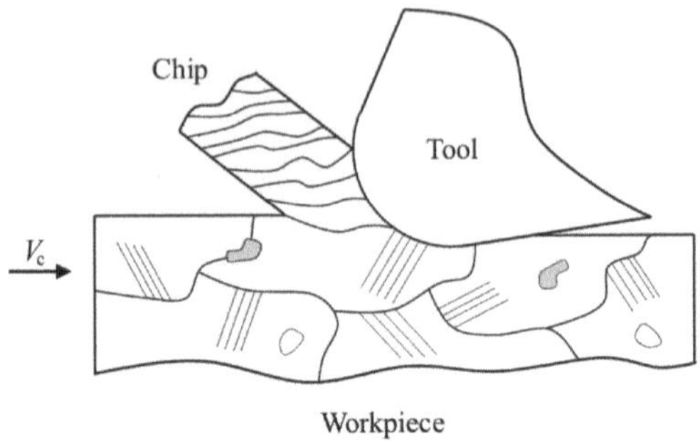

Figure 2.12 Scheme of workpiece microstructural scale improper to microcutting.

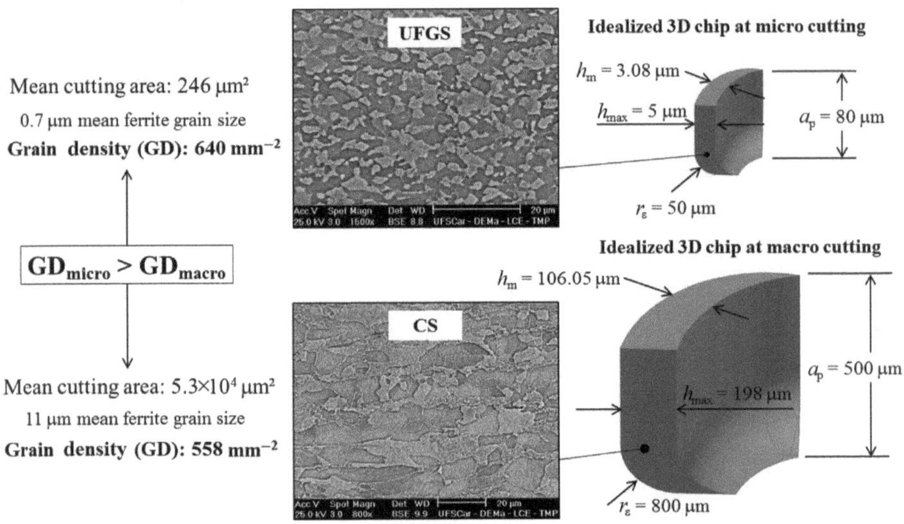

Figure 2.13 Suitable adapting between workpiece grain size and cutting scale.

According to ASTM E-112 standard, GS from 254 to 127 μm mean diameter (ASTM number 1 to 3) are coarse, 90 to 45 μm diameter (ASTM number 4 to 6) are intermediary, and 32 to 22 μm (ASTM number 7 to 8) are fine. Assis (2013) mentions that ultrafine GS are smaller than 5 μm in mean diameter (ASTM number 12 to 14). Fine GS cannot be completely suitable to microcut given that mean GS is still near or larger than cutting parameters related directly to chip formation process (e.g., chip cross sectional area). Therefore, some adjustment of the workpiece microstructure scale to the machining scale is necessary. According to Assis (2013), mean GS between 1 and 5 μm or smaller are sufficient to promote isotropic response for microcutting, so that workpiece material can behaves as a monophasic one (continuous and homogeneous). The difference between two workpiece GS can be seen in Fig. 2.14.

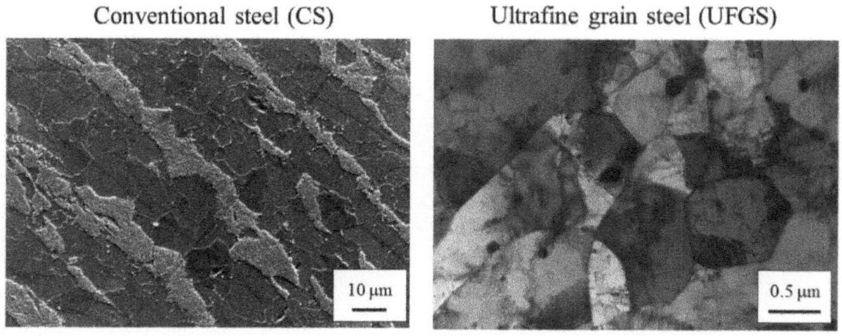

Figure 2.14 (SEM) and images of the COSAR 60 steel with 11 μm (left) and 0.7 μm (right) mean ferritic grain sizes, respectively.

Workpiece GS can be refined by producing subgrain or grain through severe plastic deformation (SPD) at room temperature, strain-induced phase transformations, and dynamic recrystallization of ferrite at warm processing (Silva Neto & Balancin, 2007). Several routes have been developed for GS refinement mainly by using experimental simulations in laboratories. Xianghua et al. (2009) declare that most proposed techniques to refine GS are based on severe plastic deformation such as equal channel angular pressing (ECAP), accumulative roll-bonding (ARB), and severe torsional straining (STS).

Historically, the development of steels, with improved properties of strength and toughness, has been reached by cold rolling and postannealing, in order to produce refined grains. Despite these methods produce nanometric or submicrometric GS, these techniques are still not economically viable for mass production of ultrafine-grained steels because of the need for special devices to generate high deformations (Xianghua et al., 2009).

Nevertheless, the grain refinement processes used to produce downsized components (e.g., ECAP) can be highly useful for products that need some microcutting operation.

A disadvantage is that the techniques based on severe plastic deformation cause residual stresses inside workpiece material, even after a great percentage of internal energy is dissipated to the grain refine process. Thus, subsequent stress relief treatments are usually carried out to assure the practical application of the component (Park et al., 2000).

Several applications for this material grade have been addressed in the automotive industry regarding components for steering wheel, propulsion shaft, camshafts, crankshafts, dampers, universal joints, axles, impact bars, drive shafts, and gear racks. Micrograin aluminum-ferrous alloys are very attractive for applications in automotive and aeronautic industries (Stolyarov et al., 2003).

Ultrafine-grained materials besides being more convenient for microcutting applications, they have both higher strength and toughness. The strength inscrease is justified because the smaller the GS, the larger is the grain boundary area which creates barriers to the sliding of dislocations. The toughness increase, on the other hand, is due to the distribution of stress on the grain boundaries being more uniform (Hertzberg, 1995).

Another relevant issue is related to the elastoplastic deformation mechanism. In coarse grains, their available volume to dislocation movement is larger than that in fine grains, and larger plastic deformation is reached. In fine grains their restrict volume needs higher energy levels to deformation. Thus, elastoplastic responses of the workpiece material due to interaction with tool cutting edge are distinct in conventional cutting and mostly microcutting for which is added the size effect of the process.

2.5 Surface integrity

The term "surface integrity" was quoted by Field and Kahles for the first time in 1964. It is defined as the set of changes in the workpiece surface and subsurface caused by manufacturing process. Thus, the surface changes were initially classified by mechanical, metallurgical, and chemical, among other kinds of transformations. In 1971, at the 21st General Assembly of "College International pour la Recherche en Productique," surface integrity officially became recognized as a research field (Becker, Santos, & Sales, 2005).

According to Field, Kahles, and Koster (1989), the knowledge of surface integrity describes and controls possible surface changes undergone by the part during its manufacturing process, including its effects on the properties and performance of materials in service. They also report the importance of studying the surface integrity citing as an example the fatigue failure, which usually has nucleated cracks close to or on the component surface.

For Chevrier et al. (2003) surface integrity measures the quality of the machined surface by interpreting the structure of the material surface and subsurface and by analyzing the influence of cutting parameters for variables such as residual stress, surface roughness and cutting force. Thus, the surface integrity can be studied as the structure shown in Fig. 2.15.

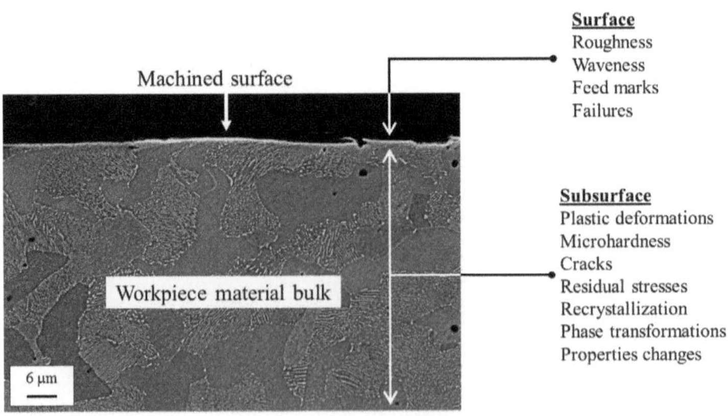

Figure 2.15 Classification of the surface integrity in a mechanical component.

A complete characterization of surface integrity includes the two aforementioned sets and additional information on fatigue strength, besides results of mechanical testing, such as tensile, rupture stress, creep, and others.

The surface analysis of a part can be divided into four different levels: roughness, waviness, feed marks and flaws. The roughness can be understood as fine irregularities, with smaller spacing, resulting from the action inherent to the cutting process caused by the tool. They are usually measured on a small interval length, called "cut off" (sample roughness length) ranging from 0.25 to 2.5 mm. The waviness are surface irregularities whose spacing is greater than the "cutoff." In general, chatter vibration, bending of the part or tool due to force or cutting temperature represents the sources and causes undesirable waviness.

Feed marks, as its name implies, indicates the relative motion relation between the workpiece and the tool during the machining process. They are marks that denote the predominant directions of surface irregularities. The flaws are unintended interruptions, unexpected and undesirable in the typical topography of the surface of a workpiece. Usually arise through defects in the workpiece material, such as inclusions, voids or cracks.

In relation to the topography of the machined surface, two very common changes observed in these evaluations are the phenomena known as material lateral flow and cavity opening (Becker et al., 2005). The material lateral flow, also known as "side flow" is the part material deformation in the opposite direction to the tool feed during chip formation (shear).

However, cavities or open grain as well known is characterized by the formation of cavities distributed over the surface and typically occurs in the machining of brittle materials. These phenomena can be observed in Fig. 2.16.

When concerning to subsurface analysis, the plastic deformation is a common alteration due to the machining process, especially in roughing operations, in which cutting parameters employed are more severe. Usually it can be identified by the appearance of elongated metallurgical structure of the component in the direction of the deformation flow. The plastically deformed burrs may be generated by built-up-edge fragments. Microhardness can be easily increased due to work hardening or more hard phases, or reduced, due to alterations in more ductile phases, depending on the temperature reached during machining. The macro or microcracks have a

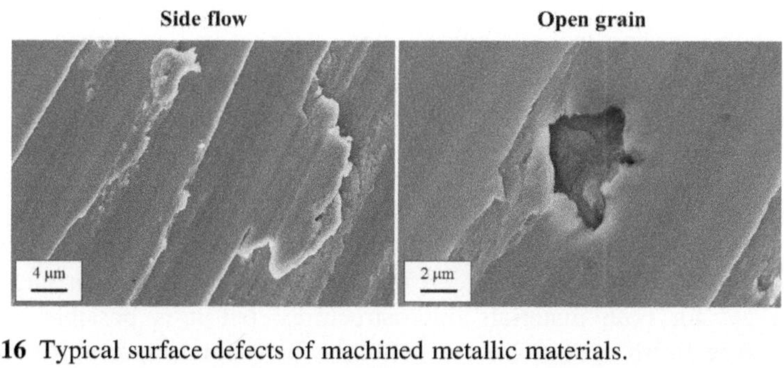

Figure 2.16 Typical surface defects of machined metallic materials.

more latent occurrence when machining brittle materials, precisely because of the absence of plastic deformation capacity. The microcracks, in particular, are extremely harmful to fatigue properties and stress corrosion, as are the initiators elements of such mechanisms, and should therefore be avoided.

Residual stress may be one of the most common variables investigated within the list of indicators of surface integrity. Stress may occur in tension or compression, high or low magnitude, shallow or deep depth. The compressive stress, as typically occurs in finishing operations (grinding and turning) of hardened parts, it is considered beneficial for increasing fatigue life. According Field et al. (1989), the residual stresses tend to be compressive for milling processes.

Recrystallization can take place in the material microstructure within the outmost surface layer during machining operation, when the workpiece reaches temperatures above the recrystallization temperature, allowing the material to return to its original state before the deformation. Phase transformations happen when the temperatures developed in the machining exceed the levels compatible with the transformations. A more common example concerns the machining of hardened steels, wherein the fragile untempered martensite can result from the phase transformation. This hardened layer on the subsurface of the part is called white layer because it does not react to chemical etching used for the detection of the microstructure.

When using microcutting, the effect of the tool edge radius is more pronounced in both, to the chip formation mechanism and to surface integrity of the part. At microscale, particularly surface damage is more likely to be probed, especially when the microstructure scale is not in the same range of the cutting conditions (Rodrigues et al., 2012). Table 2.3 (Section 2.5.5) will summarize the main defects in the generation of surface via microcutting processes.

In the following subsections various results, from the literature, on the effect of workpiece material GS in the main surface integrity variables in micromilling operation will be presented and discussed. New tool edges were used for each cutting condition to eliminate possible tool wear effects on results of surface integrity. In order to improve the knowledge on these effects some results are also included for milling operation in macroscale and variation of cutting parameters for the purpose of comparison.

2.5.1 Burr formation

Fig. 2.17 shows the top view of micromilled channels in low carbon steel named COSAR 60 (similar to AISI 1016 with microadditions) with GS of 11 and 0.7 μm, employing cutting speed of 135 m/min without cutting fluid, 800 μm tool diameter and 5 μm cutting edge radius, 32 μm depth of cut, and feed per tooth of 0.5 μm/tooth ($f_z < r_e$), 5 μm ($f_z = r_e$) and 10 μm/tooth ($f_z > r_e$). The arrows indicate the feed direction and tool rotation.

Based upon the Vogler et al. (2004), the minimum cutting thickness was not reached with the lowest tool feed ($h_{min} \approx 1.5$ μm), resulting in large burr in up-milling cut for both materials microstructures, but it is possible to observe that the surface finish of the channel bottom is better and more uniform, as well

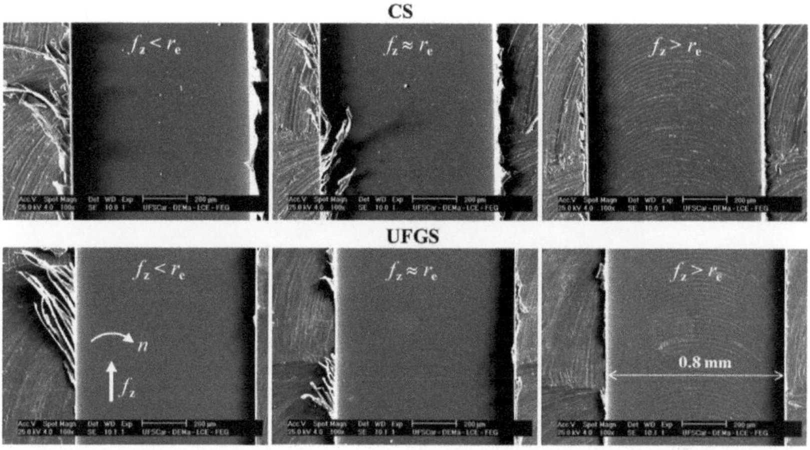

Figure 2.17 Burr formation in micromilling of the COSAR 60 steel: CS (upper) and UFGS (lower) (Assis, 2013).

as the thickness of the needle burrs for the burr formed with the ultrafine grain steel (UFGS) which may be attributed to its microstructure with smaller GS. The thickness of the needle type burrs can be derived from the GS or part material grain clusters (Santos, 2016).

When the feed per tooth is increased to the similar size and larger than cutting edge radius, the size of burrs reduced, but this effect is more prominent in UFGS. The feed with the same size to the cutting edge radius is not rather sufficient to promote the reduction in burrs' size, showing that the edge radius still exerted a deleterious effect on the chip formation, even for the UFGS. By overcoming the h_{min} and the tool edge radius, cutting thickness and consequently, chip cross section area were sufficient to form the chip and to attenuate burr formation, especially in UFGS, which presents lower plasticity than conventional steel (CS).

Micromilling tests to cut microchannels in AISI 1045 steel (12 μm and 43 μm ferrite and pearlite GS, respectively) were performed varying the end milling scale, i.e., macro and micromachining. The cutting speed used in the tests was $v_c = 60$ m/min, dry machining, with two different diameters of the cutter, i.e., 0.8 mm (micromilling) and 2.0 mm (macromilling) both with 2.7 μm edge radius. The feed per tooth and depth of cut were varied in order to verify the burr formation and the bottom finishing of the milled channel. The length machined by each tool was 13 mm. Figs. 2.18 and 2.19 show the workpiece microstructure and machined channels (Moreira, 2012).

Fig. 2.19 shows that the larger diameter tool generated needle-like burrs when cutting the workpiece with the smallest feed per tooth and depth of cut, and very small burrs for higher feed and depth of cut. The condition of maximum feed per tooth and minimum depth of cut generated a burrs-free, since the microstructure scale was more compatible with machining condition, i.e., the GD (Grain Density)

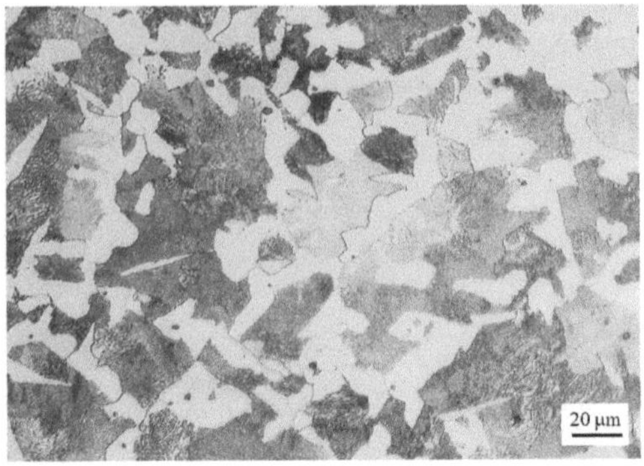

Figure 2.18 Microstructure of AISI 1045 steel for macro- and micro-milling.

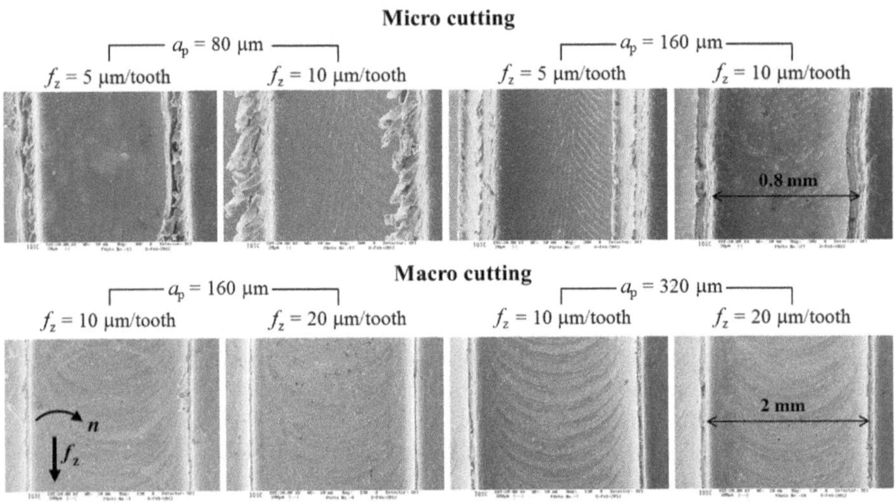

Figure 2.19 Top view of micro- (upper) and macro- (lower) end milling of AISI 1045 steel.

is 18 ferrite grains/mm^2 and 1.4 pearlite grain/mm^2. Moreover, the channel bottom topography presented well-defined feed marks.

In micromilling test, burrs formation was much more significant, particularly for smaller depth of cut. In this case, the microcutting in a workpiece of coarse microstructure amplified the anisotropic response, i.e., the GD was only 2.3 ferrite grains/mm^2 and 0.2 pearlite grain/mm^2. The anisotropic effect, in this case, is more clearly observed in the surface texture of the channel bottom as well as waviness through the tool path. When comparing the same cutting conditions for the two tool diameters, it appears that the macromilling generated better channels form and less finishing errors. The size effect of the geometric relationship between feed per tooth and edge radius of the tool is not decisive, feed per tooth was always larger than h_{min} and r_e. The results showed that there is a strong

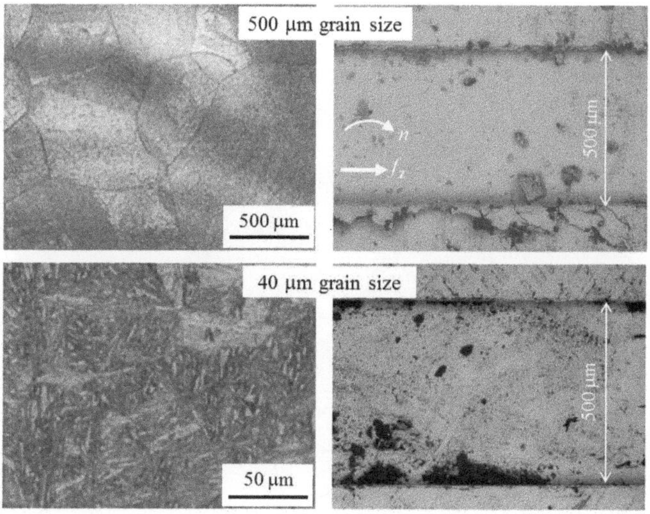

Figure 2.20 Microstructures for different GS (left) and top view of the channel micro milled in AISI H13 steel with coarse and fine grain sizes (right).

relationship between the tool diameter and GS related to burrs formation on the top of the walls generated during machining of channels. When comparing with former results it is possible to assert that microstructure has a strong effect on defect formation when the scale of the material removal process is decreased.

Lauro et al. (2014) developed a study of the effect of austenite GS (40 to 500 μm) in the channel surface finish and the formation of burrs at the micromilling AISI H13 Steel (45 HRC). It was used depth of cut of 25 μm, 0.5 mm cutter diameter, 2.77 μm edge radius, 77 m/min cutting speed and 1 μm/tooth feed per tooth. Fig. 2.20 presents both microstructures and micromilled channels.

The results presented in Fig. 2.20 show that material with larger GS generated burrs of the primary type for down-milling, due to the adopted feed per tooth being close to the minimum cutting thickness ($f_z/r_e = 0.36$). This effect of cutting edge radius (size effect) in the material with finer grains was significantly mitigated since there was practically no burr formation even in the down-milling side where the generation of burr is more common, since the chip thickness decreases to zero.

Although both materials present high hardness values added to the size effect and anisotropy, large grains may experience higher strains, given that larger grains allows longer distances for dislocation movement, which favors the burr formation and maximize its size.

2.5.2 Chip formation

Assis (2013) conducted several macro- and micro-milling tests under dry cutting condition in COSAR steel UFGS (0.7 μm) and CS (11 μm) to evaluate the effect of GS and the ratio f_z/r_e on chip formation. Macromilling adopted cutting speed

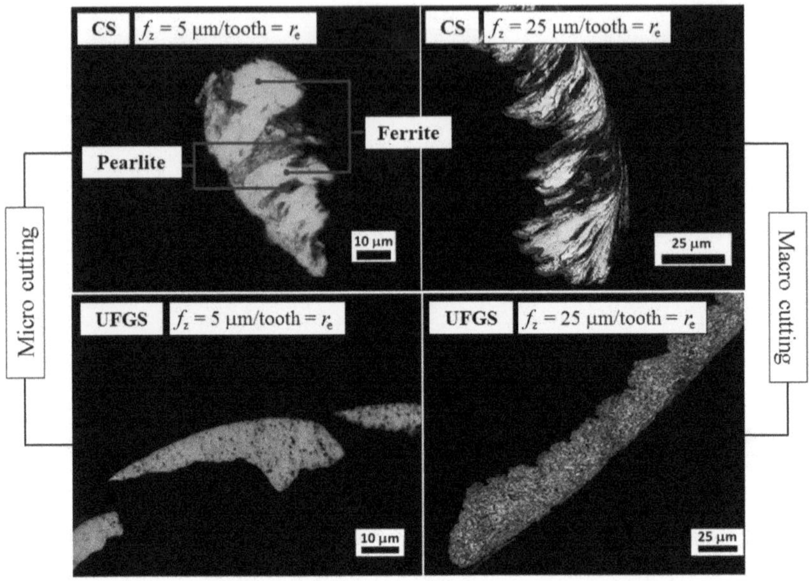

Figure 2.21 Chips of CS (upper) and UFGS material (lower) produced by micro-(left) and macro-(right) milling for which the feeds per tooth are equal to tool edge radius.

$v_c = 800$ m/min, depth of cut $a_p = 0.5$ mm, radial depth of cut $a_e = 2$ mm, and two feed conditions 0.025 mm/tooth ($f_z = r_e$) and 0.300 mm ($f_z > r_e$), two flutes cutter with 16 mm diameter (PVD-TiN), $r_e = 25$ μm. Micromilling adopted cutting speed $v_c = 135$ m/min, depth of cut $a_p = 0.032$ mm, radial depth of cut $a_e = 0.130$ mm, and two feed conditions 0.005 mm ($f_z = r_e$) and 0.018 mm ($f_z > r_e$), two flutes cutter with 0.8 mm diameter (PVD-TiN), $r_e = 5$ μm. The machined lengths were 50 mm and 5.5 mm in macro- and micro-milling, respectively. Fig. 2.21 shows the chips produced by tests.

Images from optical microscope depicts the morphology of chip formed with $f_z \approx r_e$ (micro- and macro-milling). The images show elongated ferrite grain between pearlite grains for the CS material. The change in chip formation mechanism in the CS sample has occurred due to the presence of a harder phase (pearlite) and more ductile (ferrite). When machining the UFGS sample, the chip revealed juxtaposition of lamellae, tendency to form segmented chip (serrated) and a little deformed microstructure due to the reduced GS microstructure and consequently, demanding higher energy amount to deform them. This formation process occurred despite of a more pronounced size scale, since the machining took place for $f_z = r_e$ condition.

By employing $f_z > r_e$ condition, both in the micro- and macro-milling, as shown in Fig. 2.22, it is possible to observe a pattern similar to the chip formation mechanism when the materials were milled with former cutting condition ($f_z = r_e$). Only exception is made for the trend of segmenting for $f_z = r_e$ probably occurred due to the more significant effect of tool edge radius on the chip deformation of the UFGS material.

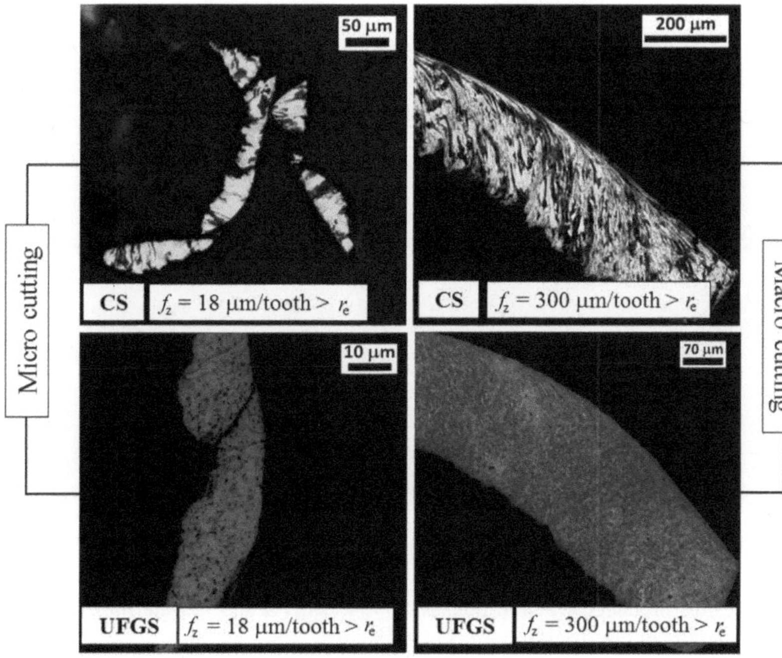

Figure 2.22 Chips of CS (upper) and UFGS material (lower) produced by micro-(left) and macro-(right) milling for which the feeds per tooth are greater than tool edge radius.

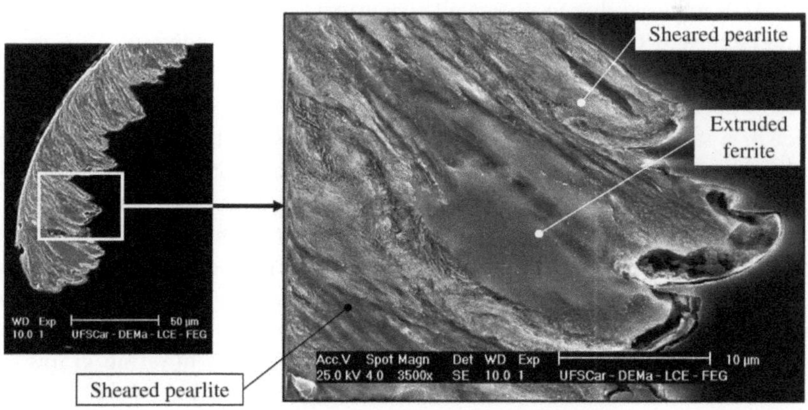

Figure 2.23 Quasi-shear-extrusion chip observed by Assis (2013).

The chip formed in the micro and macroscales from the CS sample, using $f_z = r_e$ condition is called *quasi-shear-extrusion* and has been observed by Simoneau, Ng, and Elbestawi (2006). Fig. 2.23 shows in detail the ferrite grain extruded by two perlite grains (sheared deformation).

Figure 2.24 Chips produced when milling AISI 1045 steel at macroscale for different feeds per tooth and depths of cut (tool tip radius $r_\varepsilon = 150$ μm).

The formation of this type of chip has not observed in UFGS sample, as the metallurgical modification provided a homogeneous microstructure with the presence of a single phase (ferrite). The phenomenon did not occur in the machining $f_z > r_e$ for CS material due to mitigated size effect.

The relationship between feed per tooth and tool edge radius of 12:1 in macromachining and about 4:1 in micromachining shows that for this condition the microstructure behaves homogeneously during material removal, irrespective of the presence of a second phase (with difference in hardness). The homogeneous behavior of the microstructure can also be related to the GD already mentioned. Macromilling of the CS sample with $f_z = r_e = 25$ μm resulted in GD = 45.5 ferrite grains/mm^2, while micromilling of the UFGS part with $f_z = r_e = 5$ μm presented GD = 162.8 ferrite grains/mm^2. Besides, GD is only 0.7 ferrite grains/mm^2 when milling CS material at microscale with $f_z = r_e = 5$ μm. These results show how the relationship between cutting and microstructure scales can become the chip removal mechanism anisotropic.

Finally, the microstructural uniformity of the chip with ultrafine grains can be beneficial to the cutting stability, whereas the cyclic extrusion of the ferrite phase in larger grains may cause some vibration during chip formation, which can affect the surface integrity of the machined part.

Figs. 2.24 and 2.25 show images of chips produced in macro- and micro-milling of AISI 1045 steel. Oliveira et al. (2015) showed that the chip formation mechanism depends on the machining scale, especially when associated with a microstructural scale of the workpiece not adjusted to machining in microscale. The machining conditions and the workpiece material were the same as shown in Fig. 2.19.

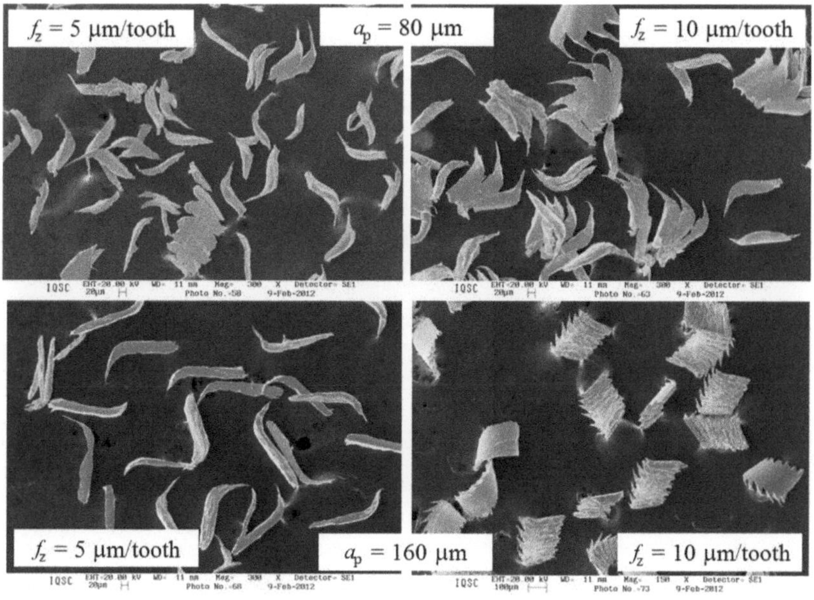

Figure 2.25 Chips produced when milling AISI 1045 steel at micro-scale for different feeds per tooth and depths of cut (tool tip radius $r_\varepsilon = 50$ μm).

It is note that chips resulting from macromilling are helical for smaller a_p/r_ε ratio and curved for larger a_p/r_ε ratio. The chips presented a defined morphology from its formation, with fairly smooth part-tool interface and free surface with uniform shear fronts and lamellae.

However, Fig. 2.25 shows that the chips produced in micromilling are distinct from those formed in macromilling. This morphological difference stems from its distinctive deformation mechanism involved by tool geometry-part material.

The chips had round and thin ends, resulting from the interaction r_ε and a_p, but the main difference lies in the chip segmentation into smaller lamellae characteristic lengths. During chip formation, there is an uneven deformation along its cross section, tending to open into a helicoidal and spliting apart in its thinnest portion. These deformations takes place in defined sections of the chip when its thickness becomes thinner around the tool cutting edge radius.

In conclusion, it can be seen that a reduction of workpiece material GS is key for micromachining process in order to have an isotropic, continuous and homogeneous, especially when analyzing the chip formation process, which reflects on the quality of micromachined product.

2.5.3 Roughness

Surface finish of microparts is one of the most important technical requirements for micromachining. Due to very small dimensions of some functional parts,

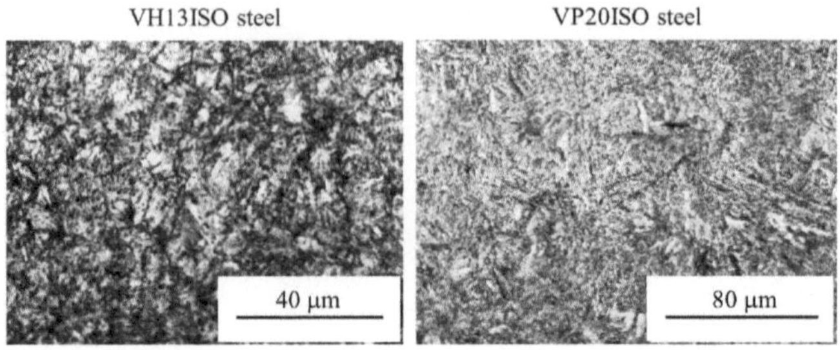

Figure 2.26 Martensitic microstructure of VH13ISO (left) and VP20ISO (right).

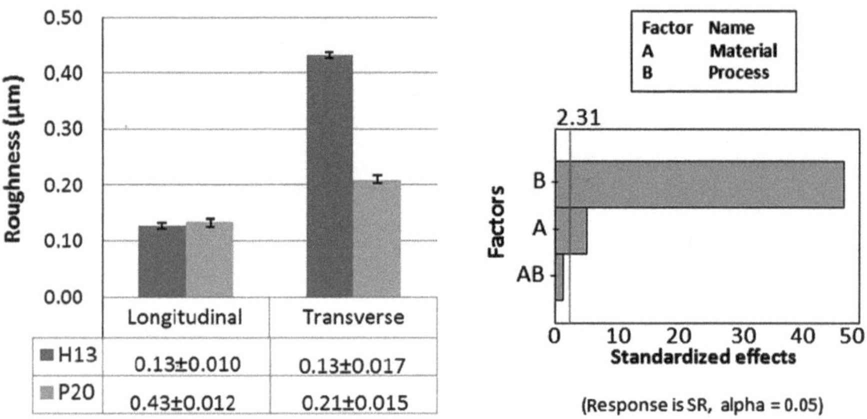

Figure 2.27 Longitudinal and transverse Ra roughness for VH13ISO and VP20ISO steels (left) and Pareto chart showing the significance of workpiece material on roughness (right).

subsequent finishing processes are more difficult to be performed. Therefore, the final roughness usually is achieved by microcutting process.

Bodziak et al. (2014) investigated the effect of microend milling process on surface integrity of tool steels used for molds and dies manufacturing. VP20ISO (30 HRC) and VH13ISO (45 HRC) steels with coarser GS (Fig. 2.26) were micromachined in dry condition by using endmill of 0.7 mm diameter, 2 flutes, tungsten carbide substrate, TiSiN coating, 10 μm/tooth feed per tooth, 60 μm depth of cut, 0.12 mm axial depth of cut, 70 and 60 m/min cutting speeds, respectively. The tool path was linear at transverse and longitudinal directions, with 35 mm long.

This microstructure is more difficult to be microcut due to the coarser austenitic GS that originated the martensitic microstructure which is formed by needles-like fine laths.

VP20ISO steel presented smaller R_a roughness values at transverse direction attributed to the smaller hardness and abrasiveness since VH13ISO steel contains greater chromium percentages as well as concentrations of silicon and vanadium. Additionally, the roughness values at transverse and longitudinal directions were distinct indicating the anisotropic effect, as seen in Fig. 2.27.

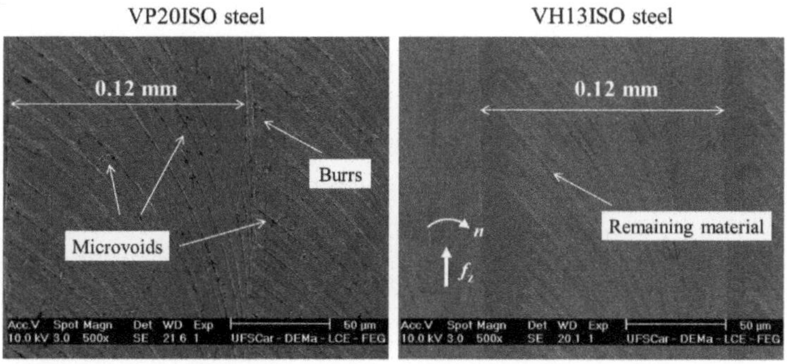

Figure 2.28 Top view of the micro-milled VP20ISO (left) and VH13ISO (right) steels.

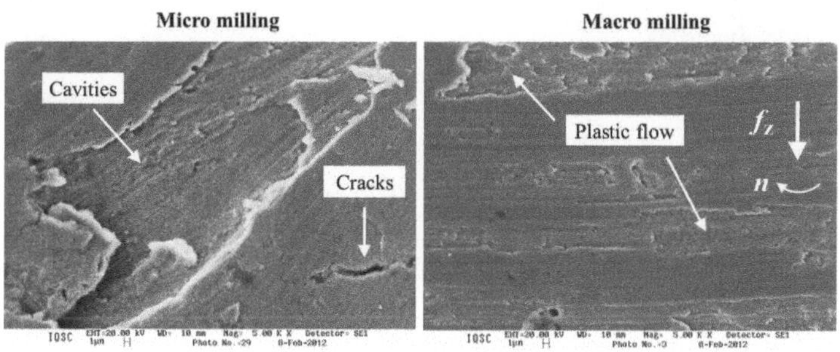

Figure 2.29 Surface damages when micro-(left) and macro-(right) milling AISI 1045 steel (10 μm/tooth tool feed, 160 μm depth of cut, 60 m/min cutting speed).

Despite roughness of VP20ISO steel is by about 50% smaller than that from VH13ISO steel, damages over machined topography were found because the size effect is more significant even if roughness value is very small. Fig. 2.28 shows that micromilled surface of VP20ISO steel presented burrs and protuding feed marks whereas VH13ISO steel did not produce any surface damages or flaws. These damages may be attributed the larger plastic deformation capacity presented by VP20ISO steel.

Similar results were found when milling the AISI 1045 steel at the same microcutting conditions presented in Fig. 2.19. Fig. 2.29 shows that microend milling caused more significant damages over cut surface than macroend milling when applying the same depth of cut and tool feed. The main identified damages were cracks and open grains which can be addressed to ferrite and pearlite grains.

Finally, Assis (2013) verified that smaller workpiece GS decreased the part roughness when feed per tooth was lesser, equal and greater than tool cutting edge radius size in microscale end milling (Fig. 2.30). The same cutting conditions presented in Fig. 2.21 were used except for cutting speed (25 m/min or 10,000 rpm).

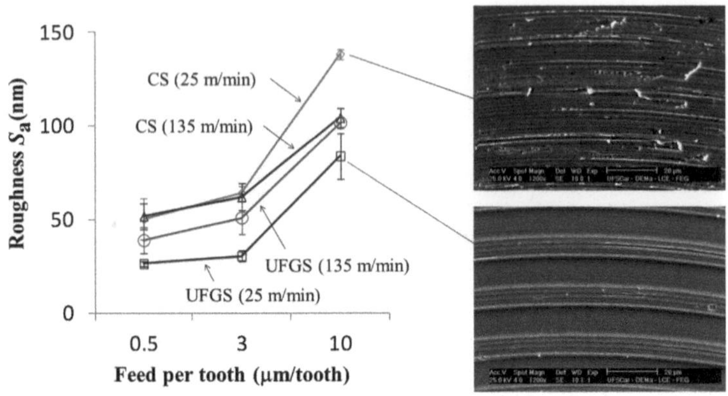

Figure 2.30 Tridimensional roughness (S_a) and surface damages when micromilling COSAR 60 CS and UFGS materials.

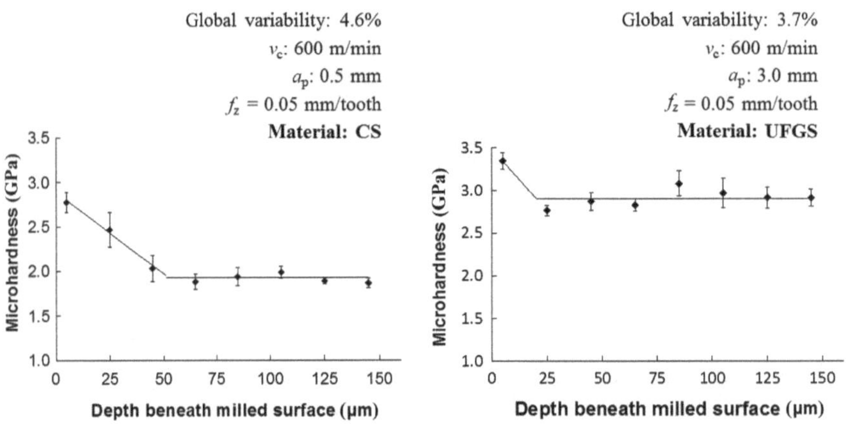

Figure 2.31 Microhardness profile measured in the section perpendicular to macromilled surface for CS (left) and UFGS (right) materials.

In terms of qualitative roughness evaluation, damages found in micromilled surface such as grooves, smeared materials, prows and microvoids were identified mainly for larger GS (images highlighted in Fig. 2.30). These defects were probed for all feeds per tooth applied and they took place mainly in workpiece with low hardness (CS) due to anisotropy effect from material microstructure GS.

2.5.4 Microhardness

Surface and subsurface microhardness of the workpiece can also be affected depending on cutting parameters, part thermomechanical properties and tool cutting edge microgeometry. Fig. 2.31 shows the microhardness gradient measured

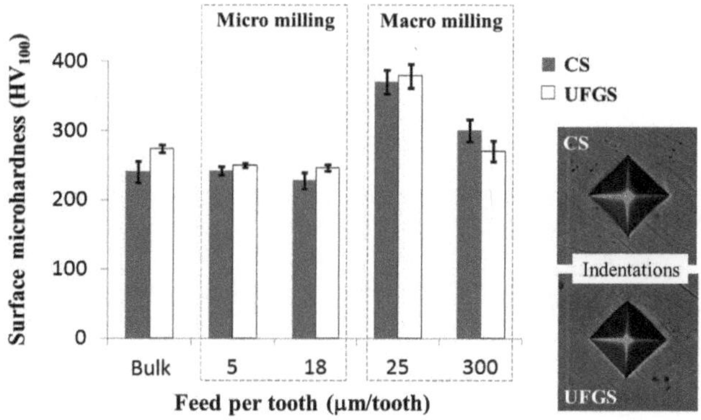

Figure 2.32 Hardness variation regarding to workpiece grain size, milling scale and feed per tooth.

beneath the piece surface after milling in dry condition COSAR 60 steel with 11 μm CS and 1.7 μm UFGS GS at macroscale cutting by using 25 mm tool diameter, 2 tungsten carbide inserts coated of Al_2O_3, 600 m/min cutting speed, 0.05 mm/tooth feed per tooth, 0.5 mm and 3.0 mm depths of cut, 19 mm radial depth of cut and 65 mm cutting length.

According to Rodrigues et al. (2012), despite the microhardness close to machined surface was higher for UFGS material, its relative increase from the bulk microhardness was lower. This result indicated that UFGS tends to sustain its bulk microhardness at low depth from the outmost surface. Even at high cutting speed, larger depth of cut and macroscale milling the material had a small increment of microhardness and a shallow affected layer. The CS material presented deeper hardened layer due to its plasticity especially at high cutting speed.

Assis (2013) has also investigated the influence of the workpiece GS and milling scale on part surface hardness. By considering the same workpiece materials and cutting conditions quoted in Fig. 2.17, the author obtained the results presented in Fig. 2.32.

It is observed that no microhardness variation was identified for CS material when micromilled. However, a small decrease (9%) of surface microhardness for UFGS material took place for both tool feeds. On the other hand, a microhardness increase of 54% and 38% for both CS and UFGS samples, respectively, when macromilling was applied with $f_z \approx r_e$. The squeezing effect of the tool cutting edge radius was more evident for macromilling. Under macromilling at higher tool feed condition, CS sample presented an increase of up to 25% in microhardeness while UFGS remained constant statistically. The greater strength of the UFGS material when compared to that of CS material hampers its cold hardening and favors its workpiece surface integrity. This work hardening of the CS material is justified by significant plastic deformation of its microstructure next to machined surface by analyzing Fig. 2.33.

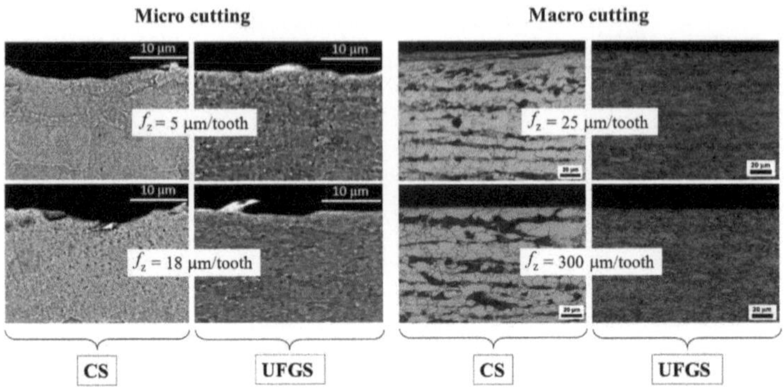

Figure 2.33 Microstructural defects when milling at macro- and micro-scales COSAR 60 CS and UFGS materials.

2.5.5 Microstructural damages

The extension of surface and subsurface damages take place usually in macro machining processes around 150 μm below the machined surface. However the affected layer depth depends strongly on cutting conditions, machining process, tool geometry and thermomechanical response of the workpiece material. The depth of defects due to microcutting trends to be lower but the damages can extend because of the size effect. Additionally, as aforementioned in Section 2.5.3, low roughness generally lower than 1 μm R_a does not exclude occurrences of defects in the surface topography such as those shown in Table 2.3.

Assis (2013) has identified a few of defects depicted in Fig. 2.33 when milling at micro- and macroscale COSAR 60 CS and UFGS materials.

The machined surfaces of the UFGS material presented less microstructural alterations when compared to CS samples. Localized defects as microburrs and cavities were probed, however without presence of severe plastic deformations. Cavities and voids more significant in quantity and size were recurrent in the CS workpiece as well as a substantial plastic deformation of microstructure near cut surface.

These damages in CS material are resulted from distinct hardnesses of ferrite (softer) and pearlite (harder) when the tool cutting edge cut these phases during machining. Moreover, grain boundaries can nucleate cracks or concentrate hard particles which contribute to generate certain damages such as open grain, cavities and cracks. All defects were more pronounced when milling at microscale.

Similar results are shown in Fig. 2.34 when milling at macro- and micro-scale AISI 1045 steel by using machining condition presented in Fig. 2.19.

The microstructure below the micromilled surface was plastically deformed by work hardening in the direction of the tool feed as well as cracks were nucleated at the pearlite-ferrite interface by reaching aproximately 5 μm depth. Cementite structure were clearly extended in the feed direction. Cracks were produced by plastic deformation (cold hardening) inside and surface of the workpiece material.

Table 2.3 Main defects and causes in micro cut surface (Assis, 2013)

Surface damage	Main features	Some causes
Cavities	Discontinuities in the machined surface.	Due to the nature of the friction condition in the contact region of the workpiece with the cutting edge.
Debris	Discontinuities in the machined surface.	Crack nucleation and propagation in front of the cutting edge, with subsequent removal of material.
Dimples	Small depressions in the surface of the workpiece.	Formation and coalescence of microvoids along the path of the fracture process that involves a considerable level of localized plastic deformation.
Severe plastic flow	Material severely deformed with flow oriented toward the cutting direction.	Plastically deformed material, belongs to the latter part of the chip.
Incomplete chip formation	Material lateral flow on the workpiece surface.	It occurs mainly when the feed per tooth is smaller than the tool edge radius.
Grooves	Scratches or asperities along the workpiece surface due to the passage of the cutting edge.	This has its origin from the burnishing caused by the tool flank surface.
Grooves of micro chips	Deeper grooves than regular grooves formed by feedrate which is also parallel to the cutting direction.	Similar to the formation of grooves but with generation microburrs that may adhere to the surface of the part.
Chatter marks	Straight grooves parallel to each other and perpendicular to the cutting direction.	Related to the tool-workpiece with increased cutting speed and tool wear.
Prows	Small bumps on the machined surface.	Formed in machining with low cutting speed; it is associated with a Built-up-edge.
Burrs	Workpiece material portion shifted to the axial direction of the tool.	Side flow, plastic bending and tear of the workpiece material due to combination between cutting conditions (e.g., tool wear) and part properties (e.g., ductility).
Elastic recovery	Elevations unrelated to surface roughness, generated by the machining and/or dynamic effects of the machine tool.	Effect of different levels of resistance to deformation among crystal grain in the polycrystalline microstructure.
Material scraps	Material adhered to the surface of the part.	Originating in debris formation, cavities or voids.
Smeared material	Material plastically deformed in the opposite direction to the tool feed.	It arises due to the increased ratio between the feed force and the cutting force.
Steps	Subsequent sensitive undulations to severe plastic flows.	Its occurrence is related to the formation of severe plastic flow.
Cracks	Microcracks that penetrate into the workpiece surface.	Change in mechanical properties between the grains, residual stress and fatigue on the part surface.
Voids	Discontinuities in the machined surface.	Differences in yield strength between the phases of the microstructure.

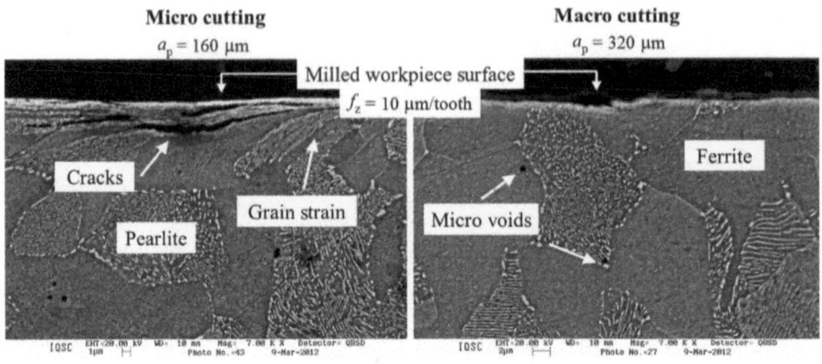

Figure 2.34 Surface damages when micro-(left) and macro-(right) milling AISI 1045 steel.

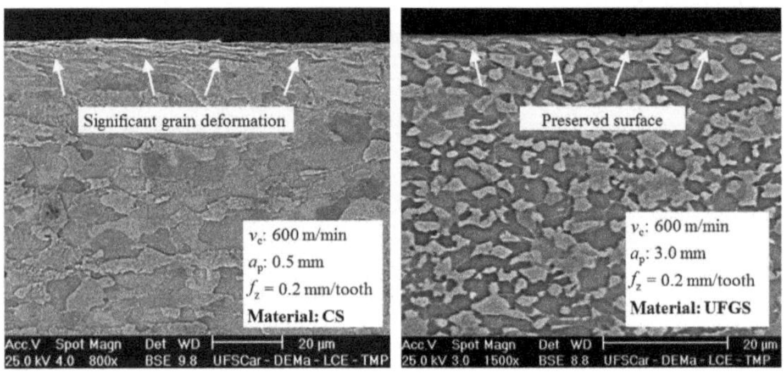

Figure 2.35 Surface damages due to macromilling of COSAR 60 CS (left) and UFGS (right).

This surface integrity was strongly impaired due to greater specific cutting energy (size effect) related to the larger grain mainly of the pearlite (43 μm size). On the other hand, macromilling preserved the workpiece surface integrity even with twofold depth of cut. Both pearlite and ferrite were not deformed and cracks were not nucleated inside microstructure. Only located microburrs were identified.

Fig. 2.35 (Assis, 2013) shows equivalent behavior of surface defects for the same workpiece materials presented in Fig. 2.31.

CS microstructure was deformed more significantly near the milled surface. This plastic response is resulted from larger GS which allow more strain within grain. Additionally, higher cutting speed contributed with grain stretching given the major shear rate for which workpiece material was submitted during machining.

However, very shallow depth of plastic deformation was found in UFGS material even when applying larger depth of cut. The higher hardness reached by grain refinement imposed strain resistance and preserved the surface integrity of machined surface. The decrease of GS increases the grain boundary area which act like barriers, hindering dislocations movement and increasing the strain resistance of material. These results are in accordance with the microhardness behavior presented in Fig. 2.31.

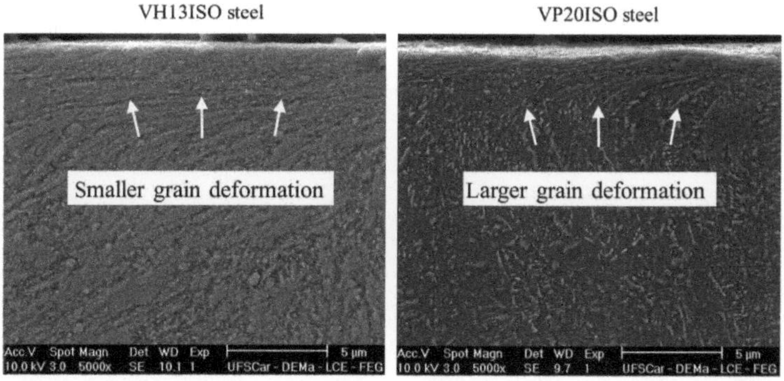

Figure 2.36 Microstructural deformation of the VH13ISO (left) and VP20ISO (right) steels near micro milled surface.

Finally, Bodziak et al. (2014) found intense microstructural deformation near micromilled surface of the VP20ISO and VH13ISO steels even with higher hardness than that of AISI 1045 and COSAR 60 steels (Fig. 2.36).

Despite the severe grain stretching other defects like cavities, cracks, voids and phase transformations were not observed. If reproduced in molds and dies machining, these functional surfaces will tend to increase the life time of the tooling once surface topography (skewness <0 and kurtosis <3) and residual stresses (compressive) are suitable. Fatigue mechanisms are greatly minimized under these surface topography conditions.

Plastic deformations were more meaningful in VP20ISO steel given its lower hardness, i.e., a certain higher ductility even with a martensitic microstructure.

2.5.6 Size effect

Fig. 2.37 shows the decrease of the specific cutting energy as a function of feed per tooth when milling slots in RSA 6061-T6 aluminum alloy at microscale cutting in transverse and longitudinal direction by using 2 flutes end mill cutter, tungsten carbide substrate, TiNAl coating, 2.7 μm cutting edge radius, 1 mm endmill diameter, 160 μm depth of cut, 60 m/min cutting speed, no cutting fluid and 13 mm cutting length.

It is observed and statistically demonstrated that tool path direction regarding workpiece extruding direction did not influenced the specific cutting energy when larger and smaller tool feeds were used, respectively. When matching part microstructural and machining scales the chip removal process behaved as isotropic, continuous and homogeneous. In addition, specific cutting energy increases more significantly from 0.6 μm/tooth tool feed because minimum cutting thickness is around 0.68–0.89 μm/tooth according to Oliveira et al. (2015). When feed per tooth is smaller than the minimum cutting thickness elastoplastic deformations are more prominent over surface and subsurface instead of shear deformation.

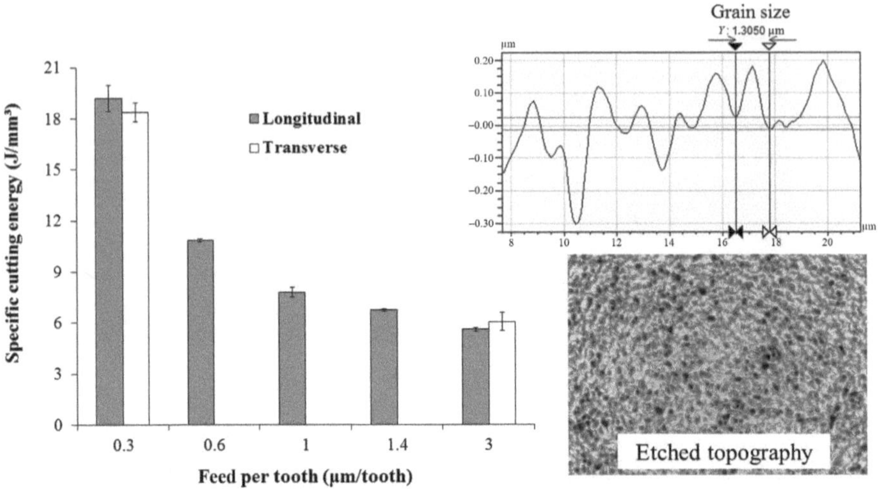

Figure 2.37 Size effect (left) when milling (slots) aluminum RSA 6061-T6 in transverse and longitudinal direction at micro-scale cutting and grain size measurement by 3D optical profiler (right).

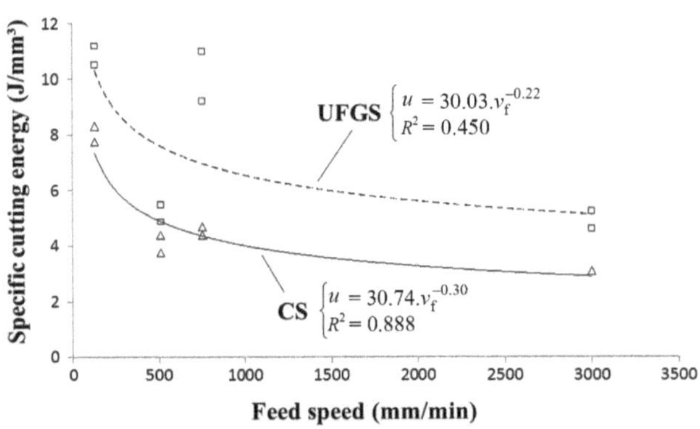

Figure 2.38 Size effect when milling COSAR 60 CS and UFGS at macroscale cutting.

Rigatti (2010) and Rigatti et al. (2013a) evaluated the relationship between workpiece GS and size effect by analyzing specific cutting energy when milling at macroscale cutting the same workpiece material and cutting conditions such as cited in Fig. 2.31. A full factorial design was implemented with 3 factors and 2 levels that is cutting speed (100 and 600 m/min), depth of cut (0.5 and 3.0 mm) and tool feed (0.05 and 0.2 mm/tooth). Radial depth of cut and cutting length were kept constant in 2.0 mm and 75 mm, respectively. Coolant and lubricant fluids were not considered.

Fig. 2.38 depicts the curves of the specific cutting energy as a function of feed speed for CS and UFGS materials. The results show that UFGS presents larger

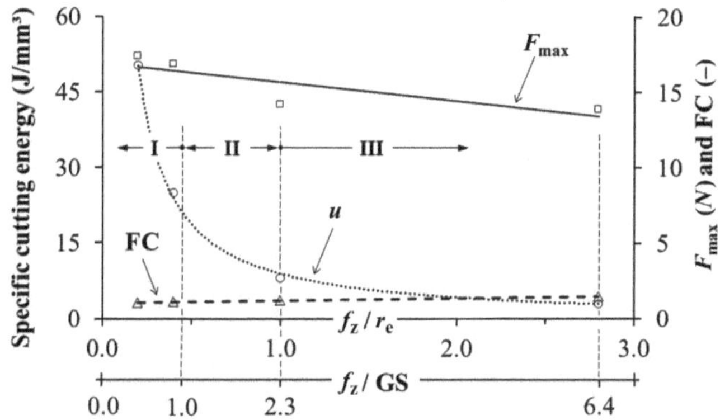

Figure 2.39 Variation of specific cutting energy (u), maximum cutting force (F_{max}) and friction coefficient (FC) as a function of the ratios f_z/r_e and f_z/GS.

values of the specific cutting energy when compared to that for CS irrespective to feed speed used. With decrease of feed speed (equivalent to feed per tooth), the size effect is characterized by an inscrease of specific cutting energy for both workpiece materials. Low amplitude of model coefficient R^2 is resulted from remaining martensite and bainite grains disperse on UFGS material matrix after thermomechanical process for grain refinement.

Although UFGS material having higher specific cutting energy, this material is less sensitive to size effect since the feed speed exponent is lower than that of CS. The multiplicative factor of the feed rate in the model equation is the specific energy to the feed speed of 1 mm/min.

Rigatti et al. (2013b) also evaluated the relationship between feed per tooth (f_z) and GS of the workpiece material on cutting force, friction at the chip-tool interface and specific cutting energy under dry macromilling of CS (COSAR 60 with 11 μm GS), 700 m/min cutting speed, 500 μm depth of cut, 2 mm radial depth of cut, and 5, 11, 25, and 70 μm/tooth feed per tooth (f_z). A two flute carbide endmill tool coated with PVD- TiNAl layer, 16 mm diameter and 25 μm edge radius was used in tests. Fig. 2.39 shows the main results.

The graphic presents two x-axes (f_z/r_e and f_z/GS) and three intervals: In (I) the plowing is the main mechanism observed from tool–material interaction where the specific cutting energy increased 1531% for $f_z <$ GS, while friction coefficient on tool rake face decreased 39% and cutting force increased 25% because of an unbalance of force generated from the occurrence of plowing and shearing of material at the cutting edge. The interval (II) between $f_z/r_e = 1$ and $f_z/\text{GS} = 1$ is indicative of the transition from a plowing-dominated to a shearing-dominated regime, commonly known as the minimum cutting thickness (h_{min}). Finally, the shearing can be considered majority when $f_z/r_e > 1$ given the decrement of cutting force and specific cutting force.

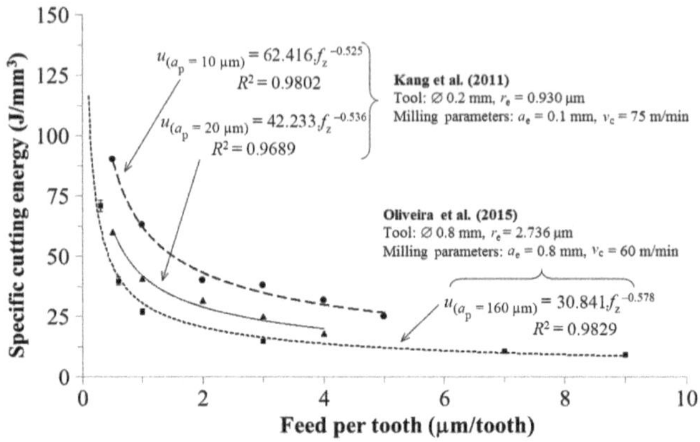

Figure 2.40 Size effect for AISI 1045 steel under different cutting conditions (Oliveira et al., 2015).

Finally, Fig. 2.40 presents a comparison among characteristic size effect curves produced by different authors and cutting conditions when milling AISI 1045 steel at microscale cutting.

For larger feed per tooth, the specific energy lies approximately between 10 and 25 J/mm^3. This magnitude is considered high for mechanical macromachining with defined geometry tool. For feed below 1 μm/tooth, the amplitudes range between 60 and 90 J/mm^3, which are equivalent to values found for abrasive processes. This makes sense, because h/r_e near or less than the unit resemble chip removal dynamics in grinding, in which the abrasive grain is rounded and cutting thicknesses are of the same order of magnitude of its radius.

Even micromilling the same workpiece material, the amplitude difference between the curves lies mainly in the depth of cut parameter. Oliveira et al. (2015) estimated lower specific cutting energies, when choosing higher depth of cut values. Meanwhile, Kang, Kim, and Seo (2011) applied depths of cut 8 and 16 times smaller, which raised the specific cutting energy.

It is worth keeping in mind that in any mechanical material removal process, to minimize energy consumption by volume of chip removed is a goal, which is equivalent to minimize the pressure on the chip cross sectional area of the tool during chip formation predominant by shear strain. When tool feed, depth of cut and cutting speed are increased up to certain limits, the specific cutting energy decreases and productivity increases, which is beneficial, but also increases the forces on the tool, the roughness of the workpiece and the cutting temperature, since chip area increase as well as the shear rate. In micromachining, these effects can be harmful, given the low rigidity and high wear rates of microend mill cutters. Therefore, it is important to find a balance when choosing the cutting parameters in micromachining.

Moreover, it is not usual to proceed micromachining near or below the tool edge radius, since under these conditions there are prevalence of elastoplastic deformation of the workpiece instead of chip removal by shear strain mechanism.

References

Afazov, S. M., Ratchev, S. M., & Segal, J. (2012). Prediction and experimental validation of micro-milling cutting forces of AISI H13 steel at hardness between 35 and 60 HRC. *International Journal of Advanced Manufacturing Technology*, *62*, 887–899.

Alting, L., Kimura, F., Hansen, H. N., & Bissaco, G. (2003). Micro engineering. *CIRP Annals*, *52*, 635–658.

Arai, R. (2008). *Estudo Teórico-experimental do efeito da flexão da ferramenta no processo de microfresamento*. Dissertation (MSc.), University of São Paulo (*in Portuguese*).

Aramcharoen, A., & Mativenga, P. T. (2009). Size effect and tool geometry in micromilling of tool steel. *Precision Engineering*, *33*, 402–407.

Assis, C. L. F. (2013). *Microfresamento de aços com grãos ultrafinos*. Thesis (PhD.), University of São Paulo (*in Portuguese*).

Becker, M., Santos, S. C., & Sales, W. F. (2005). *Integridade superficial em usinagem*, Jaboticabal, Cubo Multimídia (*in Portuguese*).

Black, J. T. (1989). Mechanics of chip formation. In: J. R. Davis (Ed.), *Metals handbook: Machining* (Vol. 16, pp. 7–18). Ohio: ASM.

Bodziak, S., et al. (2014). Surface integrity of moulds for microcomponents manufactured by micromilling and electro-discharge machining. *Journal of the Brazilian Society of Mechanical Sciences and Engineering*, *36*, 623–635.

Boothroyd, G., & Knight, W. A. (1989). *Fundamentals of machining and machine tools*. EUA: Marcel Dekker.

Brinksmeier, E., Riemer, O., & Twardy, S. (2010). Tribological behavior of micro structured surfaces for micro forming tools. *International Journal of Machine Tools and Manufacture*, *50*, 425–430.

Callister, J. R. W. D. (2007). Materials science and engineering: An introduction, John Wiley & Sons.

Câmara, M. A., Campos Rubio, J. C., Abrão, A. M., & Davim, J. P. (2012). State of the art on micromilling of materials, a review. *Journal of Materials Science and Technology*, *28*, 673–685.

Cardoso, P., & Davim, J. P. (2010). Optimization of surface roughness in micromilling. *Materials and Manufacturing Processes*, *25*, 1115–1119.

Chae, J., Park, S. S., & Freihet, T. (2006). Investigation of micro-cutting operations. *International Journal of Machine Tools and Manufacture*, *46*, 313–332.

Chen, M. J., Ni, H. B., Wang, Z. J., & Jiang, Y. (2012). Research in the modeling of burr formation process in micro-ball end milling operation on Ti-6Al-4V. *International Journal of Advanced Manufacturing Technology*, *62*, 901–912.

Chevrier, A., et al. (2003). Investigation of surface integrity in high speed end milling of a low alloyed steel. *International Journal of Machine Tools and Manufacture*, *43*(11), 1135–1142.

Chou, Y. K., & Song, H. (2004). Tool nose radius effect on finish hard turning. *Journal of Materials Processing Technology, 148*, 259–268.
Cohen, P. H. (1989). Forces, power, and stress in machining. In: J. R. Davis (Ed.), *Metals handbook: Machining* (16, pp. 13–18). Ohio: ASM.
Daves, J. R. (1989). Turning. In: J. R. Davis (Ed.), *Metals handbook: Machining* (16, pp. 135–159). Ohio: ASM.
Degarmo, E. P., Black, J. T., & Kohser, R. A. (1997). *Materials and processes in manufacturing*. EUA: Prentice-Hall.
Dib, M. H. M. (2013). *Fundamentos do fresamento: uma aplicação em microfresamento.* Dissertation (MSc.), University of São Paulo (*in Portuguese*).
Dieter, G. E. (1988). *Mechanical metallurgy*. London: McGraw-Hill.
Dornfeld, D., Min, S., & Takeuchi, Y. (2006). Recent advances in mechanical micromachining. *CIRP Annals, 55*, 745–768.
Ersoy, A., & Atici, U. (2004). Performance characteristics of circular diamond saws in cutting of different type of rocks. *Diamond and Related Materials, 13*, 22–37.
Ferreira, L. O. S., & Fernandes, J. C. B. (2003). Microdispositivos na atualidade. *Publicação da Associação Brasileira de Ciências Mecânicas, 9*, 8–10, (*in Portuguese*).
Field, M., Kahles, J. F., & Koster, W. P. (1989). Surface finish and surface integrity. In: J. R. Davis (Ed.), *Metals handbook: Machining* (16, pp. 19–36). Ohio: ASM.
Gorczyca, F. E. (1987). *Application of metal cutting theory* (p. 298). New York, NY: Industrial Press.
Groover, M. P. (2002). *Fundamentals of modern manufacturing*. USA: Wiley Publ.
Hertzberg, R. W. (1995). *Deformation and fracture mechanics of engineering materials*. New York, NY: J. Wiley.
Jackson, M. J. (2008). Micro and nanomachining. . In: J. P. Davim (Ed.), *Machining: Fundamentals and recent advances*. Portugal: Springer-Verlag London Limited.
Kalpakjian, S., & Schmid, S. R. (2000). *Manufacturing engineering and technology*. USA: Prentice-Hall.
Kang, I.-S., Kim, J.-S., & Seo, Y.-W. (2011). Investigation of cutting force behaviour considering the effect of cutting edge radius in the micro-scale milling of AISI 1045 steel. *Proceedings of the Institution of Mechanical Engineers Part B Journal of Engineering Manufacture, 225*, 163–171.
Kienzle, O. (1952). Die Bestimmung von Kräften und Leistungen an Spanenden Werkzeugen und Werkzeugmaschinen. *Zeitschrift Des Vereines Deutscher Ingenieure, 94*, 299–305.
Kim, C., Bono, M., & Ni, J. (2002). *Experimental analysis of chip formation in micromilling. North American Manufacturing Research Conference,* West Lafayette, 2002. SME: Dearborn.
King, R. I., & Hahn, R. S. (1986). *Handbook of modern grinding technology*. New York, NY: hapman and Hall.
Klocke, F., Gerschwiler, K., & Abouridouane, M. (2009). Size effects of micro drilling in steel. *Production Engineering Research and Development, 3*, 69–72.
Komatsu, T., et al. (2012). Effect of crystal grain size in stainless steel on cutting process in micromilling. *Procedia CIRP, 1*, 150–155.
Kronenberg, M. (1966). *Machining sciences and applications: Theory and practice for operation and development of machining processes*. New York, NY: Pergamon Press.
Lauro, C. H., et al. (2014). Influence of the austenitic grain size variation on the surface finishing in the micromilling process of the hardened AISI H13 steel. *Rev Matéria, 19*, 235–246.

Lee, K., & Dornfeld, D. A. (2004). A study of surface roughness in the micro-end-milling process, Berkeley. Available from <http://escholarship.org/uc/item/51r6b592>. Accessed 14.12.12.

Lekkala, R., et al. (2011). Characterization and modeling of burr formation in micro-end milling. *Precision Engineering*, *35*, 625–637.

Li, P., et al. (2010). Micromilling in thin ribs with high aspect ratios. *Journal of Micromechanics and Microengineering*, *20*, 1–10.

Liu, X., Devor, R. E., & Kapoor, S. G. (2004). The mechanics of machining at the microscale: Assessment of the current state of the science. *Journal of Manufacturing Science and Engineering*, *126*, 666–678.

Malekian, M., et al. (2012). Modeling of minimum uncut chip thickness in micro machining of aluminum. *Journal of Materials Processing Technology*, *212*, 553–559.

Malkin, S. (1989). *Grinding technology. Theory and applications of machining with abrasives*. EUA: Ellis Horwood.

Masuzawa, T. (2000). State of the art of micromachining. *CIRP Annals*, *9*, 473–488.

Melo, A. C. A., et al. (2003). *Estudo da variação da temperatura de corte no fresamento frontal*. Congresso Brasileiro de Engenharia de Fabricação, *Uberlândia, 2003*. Uberlândia: UFU, (*in Portuguese*).

Mian, J., Driver, N., & Mativenga, P. T. (2011). Identification of factors that dominate size effect in micro-machining. *International Journal of Machine Tools and Manufacture*, *51*, 383–394.

Moreira, S. R. S. (2012). *Energia específica de corte e integridade superficial no microfresamento do aço ABNT 1045*. Dissertation (MSc.), São Paulo State University (*in Portuguese*).

Oliveira, F. B., et al. (2015). Size effect and minimum chip thickness in micromilling. *International Journal of Machine Tools and Manufacture*, *89*, 39–54.

Oliveira, F. B. Estudo dos mecanismos governantes do efeito de escala na microusinagem. Dissertation (MSc.), São Paulo State University (*in Portuguese*).

Park, K. T., et al. (2000). Thermal stability and mechanical properties of ultrafine grained low carbon steel. *Materials Science and Engineering A*, *293*, 165–172.

Polini, W., & Turchetta, S. (2004). Force and specific energy in stone cutting by diamond mill. *International Journal of Machine Tools and Manufacture*, *44*, 1189–1196.

Rigatti A. M. Y. (2010) *Avaliação da força de usinagem e energia específica de corte no fresamento com alta velocidade de corte*. Dissertation (MSc.), São Paulo State University (*in Portuguese*).

Rigatti, A. M. Y., et al. (2013a). *Computational method for calculation of the specific cutting energy*. Proceedings of the 22nd International Congress of Mechanical Engineering, Ribeirão Preto, 2013. Rio de Janeiro: ABCM.

Rigatti, A. M. Y., et al. (2013b). *Material removal mechanism and cutting energy analysis at tool cutting edge scale in end milling. Proceedings of the 13th Euspen International Conference, Berlin, 2013*. Cranfield: Euspen.

Robinson, G. M., & Jackson, M. J. (2005). A review of micro and nanomachining from a materials perspective. *Journal of Materials Processing Technology*, *167*, 316–337.

Rodrigues A.R. (2005) *Estudo da geometria de arestas de corte aplicadas em usinagem com altas velocidades de corte*. Thesis (PhD.), University of São Paulo (*in Portuguese*).

Rodrigues, A. R., et al. (2012). Surface integrity analysis when milling ultrafine-grained steels. *Mat Res*, *15*, 125–130.

Salmon, S. C. (1992). *Modern grinding process technology*. EUA: McGraw-Hill.

Santos, A. G. (2016). *Microfresamento de aço inoxidável duplex UNS S32205 com ferramentas de metal duro revestidas de nitreto de titânio*. Dissertation (MSc.), Federal University at Uberlândia (*in Portuguese*).

Saxena, A. (2000). Fatigue crack growth testing. In: S. R. Lampman (Ed.), *Metals handbook: Mechanical testing and evaluation* (Vol. 8, pp. 1714–1721). Ohio: ASM.

Schmidt, J., & Tritschler, H. (2004). Micro cutting of steel. *Microsystem Technologies*, *10*, 167–174.

Schroeter, R. B. (1999). *Medição e análise de forças na usinagem de ultraprecisão de materiais de comportamento duro e frágil*. Congresso Brasileiro De Engenharia Mecânica, Águas de Lindóia, 1999. Campinas: ABCM, (*in Portuguese*).

Shaw, M. C. (1984). *Metal cutting principles*. New York, NY: Oxford Science Publications.

Shaw, M. C. (1995). *Principles of abrasive processing*. New York, NY: Oxford University Press.

Shaw, M. C. (1996). Energy conversion in cutting and grinding. *Annals of the CIRP*, 101–104.

Silva Neto, O. V., & Balancin, O. (2007). Efeito da precipitação de cementita na formação de contornos de angulo e refino de grão ferrítico em aço baixo carbono. *Acta Microscopica*, *16*, 31–37 , (*in Portuguese*).

Simoneau, A., Ng, E., & Elbestawi, M. A. (2006). Chip formation during microscale cutting of a medium carbon steel. *International Journal of Machine Tools and Manufacture*, *46*, 467–481.

Simoneau, A., Ng, E., & Elbestawi, M. A. (2007). Modeling the effects of microstructure in metal cutting. *International Journal of Machine Tools and Manufacture*, *47*, 368–375.

Sooraj, V. S., & Mathew, J. (2011). An experimental investigation on the machining haracteristics microscale end milling. *International Journal of Machine Tools and Manufacture*, *56*, 951–958.

Stolyarov, V. V., et al. (2003). Ultrafine-grained Al-5 wt.% Fe alloy processed by ECAP with backpressure. *Materials Science and Engineering A*, *357*, 159–167.

Vogler, M. P., Devor, R. E., & Kapoor, S. G. (2004). On the modeling and analysis of machining performance in micro-end milling. *Journal of Manufacturing Science and Engineering*, *126*, 685–705.

Wayman, M. L., & Juleff, G. (1999). *Historical Metallurgy*, *33*, 26–42.

Weule, H., Hüntrup, V., & Tritschler, H. (2001). Micro-cutting of steel to meet new requirements in miniaturization. *CIRP Annals*, *50*, 61–64.

Xianghua, L., et al. (2009). High performance low cost steels with ultrafine grained and mult-phased microstructure. *Science in China Series E: Technological Sciences*, *52*, 2245–2254.

Yun, H. T., et al. (2011). Ploughing detection in micromilling processes using the cutting force signal. *International Journal of Machine Tools and Manufacture*, *51*, 377–382.

Micromachining technique based on the orbital motion of the diamond tip

3

Y.D. Yan, B. Xue and X.S. Zhao
Harbin Institute of Technology, Harbin, Heilongjiang, China

3.1 Introduction

The microfluidic technique is developing rapidly with the development of the micromachining technique. The typical channel dimension on the microfluidic chip is several hundreds of micrometers. With the development of the microfluidic technique, smaller dimensions from several micrometers to nanometers are required. The existing micromachining techniques have difficulty meeting some of the requirements, such as dimensions, materials, accuracy, and complexity of the microchannels. Moreover, micro- or nanostructures also have significant importance in the fields of medicine, electronics, and optics. Therefore, it is meaningful to study novel micromachining techniques.

The micromachining technique mainly includes the lithography, the chemical reaction based the micromachining and micromechanical machining technique. The lithography technique is a batch processing method. The disadvantages are limited processing materials, difficulty to process three dimensional (3D) microstructures, and high cost (Madou, 2002). The chemical reaction based the micromachining technique can easily fabricate 3D microstructures. However its shortcomings include a difficulty to precisely control the machined depth and remove the damage of the subsurface, owing to the thermal and chemical reactions (Brousseau, Dimov, & Pham, 2010). The micromechanical machining technique can machine complex microstructures on many kinds of materials, with a perfect surface quality and form accuracy (Dornfeld, Min, & Takeuchi, 2006). Simultaneously, our previous works have verified that the mechanical machining can reach to a nanometer scale, using the atomic force microscope (AFM) based nanoscratching method. Within twenty micrometers, 3D micro/nanostructures are successfully machined (Yan et al., 2010; Yan, Sun, & Dong, 2007; Yan, Xue, Hu, & Wu, 2015). Thus, the micromechanical machining technique has attracted more attention in recent years.

The micromilling technique has already been used to machine microchannels (Vazquez, Rodriguez, Elias-Zuniga, & Ciurana, 2010). However, with a decrease in dimension, the low stiffness of the cutter and materials piled up are prone to break the cutter, owing to its high speed spindle rotation. This condition is more serious when the diameter of the cutter is less than 0.2 mm. It is very difficult to machine

Microfabrication and Precision Engineering. DOI: http://dx.doi.org/10.1016/B978-0-85709-485-8.00003-6

complex 3D microstructures or microchannels with a width of less than 0.1 mm (Tansel, Rodriguez, Trjillo, Paz, & Li, 1998). Some studies have been carried out to improve this technique, such as the manufacturing of a smaller milling cutter with a focused ion beam method, in order to machine smaller structures (Friedrich & Vasile, 1996), and the employment of vibrations to reduce the tool wear and improve the surface quality (Shen, Zhang, Xing, & Zhao, 2011). However, because the high speed spindle must be used in the conventional micromilling process, microstructures with high accuracy cannot be achieved at an affordable cost.

At present, the AFM-tip based nanomechanical machining technique is confirmed to be a potential method to machine micro- or nanostructures (Yan, Xue, et al., 2015). However the low machining efficiency of this technique restricts its application in the micro- and nanofluidic fields. Some scholars put forward some new methods to solve this problem. For example, Gozen et al. (2012a, 2012b) proposed the AFM-tip based nanomilling technique. The AFM tip is directly moved by a piezoelectric actuator in the horizontal or vertical plane with circular trajectories. Nanostructures with the width of 1 μm and the depths of less than 500 nm ere successfully machined by this method with a high efficiency. Geng et al. (2013) combined the AFM tip's reciprocal scanning and a movement of a high precision stage to develop a different novel AFM-tip based nanomilling method. Using this method, the nanochannels with the width of 5–10 μm and the depth of 200–500 nm were fabricated with a length of 1 mm. Obviously, the AFM-tip based nanomilling process is a potential method to machine the nanochannels on the nanofluidic chip. But for the microchannels with the width from 10 μm to 100 μm and the depths larger than 1 μm, this technique is not feasible.

Therefore, there is a dimensional gap between the conventional micromilling process and the AFM-tip based nanomilling process. Recently, Heamawatanachai and Bamberg (2009, 2010) presented a novel milling process based on an orbital motion of a diamond tip. This technique takes advantage of the AFM-tip based nanomilling process and overcomes the disadvantage induced by the high rotation spindle of the micromilling process. In their works, a conical diamond tip is used as the microcutter that is driven by a piezoelectric tube scanner instead of a high speed spindle. They have already showed that this technique has the ability of machining the required dimensions of the microstructures mentioned before. However, in their studies, only the conical diamond tip is used, and due to the deflections of piezoelectric tube, the posture of the tool is constantly changing along with the movement. More studies should be done to ask questions such as: (1) What's effect of the tip's shape on the machining process? (2) Is the pyramidal diamond tip more feasible for machining? The piezoelectric tube scanner is used to move the tip. The frequency of its motion can meet this requirement. However, it must be compensated, owing to the fact that the coupling effects of the movements are in three directions, and the machining dimension is restricted by the scanner's scanning range. Therefore, in this chapter, a similar setup based on the orbital motion of the diamond tip is established. The pyramidal and conical diamond tips are both employed. Their effects on the machining process are studied. A nanopiezo stage with a larger moving range is used instead of the piezoelectric tube scanner, which can keep the posture of the tool constant during the machining process. The mechanisms of the burr formation and the removal states of the material are studied in detail. Finally, the complex microstructures are machined by this technique.

3.2 Principle of micromachining using the orbital motion of the diamond tip

Figs. 3.1A and B show the schematic of the micromilling process and the tip's orbital motion based the micromachining process, respectively. The pyramidal tip is used as the example. The symbol *V* and the arrow represent the moving direction of the workpiece during machining. The micromilling process in Fig. 3.1A, shows the microcutter rotating around its center, usually driven by a high speed spindle. The section of the microchannel machined by this process is a rectangle. The width of the channel is determined by the diameter of the microcutter used. As shown in Fig. 3.1B, the pyramidal diamond tip revolves around an orbital center. This motion is realized by a 3D nanopositioning stage in this study, which will be explained in the next section. The section of the microchannel machined by this method is trapezoidal and influenced by the diamond tip shape. The width of the channel is basically determined by the radius of the orbital motion. The radius is changed according to the moving range of the nanopositioning stage. But for the micromilling process, it is impossible to machine the channel with a width less than the diameter of the microcutter. The relative motion between the tool and the workpiece in three dimensions can be realized in the same way with the conventional micromilling process. The coarse moving stages can be used. The main difference between the micromilling process and the orbital motion based micromachining process is that no high speed and high accuracy air bearing spindle is employed in the latter condition.

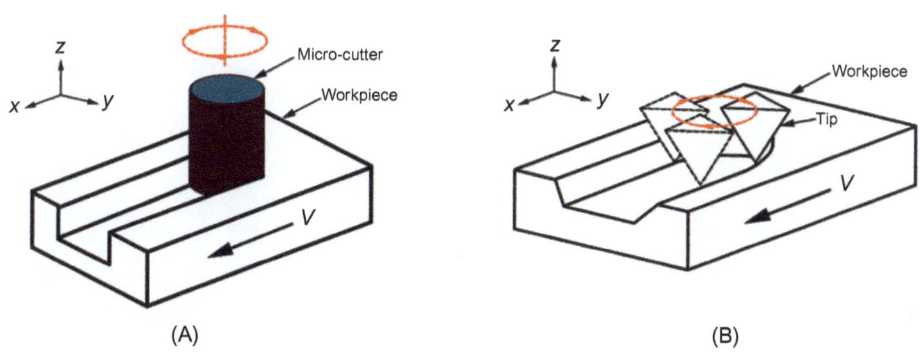

Figure 3.1 Schematic of the micromilling process (A) and the pyramidal tip's orbital motion based micromachining process (B).

3.3 Micromachining setup and test of the stage's trajectory

3.3.1 Establishment of the micromachining setup and the machining procedure

Figs. 3.2A and B show the schematic and photo of the micromachining setup based on the orbital motion of the tip. The system parts mainly consist of: a nanopiezo stage; *X*-, *Y*-, and *Z*- precision stages; the diamond tip; the aligning stage; and a force sensor. The details of each part are described as follows.

1. Nanopiezo stage. In order to make the tip move in three directions, the piezo is used as the driver, and the flexible hinge is used as the guide for the nanostage. The resistance strain gauge sensor is used to measure the displacement as the feedback signal. In this study, this stage (XP-611.XYZ) is provided by Xin Ming Tian Science and Technology Co., Ltd. China. The open-loop and close-loop travels are 120 and 100 μm, respectively. The resolution is 4 nm. The unloaded resonant frequencies in the, *X*, *Y*, and *Z* axis are 350, 220, and 250 Hz, respectively. This stage is driven by a controller through a NI USB-6259.
2. *X-Y-Z* precision stage. It includes *X*-, *Y*-, and *Z*-stage which are driven by the servo motors with the encoder to measure the displacement as the feedback signal. Its moving range is 100 mm × 100 mm × 100 mm. The repeatability is 5 μm. The resolution is 1 μm, and the maximum moving velocity is 40 mm/s. This part is provided by the Beijing Micro/Nano Optical Instrument Co., Ltd. China.
3. Diamond tip. The pyramidal and conical diamond tips are bought from the Synton-MDP company, Switzerland. The pyramidal tip is the cube corner nanoindentation tip with the face angle of 35.26°. The cone angle of the conical tip is 60°, and the tip radius is less than 1 μm. The SEM images of the used tips are presented in Fig. 3.3.
4. Aligning stages. In this system, as shown in Fig. 3.2A, there are two different aligning stages: one is the aligning stage for sample leveling and the other is for the tool angle adjusting stage, to change the cutting angle when using the pyramidal diamond tip. Both are adjusted by the precision screw. The aligning stage has two freedoms, and the moving

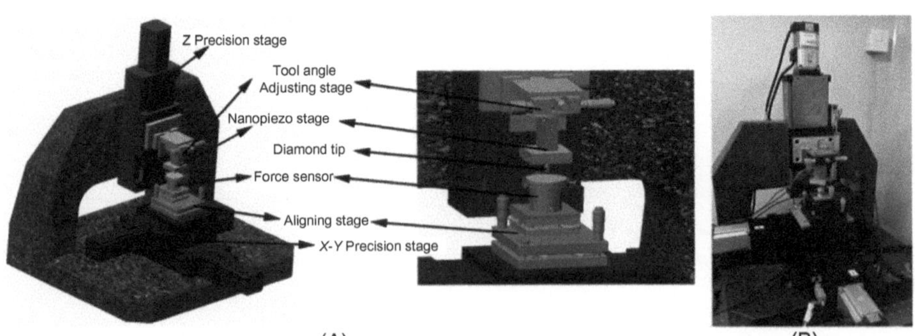

Figure 3.2 Schematic (A) and photo (B) of the micromachining system based on the orbital motion of the diamond tip.

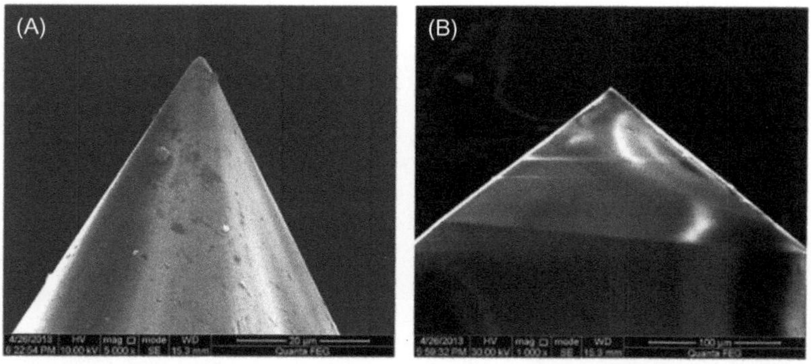

Figure 3.3 SEM images of the conical (A) and pyramidal (B) diamond tips.

range is less than $\pm 3°$. The tool angle adjusting stage uses the rolling guide with the minimum reading of 55.8″ within the range of 360°.

5. Force sensor. The force sensor is employed to measure the cutting forces. The other function of the force sensor is to level the sample surface with the aligning stage by measuring the contact forces at different locations. It is manufactured by the ATI company of USA. The model is Nano17 Ti. The measurement ranges in *X* and *Y* directions are $\pm$ 8N. The range in *Z* direction is $\pm$ 14.1N. The natural frequency is 3 KHz.

The micromachining procedure based on this setup is described as:

1. The sample is adjusted by the aligning stage to make the sample surface parallel to the horizontal plane. The aligning method is to guide the diamond tip to the sample surface with a fixed normal force, measured by the force sensor at three different locations. Simultaneously, the coordinates of the diamond tip in the *Z* direction are recognized by the nanopiezo stage. By adjusting the screw of the aligning stage, the coordinates of three points in the *Z* direction are almost the same values. Then the sample's inclination is successfully adjusted.
2. The tool angle adjusting stage changes the edge direction of the pyramidal diamond tip to achieve a better removal state. It is no need of adjusting if the conical tip is used. Then the diamond tip is driven by the nanopiezo stage with the required frequency, and the moving range is in the *X*-*Y* directions.
3. The *Z* precision stage brings the nanopiezo stage to approach the sample surface within the distance of 50 μm. After that, the nanopiezo stage in the *Z* direction brings the diamond tip to contact the sample surface by monitoring the force sensor's signal and then controls the machined depth by moving the diamond tip future.
4. Finally, the *X*-*Y* precision stage moves the sample to perform the micromachining process, and the required microstructures will be obtained.

3.3.2 Test of the trajectory of the nanopiezo stage in the orbital motion

For controlling the orbital motion of the diamond tip, it is very important to know the performance of the nanopiezo stage. In this study, the moving trajectories of the nanopiezo stage in the unload and load states are characterised respectively.

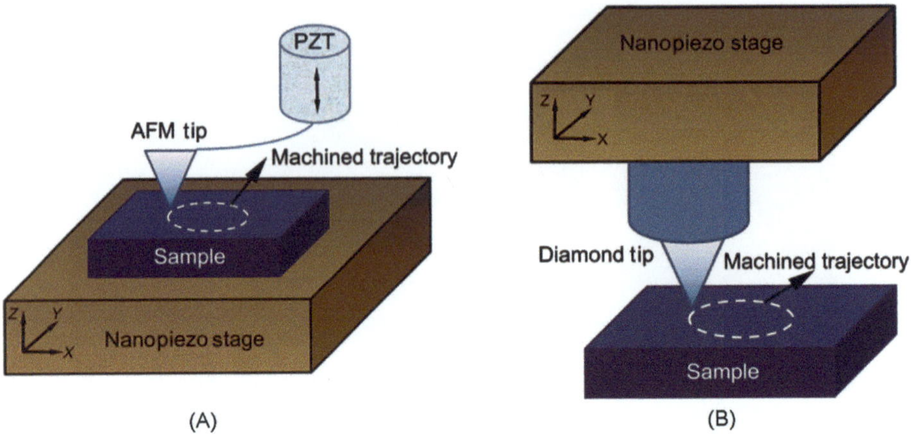

Figure 3.4 Schematic of the test of trajectories for the unload (A) and load (B) states of the nanopiezo stage based on the orbital motion.

As shown in Fig. 3.4A, the nanopiezo stage is put in an upward position and brings the polymer (polymethyl methacrylate, PMMA) plate to carry out the orbital motion. This setup is then placed under a commercial AFM head (Q-Scope250 of Ambios Company, USA). The AFM tip applies a normal load of several micro Newtons. It is enough to scratch a nanogroove on the PMMA plate. By inspecting the scratched groove, we can find the performance of the nanopiezo stage with different moving ranges and frequencies. Under this condition, the nanopiezo stage drives the sample rather than the diamond tip and no real cutting forces are applied on it. Therefore, we called this state as "the unload state."

Fig. 3.5 shows the trajectories of the stage with the diameters of 10, 30, and 50 μm and the frequencies of 80, 100, and 120 Hz. In this test, AFM works in the contact mode. The normal load is about 10 μN. Under this condition, the obvious scratches can be found on the PMMA surface. From the figure, for all of the diameters, the trajectory is perfect for the frequency of 80 Hz (Figs. 3.5A, D, and G). With the same diameter, using the lower frequency is confirmed to lead to the same conclusions by more tests. When the frequency is larger than 80 Hz (e.g., 100 Hz in Figs. 3.5B, E, and H), the circular trajectory tends to be an ellipse. With the increase of the frequency, the scratched grooves can hardly be formed on the sample surface, and the trajectories are more of a mess (Figs. 3.5C, F, and I). The reason is due to the resonance frequency of the stage and the load of the stage induced by the sample weight. Moreover, a larger moving range corresponds to a lower frequency. Since the nanopiezo stage works in the open-loop state, all of the diameters are larger than the setting values that are listed on the right row of the figure. Based on the above tests, the maximum frequency of 80 Hz and the maximum diameter of 50 μm are the good values in the unload state.

As shown in Fig. 3.4B, the nanopiezo stage is placed as it works and drives the diamond tip to carry out the orbital motion. Without the motions of the sample in the *X-Y* directions, the diamond tip only scratches a circle on the PMMA surface. This circle trajectory can reveal the performance of the Nanopiezo

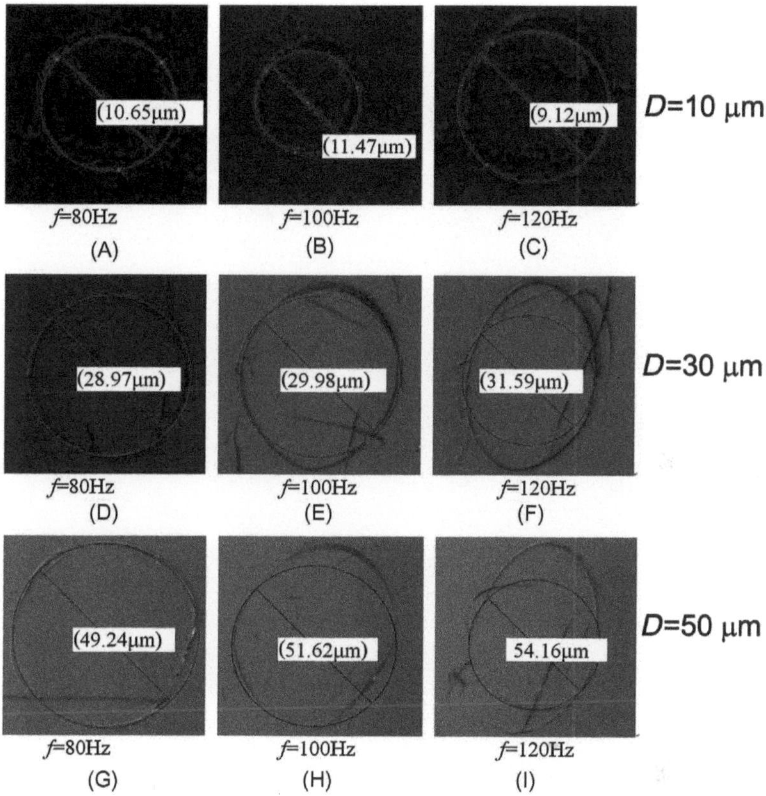

Figure 3.5 Optical images of the trajectory of the nanopiezo stage in the unload state with different revolving diameters and frequencies. (A–C) are with same diameter (10 μm) but different frequencies (80, 100, and 120 Hz). (D–F) are with the 30 μm diameter and frequencies of 80, 100, and 120 Hz. (G–I) are the 50 μm diameter and frequencies of 80, 100, and 120 Hz.

Stage with different moving ranges and the frequencies in the real machining process. Under this condition, the nanopiezo stage directly drives the diamond tip, and the real cutting forces are applied on it. Thus, we called this state as "the load state."

Fig. 3.6 shows the trajectories of the diamond tip with the diameters of 10, 30, and 50 μm and the frequencies of 40–90 Hz. In these tests, the conical diamond tip is used, owing to the easy observation of the trajectory and elimination of the tip geometry effect. The machined depth is about 1 μm. When the diameter is 10 μm (Figs. 3.6A, B, and C), an increase in the frequency leads to a bad trajectory shape. When the frequency is larger than 80 Hz, the circular groove changes to the ellipse groove. Under this condition, 70 Hz is the critical value to achieve a better result. It is the difference between the diameter of 14.49 μm in Fig. 3.6A and the setting value of 10 μm. The reason is because the measured trajectory shown in Fig. 3.6A is the envelope of the outer edge of the conical tip. It is the same condition for other tests in Fig. 3.6. For the diameter of 30 μm (Figs. 3.6D, E, and F), the critical

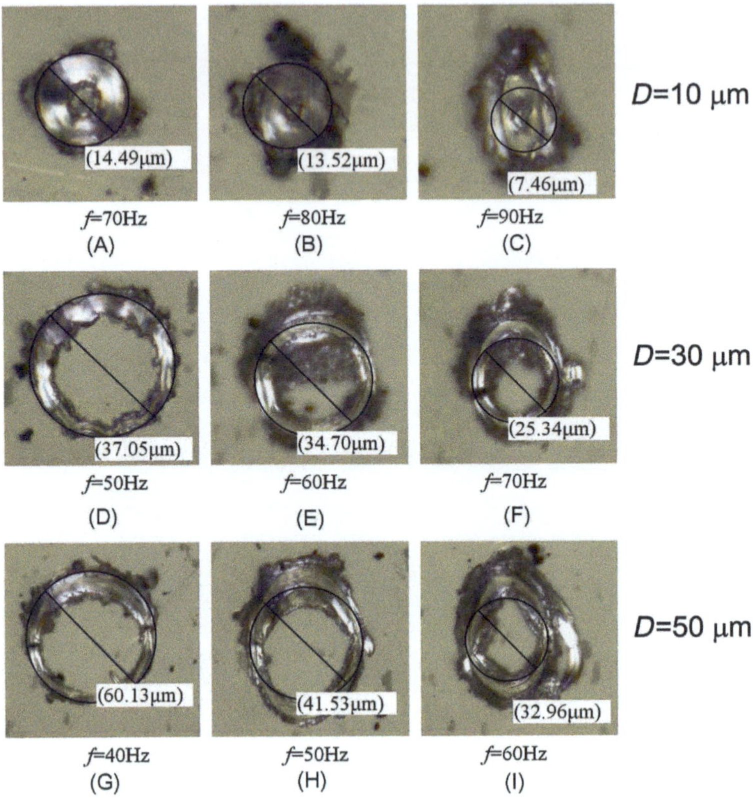

Figure 3.6 Optical images of the trajectory of the nanopiezo stage in the load state with different revolving diameters and frequencies. (A–C) are 10 μm diameter, (D–F) are 30 μm diameter, and (G–I) are 50 μm diameter.

frequency is 50 Hz. For the diameter of 50 μm (Figs. 3.6G, H, and I), the critical frequency is 40 Hz. Because under this condition, the real load is applied on the tip, and all the critical frequencies lower than 80 Hz are used in the unload state. All these critical frequencies are obtained with the machined depth of 1 μm. Results from more tests show that, with a larger machined depth of less than 4 μm, the frequency of 10 Hz is feasible for most machining conditions based on our system. Actually, such low frequency will lead to a low machining efficiency.

3.4 Micromachining mechanism using the orbital motion of the tip

3.4.1 Comparison of chip states with the conical and pyramidal tips

The pyramidal and conical diamond tips are two common tips which can be purchased from many manufacturers of the nanoindenaion device. The pyramdial

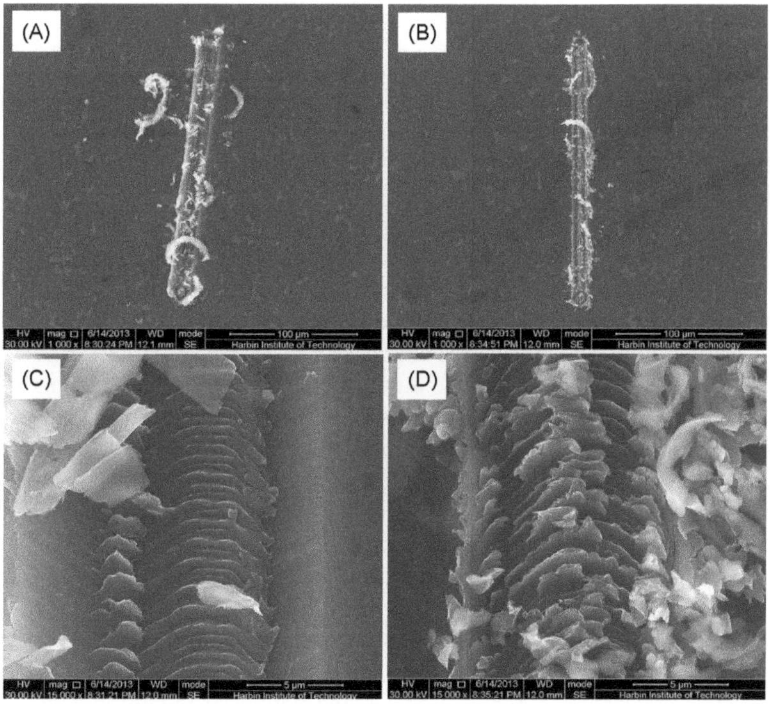

Figure 3.7 SEM images of the machined channels with the pyramdial tip ((A) and (C)) and the conical tip ((B) and (D)).

diamond tip has three edges. It may be more suitable to be used as a cutting tool; although, the conical tip has already been used to perform the micromachining process (Heamawatanachai & Bamberg, 2009). The following tests are carried out to clarify which one is better.

Using two kinds of tips, the microchannels are machined as shown in Fig. 3.7. The diameter of the tip motion is 10 μm. The frequency of the nanopiezo stage is 30 Hz. The machined depth is 1 μm. The machining velocity of the precision stage is 15 μm/s. The length of the channel is 200 μm. The sample material is aluminum alloy (2024). Figs. 3.7A and C are the SEM images of the channel and its local enlarged part, using the pyramidal diamond tip, respectively. Figs. 3.7B and D are the SEM images of the channel and its local enlarged part, using the conical diamond tip, respectively. From the figure, the following conclusions are obtained. For the pyramidal diamond tip, the long belt-type chips are achieved, indicating that the materials are efficiently removed. Thus, little burrs are found on both edges and the bottom of the channels. For the conical diamond tip, on the contrary, large amount of burrs are formed at both sides and the bottom of the groove. Few belt-type chips are formed. The difference is mainly due to the real cutting angle during the orbital motion of the tip. Actually, the conical diamond tip is suitable for the machining of brittle materials because of the large negative rake angle during machining. In this study, we mainly focus on the machining mechanism with the pyramidal diamond tip.

3.4.2 Difference between the micromilling process and this technique

As metioned in Section 3.2 in the miromilling process, the microcutter rotates with a center. In the tip orbital motion based micromachining process, the tip revolves, and the relative position between the tip and the workpiece changes with the revolution. Fig. 3.8 shows the schematics of two machining processes.

In Fig. 3.8A, the solid and dotted lines represent the original and new positions of the cutter in the micromilling process. The shadow part shows the removed materials. In the milling process, because the rake face of the cutter is a curved surface and it rotates around the point O, the rake angle is constant. With the feed of the tool, the uncut chip thickness (h) changes from zero to the maximum value and back to the zero for each edge of the cutter. For the following machining tests, the state is theoretically kept the same. In Fig. 3.8A, the solid and dotted triangles show the original and new positions of the pyramidal diamond tip, respectively. In the micromachining process, the rake angle (γ) changes with the revolution of the tip, and the uncut chip thickness varies according to different postions of the tip presented by the h_2. This process is different from the micromilling process. Fig. 3.8B shows the details of one condition of the relationship between the tip orientation and the feeding direction, in the machining process of one revolution of the pyramidal diamond tip. The term, f, shows the feeding direction and ω shows the revolving direction. In one revolving cutting process, there are three kinds of states: single-edge cutting, Courses 1 and 3, and double-edge cutting, Course 2. The solid dark triangle shows the position of the diamond tip. It can be found that the length of the Course 1 is the longest, and that of the Course 3 is the shortest. Because two cutting edges would together join in the machining process, Course 2 is more complex than Courses 1 and 2. For Course 1 and 2, with the movement of the diamond tip, the rake angle (γ) and the uncut chip thickness are constantly changing. For Course 3, the rake angle (θ) is positive and changing with the tip revolution. The shape of the prescribed removed material is presented in Fig. 3.8C, which is different from the shadow part of the milling process shown in Fig. 3.8A.

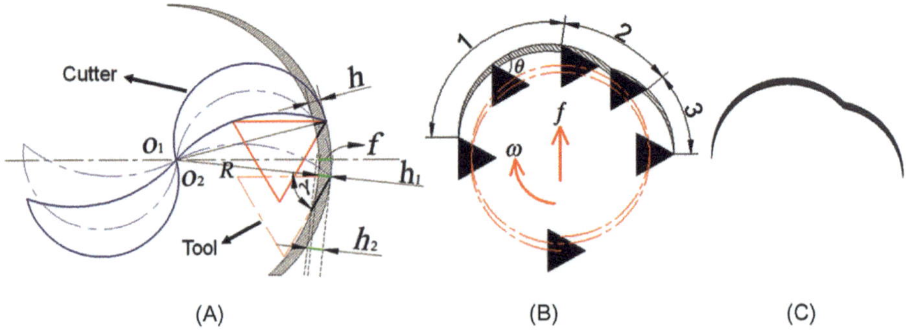

Figure 3.8 Schematics of the uncut chip thicknesses in the micromilling process (A) and the tip orbital motion based micromachining technique (B). (C) is the uncut chip thickness in one circle extracted from (B).

The tip orbital motion based micromachining is more complex. More processing experiments should be carried out to find out the influencing factors of this process.

3.4.3 Determination of the uncut chip thickness and the cutting rake angle

In order to understand the machining emchanism, the uncut chip thickness and the cutting rake angle of the tip orbital motion based micromachining should be achieved. The cutting process of the tip edge is looked on as an oblique cutting with a varied cutting rake angle, and while the tip is revolving, two or three cutting edges would participate in the cutting process. This is a complicated cutting process. For simplifying this process, a two-dimensional (2D) cutting plane as shown in Fig. 3.8A is used to calculate the uncut chip thickness and the cutting rake angle.

$$\frac{\sin(\varphi)}{oo'} = \frac{\sin(\pi - \theta_2)}{R} = \frac{\sin(\theta_1)}{o'A} \tag{3.1}$$

$$h = R - o'A \tag{3.2}$$

Fig. 3.9 shows the geometric relationship for calculation of the uncut chip thickness and the cutting rake angle in the milling process. Due to a fixed cutting rake angle, when the tool turns to the angle θ_2, the corresponding uncut chip thickness (h) can be calculated by the analytical method according to the Eqs. (3.1) and (3.2). However, in the machining process of the tip orbital motion, owing to the tip translantion movement of the revolution, the cutting rake angle continuously varies, and hence the uncut chip thickness is different from that in the milling process, especially for the pyramidal tip. As shown in Fig. 3.8A, when the center of the revolution arrives at the position O_2, the uncut chip thickness, based on the milling process, is represented by h_2, which is the distance between the intersection of the moving radius (R) and the machined surface, and the intersection of the moving radius and the unmachined surface. The amount of the feed per circle (f) is also the maximum thickness in the milling process. The actual amount of the uncut chip thickness in the tip-based trajectory machining is h_2 which is the vertical distance between a pair of parallel lines along the instantaneous velocity direction, passing through the intersection of the rake face, the machined surface, and the tip edge apex. It can be seen that at this position owing to the effect of the negative cutting rake angle (γ), h_2 is different from h_1 and has exceeded f. In addition, at the position O_1, the difference between the uncut chip thicknesses is relatively small. Therefore, the analytical method used for calculating the uncut chip thickness in the milling process is not applicable, while the numerical calculation is employed. The variation of the cutting rake angle γ during single edge cutting processes (Courses 1 and 3) is calculated by using θ (the angle between the tip's rake face and the unmachined workpiece surface) minus 90°. Course 1 and 3 in the Fig. 3.8B are mainly studied because they are not only single edge cutting processes, but also are used to machine two sides of channel.

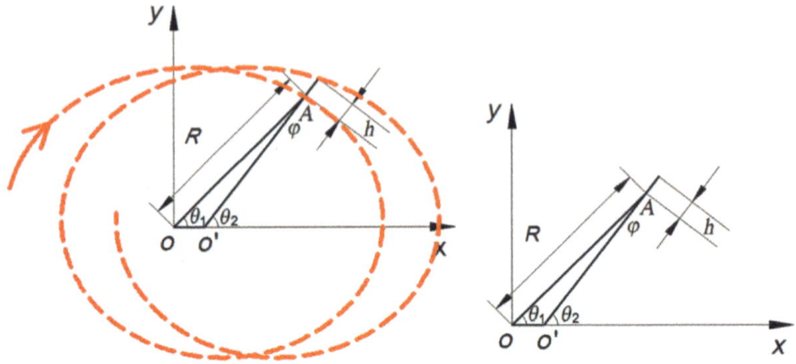

Figure 3.9 Geometric relationship for calculation of the uncut chip thickness and the cutting rake angle.

Fig. 3.10 shows the relationships between the tip revolution angle, the uncut chip thickness, and the cutting rake angle of the trajectory presented in Fig. 3.8B. The dotted line represents the uncut chip thickness of the conventional milling process. The solid and broken lines show the uncut chip thickness and the cutting rake angle for Courses 1 and 3, respectively. The processing parameters are: a 0.4 μm feed mount per revolution, 10 μm diameter of the revolution motion, 1.2 μm machined depth, and the triangle side length is 3 μm, correspondingly. The machining process is presented in Fig. 3.4B.

- *The cutting rake angle.* For Course 1, the cutting rake angle decreases from 30° to −70°, with the increase in the tip revolution angle. At the beginning of the cutting process, the tip edge cuts into the sample surface, and a positive rake angle is achieved with the tip revolution angle from 0° to 100°. For Course 3, the cutting rake angle also decreases from 15° to −30°, with the increase in the tip revolution angle. When the tip leaves from the machined structure, the maximum rake angle of −30° is presented. As mentioned above, this property is different from the conventional micromilling process, during which the rake angle is a fixed value. In generally, for plastic workpiece materials, such as aluminum, a positive rake angle is prone to lead to the cutting state and effectively remove the materials, and a negative rake angle will result in the pileup of the removed materials, forming different kinds of burrs. Therefore, based on the above analysis, using the trajectory shown in Fig. 3.8B, the microgroove with a smooth left side and a pileup, or burrs, on the right side can be obtained. This will be discussed in the following section.
- *The uncut chip thickness.* When the tip revolution angle varies from 0° to 40° and from 140° to 180°, the uncut chip thickness has little difference with that of the milling process. The uncut chip thickness increases with the tip revolution angle when it is less than 100° in Course 1. The maximum value, 0.55 μm, is larger than the feed mount per revolution of 0.4 μm. Under this condition, a negative rake angle is used.

As shown in Fig. 3.10B, when the tip revolution angle varies from 0° to around 30°, the cutting rake angle is positive, and the material is removed mainly in way of chips. After that, when the tip revolution angle changes from 30° to 100°, the rake angle becomes negative, and the uncut chip thickness is larger. Under this condition, the materials are mainly deformed by ploughing and are not prone to be removed. When the tip revolution angle is within the scope of 100°−140°, two cutting edges

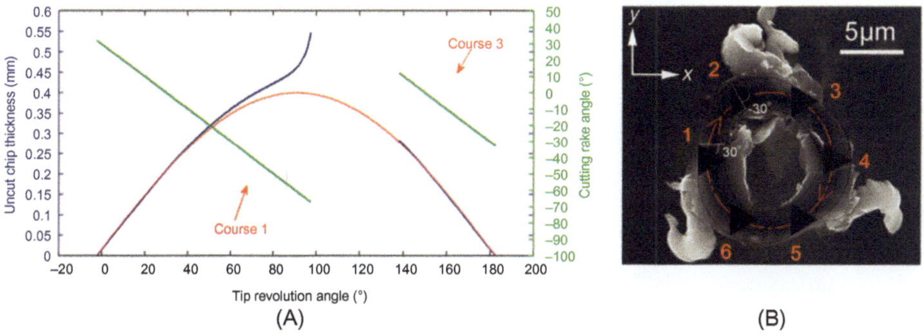

Figure 3.10 (A) Relationships between the tip revolution angle and the uncut chip thickness and the cutting rake angle of the trajectory presented in Fig. 3.8B. (B) The ring machined on the aluminum alloy surface by one revolution of the pyramidal tip which is a complicated cutting process and cannot be expressed by cutting raka angle and uncut chip thickness.

together join in the machining process which is a complicated cutting process and can not be expressed by cutting raka angle and uncut chip thickness, and hence the rake angle and the uncut chip thickness are not provided here. Under this condition, ploughing and extrusion are estimated to play the key roles as Course 2 shown in Fig. 3.8B. In the Fig. 3.10B, in the process from point 2 to 3, the belt chips are piled up. When the tip revolution angle changes from 140° to 180°, although the uncut chip thickness becomes small, the rake angle makes the materials to be removed more in type of the chips, which are pushed to the position of point 4. Fig. 3.10B shows one complete circle of machining. When the micromachining process is carried out with a feed, only the tip revolution angle from 0° to 180° is employed.

3.5 Formation mechanism and control methods of burrs

In this section, the burrs formed during machining channels, based on the orbital motion of the tip, are mainly studied on the material of Al. The types of burrs are classified by analyzing their morphologies and sizes. By comparing channels machined, respectively, by the conical tip and the pyramidal tip, the optimal machining way is proposed. Meanwhile, by optimizing the machining process and the deburring process, channels with smooth sides are obtained.

3.5.1 Burr formation during machining with the conical tip

Due to the symmetry of the conical tip along different feeding directions, the machining trajectories are the same. Therefore, the effects of feeding directions on machining channels are neglected. Fig. 3.11 shows the AFM images of the channels machined by the conical tip with different feed rates (Fig. 3.11A 2 μm/s and (B) 10 μm/s). The machined depth is set to 1.5 μm for all tests. In two conditions, burrs formed on both sides of the channel; and under the high feed rate condition, burrs are more serious, which is attributed to the increase of the uncut chip

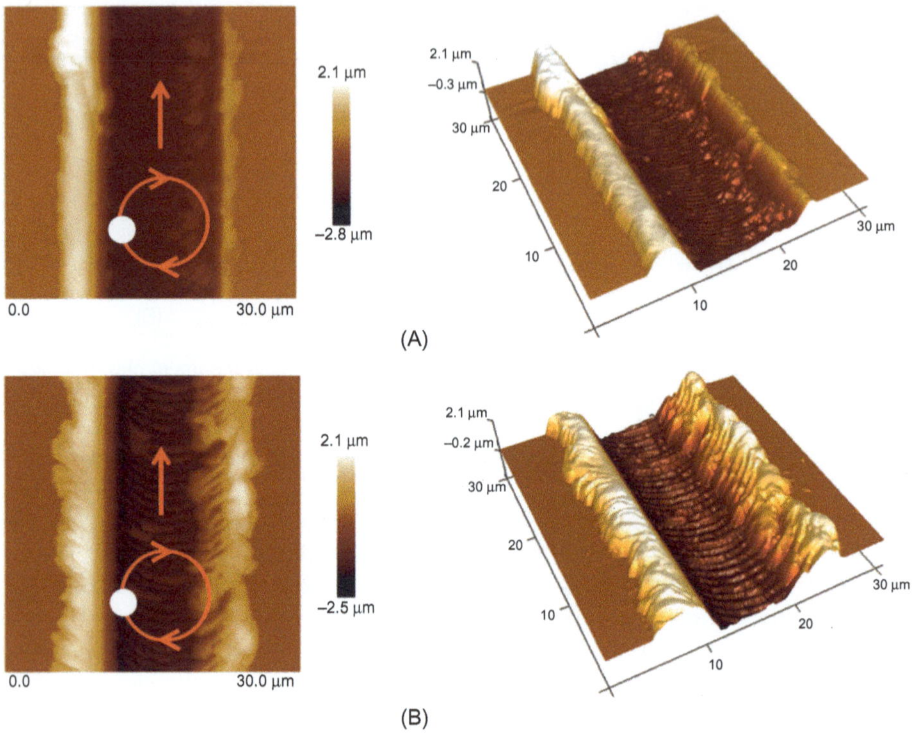

Figure 3.11 The channels machined by the orbital motion of the conical tip with different feed rates 2 μm/s (A) and 4 μm/s (B).

thickness. In Fig. 3.11A the burr on the left side (the upmilling side) is formed by the plastic deformation of materials and is larger than that at the right side (the downmilling side). At the upmilling side, the uncut chip thickness begins to increase at the zero value, owing to the influence of the circular cutting edge. There will not be chips formed until the tip revolves to a certain angle, achieving the minimum chip thickness; and hence, before forming the chip, the materials are violently extruded by the tip face forming the burr. At the downmilling side, though the uncut chip thickness decreases back to zero from the maximum value, the chip is formed with the revolving tip; and hence, the caused plastic deformation is relatively small. In the case of high feed rate in Fig. 3.11B, owing to the increase of the uncut chip thickness in each revolution, before achieving the minimum chip thickness the angle needed to revolve of the tip during each circle is decreased, which results in less extrusion between the tip face and the machined material. The burr at the upmilling side is similar to that in the milling process (De Assis, Jasinevicius, & Rodrigues, 2015) that has a wave-like shape at the channel's top. It is due to the fracture of the materials, and they are pushed by the tip to the side's top. At the right side, the burr is the residual material that is unremoved by the revolving motion. It can be found that in two feed rates there are both burrs formed at the channel's top, which are difficult to be removed by postprocessing.

3.5.2 Burr formation during machining with the pyramidal tip

Different from the symmetry shape of the conical tip, along different feeding directions, the machining trajectories of the pyramidal tip are different. Because of the triangular section of the pyramidal tip, with the circular revolution there are main four kinds of machining trajectories as show in Fig. 3.12, and each machining trajectory corresponds to a type of channel. It can be seen that in Figs. 3.12B and C, there are burrs formed on both sides of the channels; while in Figs. 3.12A and D, burrs are only formed on the downmilling side, and the upmilling side is smooth. In the figure, the lines through the centers of the triangles are the moving trajectories of the tip, and the lines through the apexes of the triangles are the machining trajectories along different directions. The shaded areas are the uncut chip thicknesses in each revolving cut. The θ is the angle between the tip rake face and the machining trajectory, which corresponds to the cutting rake angle. The differences of the machined channels and formation mechanisms of burrs are due to the variations of the uncut chip thickness and the cutting rake angle, which was studied in detail during the previous study (Xue et al., 2015). However, there is another important factor for deciding the positions of burrs—the cutting edge inclination angle. Owing to the pyramidal shape of the tip, there would be an inclination angel (λ) existing during the tip's cutting processing, and hence, the tip's cutting can be looked on as the oblique cutting as shown in Fig. 3.13A. In the oblique cutting process, the chip flow is not along the direction perpendicular to the cutting edge (V_1) but along the resultant direction (V), added with a velocity component tangential to the cutting edge (V_2). However, with different feeding directions and the tip's edges, the directions of the velocity component (V_2) would be different (upward or downward). For instance, in Fig. 3.13B, when the tip feeds forward along the arrow direction, the V_2 of the edges one and two are both upward and the edge three does not participate in cutting. In Fig. 3.13C, the V_2 of the edges one and two are both downward. During the tip's revolving cut, the direction of V_2 decides the flow direction of the chip. It can be

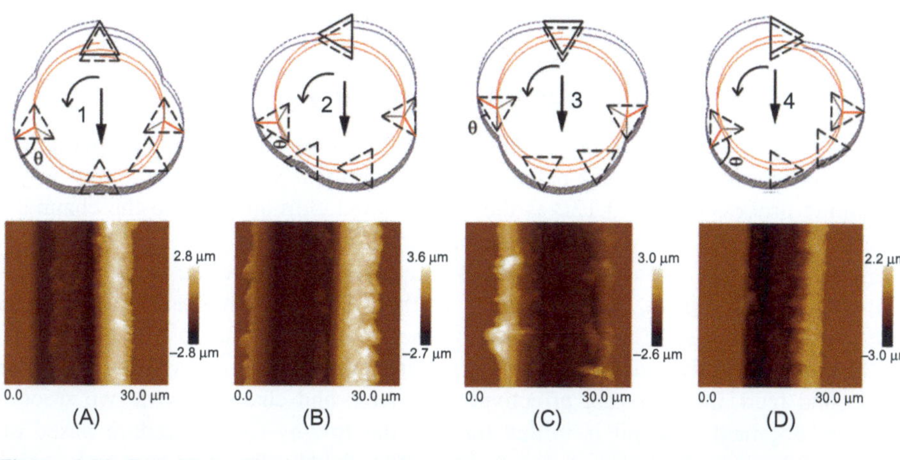

Figure 3.12 Four kinds of machining trajectories and the corresponding machined channels. (A) The channel machined by Trajectory 1. (B) The channel machined by Trajectory 2. (C) The channel machined by Trajectory 3. (D) The channel machined by Trajectory 4.

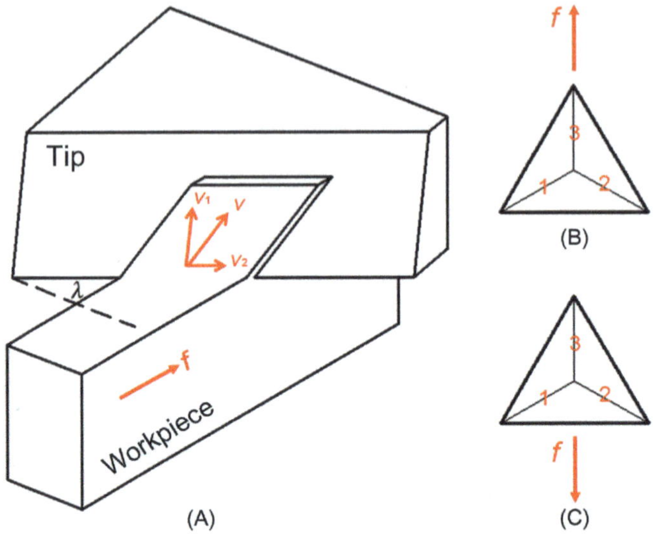

Figure 3.13 Schematic of the oblique cutting of the tip's machining process. (A) Schematic of the oblique cutting process. (B) The edge forward feeding. (C) The face forward feeding.

judged, that along the tip's moving direction, if the tip apex is in front of the cutting edge, the V_2 is upward; if the tip apex is in the back of the cutting edge, the V_2 is downward. At the right sides of Figs. 3.12A, B, and D, the burr formation results from the accumulated chips. However, at the right sides of Figs. 3.12A and B, the cutting edges (marked in bold) cause the upward V_2, and hence the produced chips flow upward to the channel's top. In Fig. 3.12D the cutting edge at the right side approximately does not have the component of V_2, which can be seen as an orthogonal cutting, and hence the chips would turn laterally pushed by the tip and attached on the channel's sidewall, which is the condition that the burr can be easily removed.

3.5.3 Methods of formation of slight burrs

From the above analysis, it can be known that the orbital motion of the pyramidal tip is more feasible to machine better channel structures. Meanwhile, the burr in the machining process of Fig. 3.12D is the accumulated chip attached to the channel's sidewall, and hence, it is more useful to optimize this process.

First, the strategy of the second feed is conducted to remove the burr at the sidewall as shown in Fig. 3.14A. According to the study (Xue et al., 2015), the feed rate was set to 4 μm/s. The burr caused by the plastic deformation of the materials can be suppressed. Based on this tip's orientation and the revolving direction after the second feed, the burrs are effectively removed, and channels with two smooth sides are obtained. The pit is milled through the row-by-row procedure based on this channel machining process, as shown in Fig. 3.14B. The burr is mainly accumulated at one side of the pit, and the other three sides are smooth. Meanwhile, the

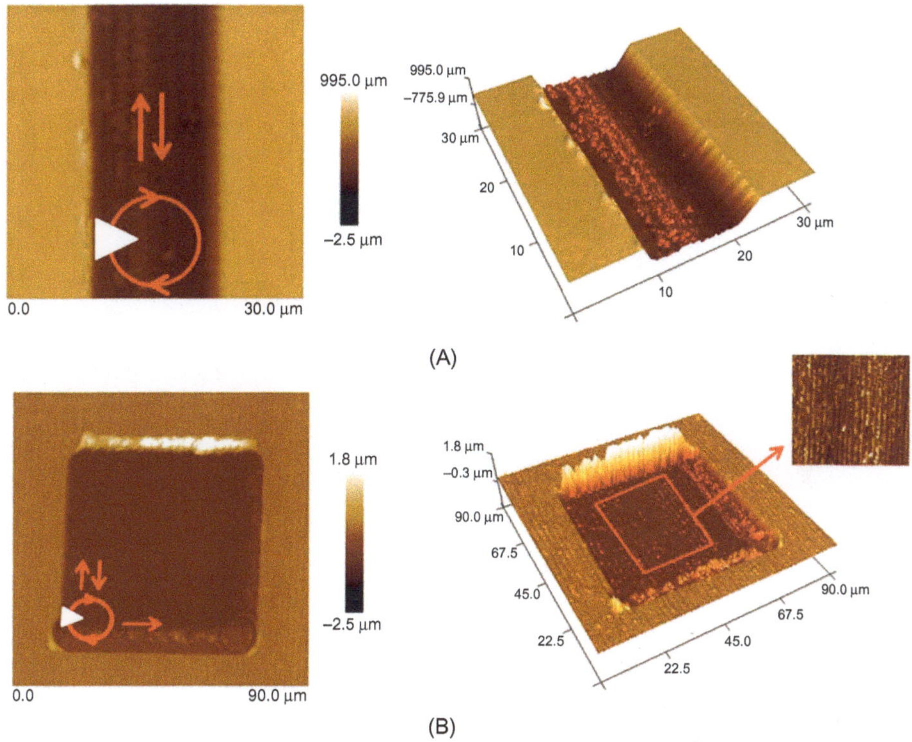

Figure 3.14 The channel (A) and pit (B) machined by tip revolving machining with the reciprocating feed.

bottom surface quality of the pit (marked in bold) is measured. The roughness, R_a and R_q, are 34.2 and 43.4 nm, respectively.

Second, different tip's trajectories can be used to reduce the burr. In the traditional milling process, owing to the limitation of the rotation of the spindle, the cutting edge of the mill can only do the spiral movement based on a circle. However, the orbital motion of the tip is based on the resultant motion of the translation of each axis, which is unable to change the tip posture and exhibits a high flexibility. When the *x*- and *y*-axes do the simple harmonic motions with the same amplitude, frequency, and out of phase, the trajectory is the circle similar to the milling condition. However, if the amplitudes of two axes are different with other parameters unchanged, the trajectories are the ellipses, which is unable to be conducted in the milling process. Therefore, this is the advantage of the orbital motion machining. When the elliptical trajectory is used, the variations of the uncut chip thickness and cutting rake angle are different from those in the circle trajectory (Yan, Geng, & Hu, 2015). A shape elliptical trajectory and a flat elliptical trajectory are designed as shown in Figs. 3.15A and B. For the shape ellipse, the long axis (along the feeding direction) is 7 μm in length, and the short axis is 5 μm. For the flat ellipse, the long axis is 5 μm, and the short axis (the feeding direction) is 3 μm. This is because

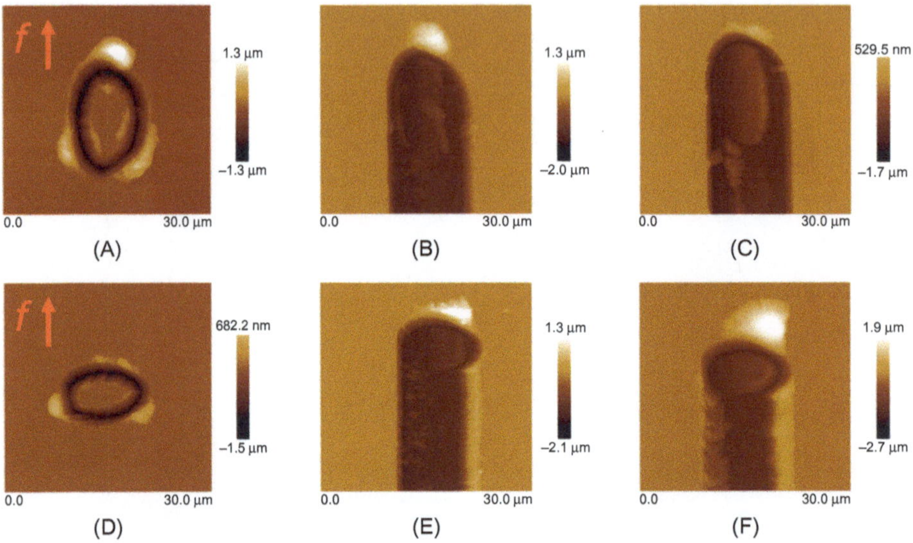

Figure 3.15 The channels machined by the shape elliptical (A) and flat elliptical (D) trajectories, respectively, with different feed rates. (B) and (C) are under the condition of shape elliptical trajectory with feed rates of 2 and 4 μm/s, respectively. (E) and (F) are under the condition of flat elliptical trajectory with feed rates of 2 and 4 μm/s, respectively.

the if the channel's width is same, the chip volume in each revolution is made the same. It can be seen that both channels in Figs. 3.15B and C (feed rates are 2 and 4 μm/s, respectively), machined by the shape ellipse, have two smooth sides. However, under the flat ellipse condition, the burr at the right side has not been removed, and the burr size increases as the feed rate increases. Because of the tapered shape of the tip, some materials are unremoved to remain at the channel's bottom (Yan, Geng, et al., 2015). In Fig. 3.15F, the left side has been undermined by the residual materials at the bottom, which is due to a larger uncut chip thickness of the flat ellipse in each revolution, causing more residual materials. Meanwhile, comparing Fig. 3.15B with Fig. 3.15C, the lower feed rate results in the smaller variation of the uncut chip thickness, which makes the channel's bottom quality better.

Third, except for approaching the workpiece surface, the *z*-axis can also be employed to provide the moving trajectory for the tip, which is used as a deburring method for the channel. By making the *z*-axis do the simple harmonic motion, accompanying the resultant motion of *x*- and *y*-axes, the tip can achieve a 3D movement. Fig. 3.16A is the 2D motion in the horizontal plane, in which the machined channel is presented in Fig. 3.12D. Fig. 3.16B is the 3D motion in space by introducing the *z*-axis motion based on the Fig. 3.16A. The altitude difference is about 500 nm along the feeding direction. According to the schematic of Fig. 3.16B, the channel was machined by once feed with the feed rate of 4 μm/s as shown in Fig. 3.16C. It can be seen that by using this 3D trajectory, the burr at the right side has been successfully removed, and the machined channel has two smooth sides, which confirms that this approach is a good deburring method. In addition, owing to

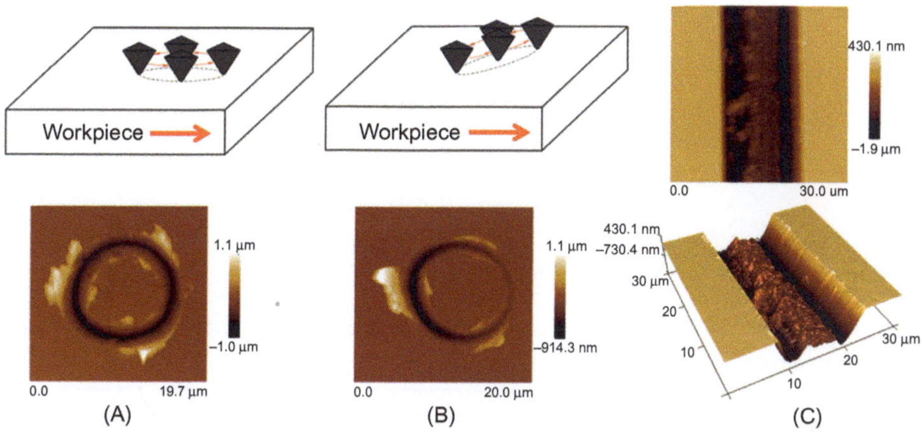

Figure 3.16 The designed 3D trajectory used for machining channel. (A) The two-dimensional trajectory. (B) The three-dimensional trajectory. (C) The channel machined by the three-dimensional trajectory.

once feed, there is the trimming process for channel's bottom compared to the condition of twice feeds, which makes the bottom quality worse than that in Fig. 3.14A. Reducing the feed rate can also improve the bottom quality.

3.6 Effects of the processing parameters and fabrication of microstructures

3.6.1 Effects of the processing parameters on machining microchannels

In this section, effects of the processing parameters including the revolving frequency, feed rate per revolution, diameter of the tip's circular moving trajectory, and the machined depth on the surface roughness (R_a) are studied. R_a of 30 µm × 6 µm area at the bottom of the machined channel is measured by AFM. As shown in Fig. 3.17, $A = 30$ and $B = 6$ µm. The sample material is the Aluminum alloy (2024) machined by the ultra-precision turning, with the original surface roughness (R_a) of 5 nm. The conical diamond tip is employed in this process. The detail settings for these processing parameters and the corresponding measured R_a are listed in Table 3.1. The revolving frequency changes from 10 to 40 Hz. The feed rate per revolution changes from 0.2 to 2 µm. The diameter of the tip's circular moving trajectory changes from 10 to 50 µm. The machined depth changes from 0.5 to 3 µm. The relationships between these processing parameters and R_a are plotted in Fig. 3.18.

As shown in Fig. 3.18A, when the revolving frequency is less than 30 Hz, R_a changes a little. When it is larger than 30 Hz, R_a abruptly increases. This value is far less than the value of 70–80 Hz presented in Fig. 3.6A. Higher frequency will bring more vibrations of the stage into the machining process. Therefore, a lower

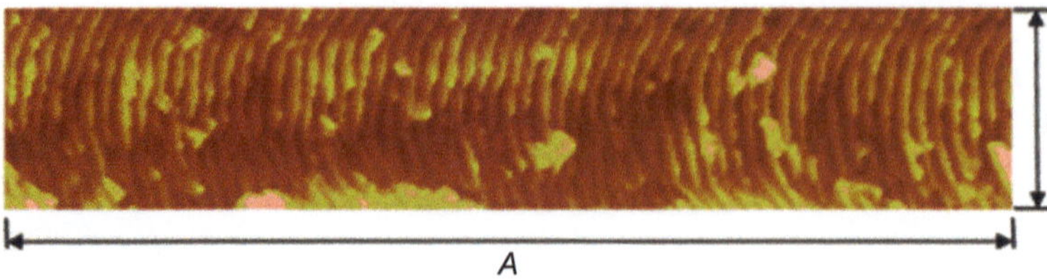

Figure 3.17 The measured example of the surface roughness (R_a) at the channel bottom.

Table 3.1 **The detail settings of the processing parameters and the surface roughness R_a**

No	Revolving frequency f(Hz)	Feed rate per revolution a (μm)	Diameter of the tip's circular moving trajectory D(μm)	Machined depth h(μm)	Surface roughness R_a(nm)
1	10	1	10	1	101.3
2	20	1	10	1	165
3	25	1	10	1	150.9
4	30	1	10	1	172.1
5	40	1	10	1	292
6	20	0.2	10	1	103.5
7	20	0.5	10	1	169
8	20	1	10	1	237
9	20	1.5	10	1	270.6
10	20	2	10	1	293.8
11	20	1	10	1	165
12	20	1	20	1	189.3
13	20	1	30	1	180.6
14	20	1	40	1	209
15	20	1	50	1	355
16	20	1	10	0.5	296.1
17	20	1	10	1	317
18	20	1	10	1.5	370
19	20	1	10	2	378.9
20	20	1	10	3	454

frequency is better for the higher surface quality. However, it will lead to the low machining efficiency. As shown in Fig. 3.18B, R_a increases with the feed rate per revolution and will reach to the saturation state when it is larger than 1.5 μm. A feed that is too large rate leads to more residual tracks, owing to the limited radius of the diamond tip as shown in Fig. 3.7. As shown in Fig. 3.18C, when the diameter of the tip's circular moving trajectory is less than 30 μm, the surface roughness (R_a) varies a little. When it is larger than 30 μm, R_a abruptly increases. With the fixed revolving frequency of 20 Hz provided in Table 3.1, a larger diameter will lead to

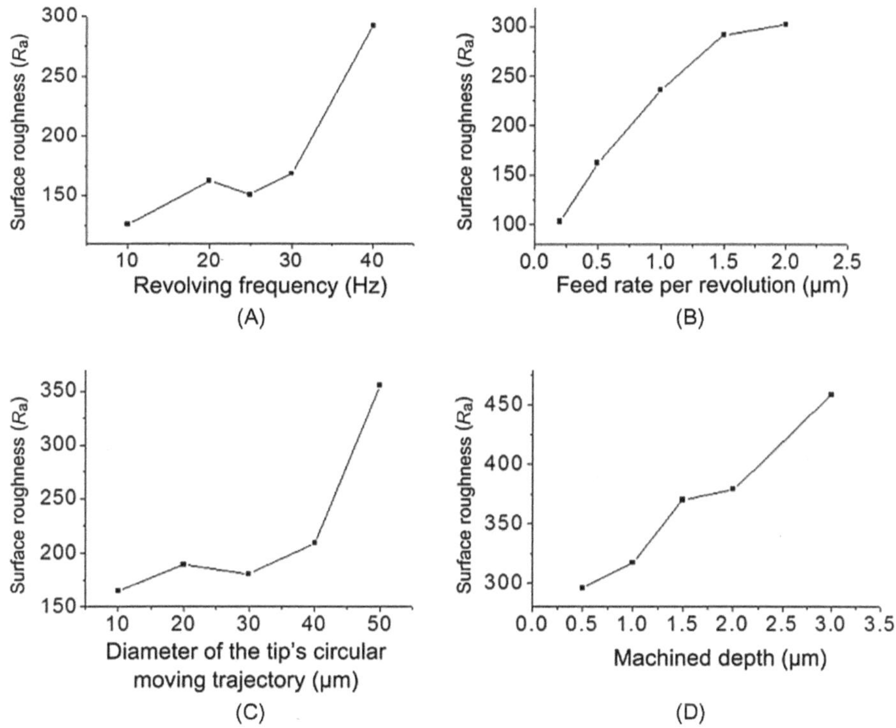

Figure 3.18 Relationships between the processing parameters and the surface roughness. (A) Revolving frequency versus R_a. (B) Feed rate per revolution versus R_a. (C) Diameter of the moving trajetory versus R_a. (D) Machined depth versus R_a.

the instability of the nanopiezo stage, which results in a bad surface quality. As shown in Fig. 3.18D, R_a increases with the machined depth. The large cutting forces induced by the large machined depths lead to the large deformations of the materials. Then the surface quality will become worse.

Therefore, in order to get a smooth machined surface, the processing parameters are: the revolving frequency is 10 Hz, the feed rate per revolution is less than 1 μm, the diameter of the tip's circular moving trajectory is 10 μm, and the machined depth is 1 μm. These optimal parameters do also work in the condition of the pyramidal tip.

3.6.2 Effect of the feed on machining microstructures

Using this technique, it is very easy to machine a required microchannel (e.g., a channel with 10 μm in width). But if it is desired to machine large scale microstructures this way, the other feed (in Y axis) perpendicular to the feed direction X as presented in Fig. 3.1B is needed. In this work, we provided two approaches to realize 2D microstructures for getting better surface quality, as shown in Fig. 3.19. The dotted circles and the sequence numbers of one, two, three, four, and five show the tip's position during the machining process. The arrows show the tip's moving direction. Fig. 3.19A shows the reciprocal feed mode. Point 1 shows the original position of the tip. First,

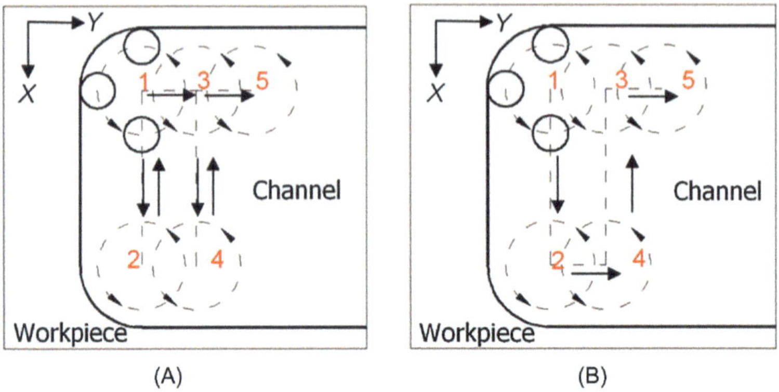

Figure 3.19 Schematics of two approaches to machine a wider microchannels. (A) The channel is machined by the reciprocal feed mode. (B) The channel is machined by the side feed mode.

the tip moves from the point 1 to the point 2 and then returns to the point 1 along the same trace. Second, the tip goes to the point 3 with a feed. Next, the tip continues to go to the point 4 and then return to the point 3. All the grooves are machined twice. Fig. 3.19B shows the side feed mode. First, the tip moves from the point 1 to the point 2 and then moves to the point 4 with a feed. Second, it continues to move to the point 3 and then feeds to the point 5. During this process, each groove is machined for only once. For both conditions, the width of the required microchannel is determined by the length of each machining in the X direction, and the length is obtained by the feed multiplied by the number of the feed in the Y direction. From the machining processes presented in Fig. 3.19, it can be found that the tip shape will play a key role in the machined surface quality. In the present study, a symmetrical conical diamond tip and an unsymmetrical pyramidal diamond tip are both used in the machining processes. Therefore, effects of two kinds of tips are studied as follows.

In order to study the effects of different modes and shapes of the tip on the surface quality, the following tests are carried out using the machining parameters: the revolving frequency is 10 Hz, the diameter of the tip's circular moving trajectory is 10 μm, and the machined depth is 1 μm. The feed rate per revolution is 1.0, 2, 3, 4, 5, and 6 μm. The conical and pyramidal diamond tips are employed. The surface roughness (R_a) data are obtained from the area 20 μm × 20 μm from the machined channels by AFM.

Figs. 3.20A and B show the relationships between the feed (f) and R_a, with the side feed mode and the reciprocal feed mode and using the conical diamond tip, respectively. When the feed is 1 μm in both modes, the surface roughness (R_a) is the smallest (about 32 nm). With the same machining parameters, the high efficiency in the side feed mode can be achieved by comparing it with the reciprocal feed mode, as shown in Fig. 3.19. Therefore, when the feed of 1 μm is used, the side feed mode is better. After that, the surface roughness increases with the feed in both the side feed and reciprocal feed modes. However, a perfect surface is obtained in the reciprocal feed mode. About half values of R_a (Fig. 3.20A) can be achieved for this mode. Thus, for the larger feed, the reciprocal feed mode should be used for the better surface quality.

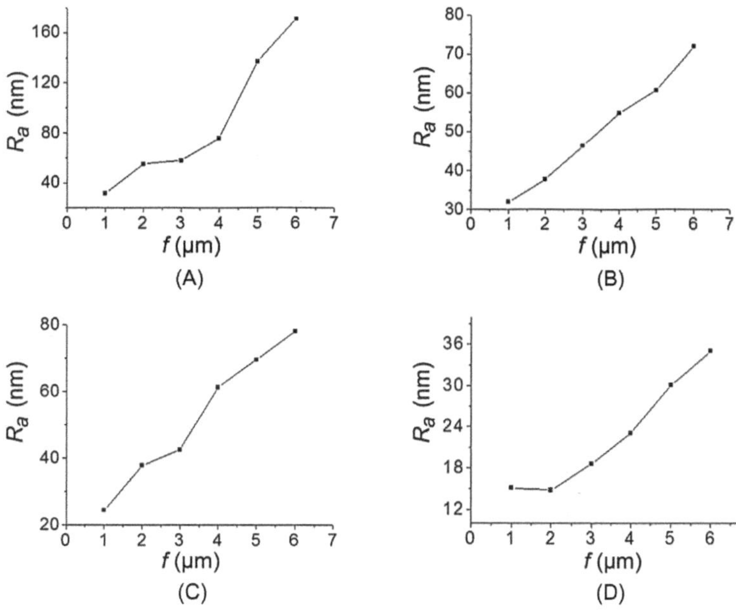

Figure 3.20 Relationships between the feed *(f)* and R_a with the reciprocal feed mode and the side feed mode using the conical and the pyramidal diamond tips. (A) and (B) are the conditions of the conical tip, respectively using the side feed mode and the reciprocal feed mode. (C) and (D) are the conditions of the pyramidal tip respectively using the side feed mode and the reciprocal feed mode.

Figs. 3.20C and D show the relationships between the feed *(f)* and R_a, with the side feed mode and reciprocal feed, mode using the pyramidal diamond tip, respectively. When the feed is 1 μm in the side feed mode, R_a is about 24.4 nm. In the reciprocal feed mode, when the feed is in the range of 1–2 μm, R_a is around 15 nm. Although the side feed mode owns a higher machining efficiency, this mode is not feasible to be used, owing to the poor surface quality. Therefore, the feed of 2 μm and the reciprocal feed mode are used for achieving a better surface quality with the pyramidal diamond tip. In addition, it is comparable for the conical diamond tip state, that R_a in the reciprocal feed mode is the half of the value obtained in the side feed mode. A larger feed will also result in a bad surface quality.

Moreover, the surface quality achieved by the conical tip is worse than that obtained by the pyramidal tip for both feed modes as shown in Fig. 3.20. Fig. 3.21 shows the AFM images of the local area machined in the side and reciprocal feed modes, using the pyramidal diamond tip. For the side feed mode presented in Fig. 3.21A, the formed burrs are increased and remained in the machined surface. For the reciprocal feed mode, the machined surface is polished by the tip for the second time when the tip retraces from the point 1 to the point 2, as presented in Fig. 3.19A. Therefore, the surface is much smoother under this condition. The corresponding R_a values become smaller, as shown in Fig. 3.20.

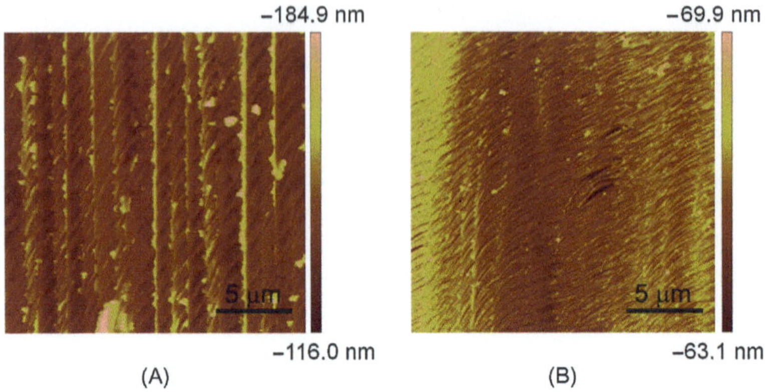

Figure 3.21 AFM images of local area machined in the side (A) and reciprocal (B) feed modes using the pyramidal diamond tip.

3.6.3 Fabrication of typical microstructures

Machining microchannels with the width less than 100 μm using this technique is an important advantage compared to the conventional micromilling process. These types of microchannels will be employed in the microfluidic application. Therefore, in this section, some complex 2D microstructures like microchannels are machined by this method. The conical diamond tip is used for the uniform width of the channels and is unaffected by the different feeding directions. The revolving frequency is 10 Hz. The diameter of the tip's trajectory is 5 μm. The feed rate per revolution is 1.0 μm. The samples are the injected mold polymer PMMA plate.

Figs. 3.22A and B show the hexagon and circular microchannels, respectively. The length of the hexagon microchannel is 50 μm. This channel is closed very well, indicating that the position accuracy of the *X-Y* precision stage is enough for this process. The diameter of the circular microchannel is 100 μm. The start point does not coincide with the end point under this condition. The accumulated error of the precision stage will play a role in the machined results, which can be reduced by a compensation process. By using the AFM to measure the machined surface, the surface roughness (R_a) 30–50 nm is obtained. In the microfluidic field, a straight channel connects to the circular channel. As shown in Fig. 3.22C, two straight microchannels, with the angle of 30°, connect to a circular microchannel, with the central angle of 150°. The lengths of two straight channels are 200 and 120 μm, and the radius of the circular microchannel is 40 μm. Using the same machining process, a microstructure letter "1" consisting of a circular microchannel with the central angle of 30° and a straight microchannel with the length of 120 μm is presented in Fig. 3.22D.

To confirm this approach further, using the same machining parameters, two kinds of Y-type microchannels, which are often used in the microfluidic fields, are machined. Figs. 3.23A and D show the dimensions of the microchannels. The unit used in the figure is the micron. Figs. 3.23B and D give the optical images of the machined channels. The Figs. 3.23C and F are the AFM images of the intersection area. The surface roughness (R_a) of the area at the bottom of the microchannel can reach up to 37.8 nm.

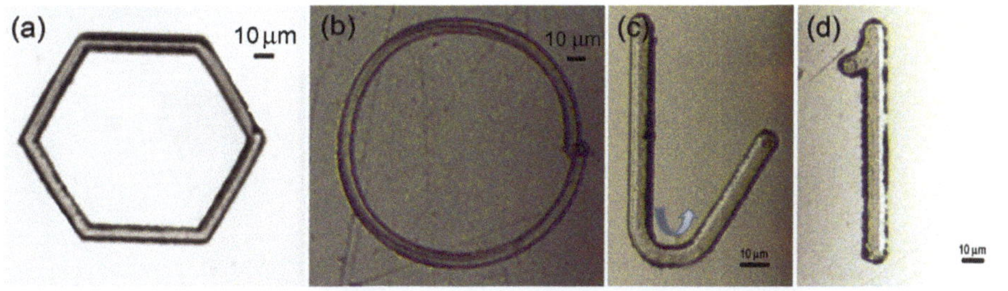

Figure 3.22 Typical 2D microchannels machined by this technique. (A) The hexagon microchannel. (B) The circular microchannel. (C) The curve microchannel. (D) The structure of letter "1".

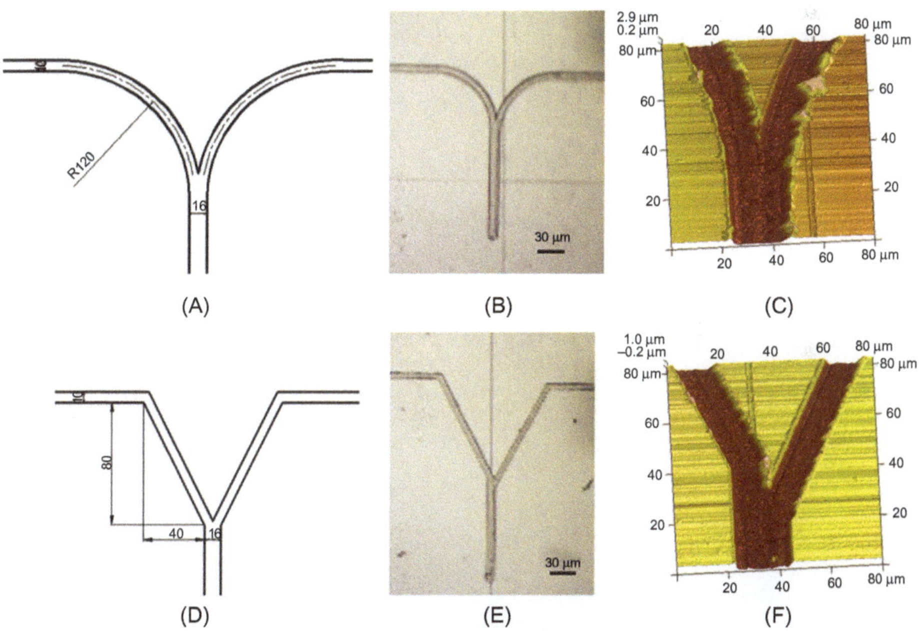

Figure 3.23 Fabrication of two kinds of Y-type microchannels. (A) and (D) show the dimensions of Y-type microchannels. (B) and (E) are the optical images of the machined channels. (C) and (F) are the AFM images.

Moreover, the microstructure array can also be machined by this method. Fig. 3.24 shows a local area of the array with the square length of 10 μm and the distance between adjacent channels of 10 μm. The number of the channel is 10 for orthogonal directions. Owing to the employment of the conical diamond tip, as shown in Fig. 3.24B, some materials are not effectively removed and are piled up on the sides of the squares. The pyramidal diamond tip may result in a better result with optimized machining parameters. But the machining process is more complex.

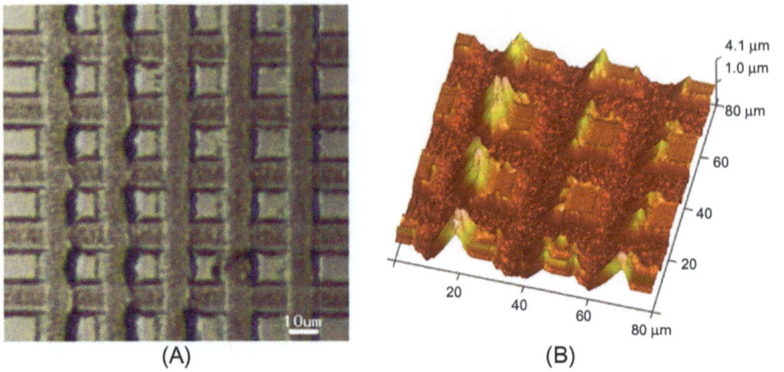

Figure 3.24 Optical (A) and AFM (B) images of a microsquare array.

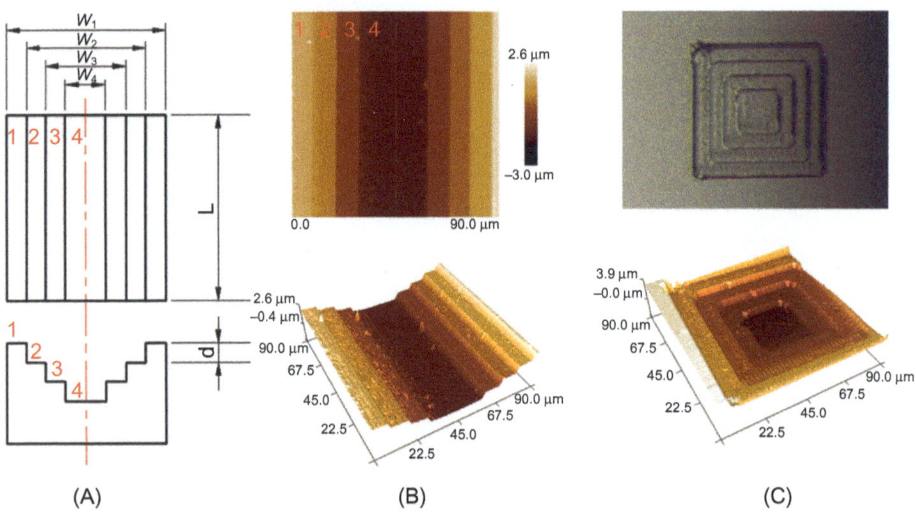

Figure 3.25 3D microstructures machined respectively on the aluminum alloy and PMMA surfaces. (A) Schematic of the machining strategy for the 3D microstructure. (B) The 3D structure machined on aluminum alloy surface. (C) The 3D structure machined on PMMA surface.

Using the schematic of Fig. 3.19, the layer by layer technique can be used to machine 3D microstructures by changing the dimensions of the micropits. Each pit is machined row by row, and each row is based on the machining method of the channel. As shown in Fig. 3.25A, the surface 1 is first milled with the dimension of width W_1 and length L. After finishing surface 1, the tip engages downward to the prescribed depth (D) and then begins to machine surface 2 with the dimension of W_2 and L using the same center line as the surface 1. And so on, for the surfaces 3 and 4. To obtain the structure with the smooth edges and a better surface bottom quality, the optimal machining method for the channel (as above studied in Fig. 3.14) has been adopted for machining aluminum alloy. Fig. 3.25B shows that the ladder structure is successfully machined. It can be seen that the structure has smooth steps with different depths. The distance between two layers is set to 1 μm. Owing to the high stiffness of the actuator, the machined depth can be well controlled. In addition, due to the difference of the material properties, under the condition of machining PMMA, the best way (Fig. 3.12D) of machining aluminum alloy does no longer work, while the machining way in Fig. 3.12A is the best case for PMMA which can produce the channel with two smooth sides. Therefore, based on the machining process of the channel, the 3D structures are machined on the PMMA surface as shown in Fig. 3.25C. The optical image in Fig. 3.25C shows the structure is made up of square pits with different dimensions. From the above analysis, while machining the pit on the aluminum alloy surface, the burr will occur at the top side (Fig. 3.14D) and other three sides are smooth. However, it can be found that comparing with the pit machined on the aluminum alloy surface, the pit on the PMMA surface in Fig. 3.25C has four smooth sides, which is mainly due to the brittleness property of PMMA material.

3.7 Summary and future works

The micromachining technique is based on the tip orbital motion, which is a new technology derived from the micromilling process and the AFM-tip based nanoscratching technique. By the trajectory movement of the diamond tip combined with the feed movement of the precision stage, it is able to machine microstructures like the microchannel, the microstep structure, and so forth on wide-range material surfaces. In this chapter, the principle of this micromachining technique is presented, and then it is compared with the conventional micromilling process. Based on this idea, a micromachining setup is established using a nanopiezo stage to perform the orbital motion. Based on analysis of the machining mechanism, the processing experiments on different materials, including polymers and metals with the conical and pyramidal diamond tips, are carried out. The burrs formation and control approaches are studied. 2D and 3D microstructures are successfully machined, which indicates that this technique will be used in the microfluidic fields in the near future.

However, this technique is still a new mechanical machining method to machine microstructures. There are many important points that are needed to be clarified and studied in the future.

1. *Improve the machining accuracy*. Based on the studies in this chapter, results show that a high machining accuracy is obtained with a pyramidal diamond tip rather than the conical tip, which is also used by (Heamawatanachai & Bamberg, 2009, 2010). In order to improve the machining accuracy further, the pyramidal tip or an even more complex tip shape should be used and be designed specifically. This work must be supported by many experiments in the future. Moreover, using this kind of diamond tip, the machining scale is still at the micrometer range. If one wants to reach to the nanometer range, an AFM tip with the tip radius of less than 10 nm may be used as the cutting tool. Also, the orbital motion is still employed. Similar studies have been performed by Gozen et al. in 2010

(Gozen & Ozdoganlar, 2012a, 2012b). This machining process is defined as nanomilling, a rotating tip based nanomechanical manufacturing process. Actually, the technique presented in this chapter and Heamawatanachai's studies **(2009; 2010)** can be defined as a novel micromilling process.

2. *Improve the machining efficiency.* Because no high rotation spindle is used in this work, a relative low machining efficiency becomes an obstacle in order to apply this technique to the industrial applications. Using a piezo tube can make the tip revolve more quickly (Heamawatanachai & Bamberg, 2009). However, a small moving range and a compensation process for the *Z*-axis also lead to a low machining efficiency for the large scale microstructures. In our works, a nanopizeo stage is used, and a large machining scale can be easily obtained. However, the tip cannot move as fast, which results in a low machining efficiency. Therefore, in future studies of this technique, a large moving range and a relative high moving frequency are both required for the moving parts. Some new techniques in the motion parts will push this technique ahead.
3. *Expand the workpiece materials.* In this chapter, only the Al alloy and the polymer PMMA are used. In the previous studies, brittle and ductile materials including silicon, Ti, and Al alloy are machined with the conical diamond tip (Heamawatanachai & Bamberg, 2009). By using the pyramidal diamond tip and different kinds of trajectories, a better surface quality may be achieved. Therefore, in future studies, more types of materials should be explored further, based on these results. Moreover, previous studies also found that there is a severe tip wear when machining Ti, and different kinds of burrs and cracks exist when machining some materials. The ultrasonic vibration assisted method may make the machining quality better on more sample material surfaces.
4. *Machine more complex microstructures.* To date, only 2D microchannels structures and simple 3D microstructures have already been demonstrated by this technique. As a machining method similar to the micromilling process, complex microstructures can be machined by this method. This must depend on the insight into the machining mechanism of this technique and the detailed studies of the optimization of the processing parameters.
5. *Applications of the machined microstructures.* Previous works have confirmed that this technique is a low cost, highly accurate, and efficient mechanical machining method to machine microstructures less than 100 μm. The next important study of this technique is to use these microstructures in real applications. Currently, our group is doing this kind of work by applying the machined microchannel as the key part of the microfluidic chip. Moreover, many complex microstructures can be machined by this way. How to use them to demonstrate their effects on the properties such as the hydrophobic property and the optical characteristic, will greatly expand the application range of this technique.

Acknowledgments

The authors gratefully acknowledge the financial support from the Foundations for National Natural Science Foundation of China (51521003, 51222504), Self-Planned Task (SKLRS201606B) of State Key Laboratory of Robotics and System (HIT), the Program for New Century Excellent Talents in University (NCET-11-0812), the Fundamental Research Funds for the Central Universities (HIT.BRETIV.2013.08), and the National Program for Support of Top-notch Young Professors. The authors also gratefully acknowledge the works of Master Qiong Cai and Bachelor Dongze Wu.

References

Brousseau, E. B., Dimov, S. S., & Pham, D. T. (2010). Some recent advances in multi-material micro- and nano-manufacturing. *International Journal of Advanced Manufacturing Technology*, *47*, 161–180.

De Assis, C. L. F., Jasinevicius, R. G., & Rodrigues, A. R. (2015). Micro end-milling of channels using ultrafine-grained low-carbon steel. *International Journal of Advanced Manufacturing Technology*, *77*(5–8), 1155–1165.

Dornfeld, D., Min, S., & Takeuchi, Y. (2006). Recent advances in mechanical micromachining. *CIRP Annals - Manufacturing Technology*, *55*, 745–768.

Friedrich, C. R., & Vasile, M. J. (1996). Development of the micromilling process for high-aspect-ratio microstructures. *Journal of Microelectromechanical Systems*, *5*, 33–38.

Geng, Y. Q., Yan, Y. D., Zhao, X. S., Hu, Z. J., Liang, Y. C., Sun, T., & Dong, S. (2013). Fabrication of millimeter scale nanochannels using the AFM tip-based nanomachining method. *Applied Surface Science*, *266*, 386–394.

Gozen, B. A., & Ozdoganlar, O. B. (2012a). Design and evaluation of a mechanical nanomanufacturing system for nanomilling. *Precision Engineering – Journal of the International Societies for Precision Engineering and Nanotechnology*, *36*, 19–30.

Gozen, B. A., & Ozdoganlar, O. B. (2012b). A method for open-loop control of dynamic motions of piezo-stack actuators. *Sensors and Actuators A: Physical*, *184*, 160–172.

Heamawatanachai, S., & Bamberg, E. (2009). Design and characterization of a PZT driven micromachining tool based on single-point tool tip geometry. *Precision Engineering – Journal of the International Societies for Precision Engineering and Nanotechnology*, *33*, 387–394.

Heamawatanachai, S., & Bamberg, E. (2010). Cutting force model of orbital single-point micromachining tool. *International Journal of Machine Tools and Manufacture*, *50*, 815–823.

Madou, M. J. (2002). *Fundamentals of microfabrication: The science of miniaturization* (2nd ed.). Boca Raton, Florida: CRC Press.

Shen, X. H., Zhang, J. H., Xing, D. L., & Zhao, Y. F. (2011). A study of surface roughness-variation in ultrasonic vibration-assisted milling. *International Journal of Advanced Manufacturing Technology*, *58*, 553–561.

Tansel, I., Rodriguez, O., Trjillo, M., Paz, E., & Li, W. (1998). Micro-end-milling-I. wear and breakage. *International Journal of Machine Tools and Manufacture*, *38*, 1419–1436.

Vazquez, E., Rodriguez, C. A., Elias-Zuniga, A., & Ciurana, J. (2010). An experimental analysis of process parameters to manufacture metallic micro-channels by micromilling. *International Journal of Advanced Manufacturing Technology*, *51*(9–12), 945–955.

Xue, B., Yan, Y. D., Li, J. R., Yu, B. W., Hu, Z. J., Zhao, X. S., & Cai, Q. (2015). Study on the micro-machining process with a micro three-sided pyramidal tip and the circular machining trajectory. *Journal of Materials Processing Technology*, *217*, 122–130.

Yan, Y. D., Geng, Y. Q., & Hu, Z. J. (2015). Recent advances in AFM tip-based nanomechanical machining. *International Journal of Machine Tools and Manufacture*, *99*, 1–18.

Yan, Y. D., Hu, Z. J., Zhao, X. S., Sun, T., Dong, S., & Li, X. D. (2010). Top-down nanomechanical machining of three-dimensional nanostructures by atomic force microscopy. *Small*, *6*(6), 724–728.

Yan, Y. D., Sun, T., & Dong, S. (2007). Investigation on AFM-based micro/nano CNC machining system. *International Journal of Machine Tools and Manufacture*, *47*(11), 1651–1659.

Yan, Y. D., Xue, B., Hu, Z. J., & Wu, D. Z. (2015). Machining slight burr formed micro-channels with different moving trajectories of a pyramidal diamond tip. *International Journal of Advanced Manufacturing Technology*, accepted.

Microelectrical discharge machining of Ti-6Al-4V: implementation of innovative machining strategies

4

G. Kibria[1] *and B. Bhattacharyya*[2]
[1]Aliah University, Kolkata, West Bengal, India, [2]Jadavpur University, Kolkata, West Bengal, India

4.1 Introduction

In last 15 years, the demand for microproducts and miniaturized components has increased rapidly in various microengineering applications in automotive, avionics, biotechnology, communications, optics, and electronics industries. To fulfill these demands and ultimately meet the requirements of product miniaturization, micromachining processes—both conventional and nonconventional—are efficiently and effectively utilized. These processes also play a vital role in micromanufacturing and microfabricating miniaturized products and microsystems (Venkatesh & Izman, 2007, Yu, Rajurkar, & Shen, 2002). Among various micromachining technologies, microelectro discharge machining (micro-EDM) is one of the important and cost-effective thermoelectric type micromachining method capable of machining electrically conductive materials regardless of their hardness and strength. During the machining process, there is no direct contact between the tool electrode and the workpiece; therefore, machining errors due to deformation of the tool electrode as well as chatters, vibration-related inaccuracies, and mechanical stresses are absent (Ho & Newman, 2003).

In the Union Soviet Socialist Republics (USSR) in the 1940s, material removal (erosion) by a series of spark discharges in a controlled manner was first reported by two scientists, Doctors B.R. and N.I. Lazarenko, for stock removal from the workpiece (Ho & Newman, 2003). They invented a simple servo controller and successfully maintained the inter-electrode gap (IEG) between the tool and the workpiece. They also investigated the destructive effects of discharges and the mechanism of material removal to machine the desired shape by properly controlling the repetitive discharges. A number of research investigations were then carried out by many scientists for more enhancements related to discharge phenomena, spark gap control, and feed mechanism, as well as for better control of the erosion technique by employing computer numerical control, adaptive control mechanism, improved machine intelligence, and better flushing conditions. In 1968, Kurafuji and Masuzawa demonstrated the micro-EDM process and achieved a miniature

Microfabrication and Precision Engineering. DOI: http://dx.doi.org/10.1016/B978-0-85709-485-8.00004-8

microhole by micro-EDM drilling in a 50 μm thick plate of carbide material (Kurafuji & Masuzawa, 1968). Considerable research and development of EDM and micro-EDM processes have been done throughout the last 20 years to achieve high-precision machined features, improved material removal rate (MRR), and reduced tool wear rate (TWR).

4.2 Principle of electrical discharge machining

Electrical discharge machining is one of the thermo-erosive processes in which controlled spatially and temporally separated pulsed discharges are used to machine electrically conductive materials irrespective of their mechanical, thermo-physical, and chemical properties (Kunieda, Lauwers, Rajurkar, & Schumacher, 2005). Series of discrete sparks are generated between the shaped tool electrode and the workpiece submerged in a dielectric fluid, namely, kerosene, EDM oil, deionized water, and paraffin oil, among others. Between the tool electrode and the workpiece surface, very small IEG is kept through which the dielectric fluid is passed (D'Urso & Merla, 2014). The schematic of the basic EDM process is shown in Fig. 4.1. The instantaneous location of discharge underneath the tool face is determined by the smallest gap between the tool and the workpiece. The machining process consists of successively removing very small volumes of work material—partly molten and partly vaporized—during discharge when the localized temperature at the discharge spot exceeds the melting and vaporization temperature of the workpiece material (König & Klocke, 1997).

Fig. 4.2 illustrates the sequence of phenomena occurring during a particular discharge at the IEG. When a pulsed DC voltage is applied to the electrode and workpiece, a strong electrical field is produced at the point where there is minimum IEG due to surface microirregularities. The microscopic contaminants suspended in the dielectric fluid start to drift and align at the strongest point of the field because of the electromagnetic field. These contaminants, along with other particles, construct the conductive bridge across the IEG, typical spark gap distance varies from 10 to 100 μm. As the voltage between the electrode and

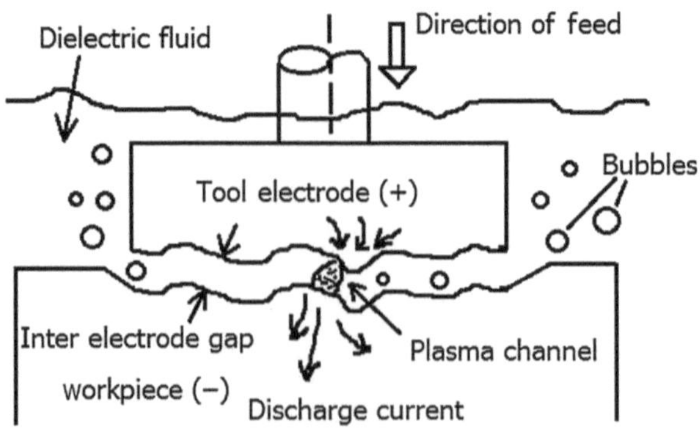

Figure 4.1 Schematic view of concept of EDM process.

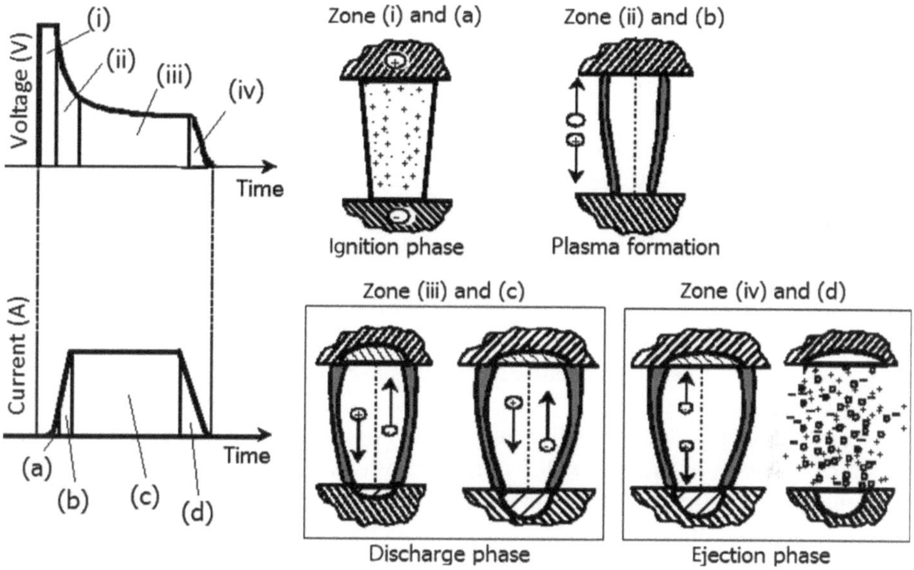

Figure 4.2 Schematic view of voltage and current trends of pulse along with the phases of an electrical discharge.

workpiece increases at the beginning of the pulse, the surface temperature of the workpiece material increases. A certain amount of the dielectric fluid and charged particles of the conductive bridge vaporizes and ionizes, thereby forming a plasma channel. When the potential difference across the spark gap sharply falls, voltage breakdown occurs. At this time, the plasma channel starts to conduct the applied current whose magnitude rises instantaneously. The abrupt increase in current results in an instantaneous increase in localized temperature and pressure in the plasma channel. The extremely high temperature of the discharge melts and vaporizes a small amount of material from the surfaces of both the electrode and workpiece at the points of discharge. The vaporization of a very small amount of materials from both electrodes as well as dielectric fluid generates gaseous bubbles in the plasma channel. These bubbles rapidly expand outward radially from their point of origin. At the end of the discharge, the supply of electrical pulse is terminated. This sudden termination of the pulsed power results in a collapse of plasma channel and consequently the vapour bubble under the influence of pressure imposed by dielectric fluid from the environment. The violent in-rush of a relatively cool dielectric fluid results in an explosive expulsion of molten materials both from the tool electrode and the workpiece surfaces, resulting in the formation of a small crater, depending on the current density parameter. The molten materials resolidified because of rapid cooling and took the shape of small spherical particles and formed debris. These debris are expelled from the IEG when fresh dielectric fluid covers the IEG for repeating the discharge cycle.

4.3 Overview of micro-EDM

Micro-EDM is the micromachining process of electrically conductive materials utilizing the EDM technology. The dimension of material removal or unit removal (UR) per discharge is in microdomain so that microsize craters can be produced on the workpiece surface by precisely controlling the sparks that occur between the microsize tool electrode and the workpiece in the presence of dielectric fluid at the IEG (Lim, Wong, Rahman, & Edwin Lee, 2003). The basic mechanism of micro-EDM is similar to that of the EDM. However, differences such as utilizing micro-sized tool, the amount of discharge energy, and the *X*-*Y*-*Z* axes resolution make the micro-EDM process more precise and reliable and capable of microfeatures generation (Masuzawa, 2000). In micro-EDM, very high-frequency pulses (>200 Hz), small discharge energies (10^{-6}–10^{-7} J), and potential difference (40–100 V) between the electrodes are applied to achieve high-accuracy features and surface finish (roughness as much as 0.1 μm) (Gentili, Tabaglio, & Aggogeri, 2005). Utilizing improved pulse generator and precise servo feed system, the microtool electrode can be moved at micron rate to maintain the required IEG and also to retract the microtool if the servo feed senses any short-circuit between the electrodes. Another important component of the micro-EDM system is the dielectric circulation unit. The required amount of flushing pressure and the type of flushing are used to deliver the fresh dielectric in the machining zone and to ensure that the debris is ejected from the IEG.

4.4 Differences between EDM and micro-EDM

Although the physical principle of micro-EDM with respect to the material erosion technique is similar to conventional EDM, there are a number of differences between these two processes, particularly in carrying out micromachining operations. These differences are mainly in terms of the dimension of the tool used, utilization of power supply for scaling down the discharge energy, resolution of axes travels, IEG control, and employing improved flushing techniques.

The significant differences between EDM and micro-EDM are as follows:

- The dimensional radius of the plasma channel generated during the spark is the most important difference between EDM and micro-EDM. As the microtool size in micro-EDM is small, the size of the plasma channel is comparable with the tool dimension. However, for EDM, compared with tool size, the plasma channel is much smaller (Katz & Tibbles, 2005).
- The maximum peak energy applied in micro-EDM is limited to control the material removed per discharge (unit removal, UR) during machining. Consequently, the dimension of the crater size is also smaller. However, for conventional EDM, the UR is comparatively high.
- The maximum applicable energy in micro-EDM limited as excessive discharge energy can produce microtool rupture or burnt. However, for EDM, the discharge energy is not a limited factor as long as tool can sustain the applied energy of sparks.

- The flushing pressure applied in micro-EDM is limited as the dielectric fluid pressure acting on the microsized electrode can cause tool deflection as the electrode stiffness is lower. However, for conventional EDM, these problems can be avoided as the tool dimension is large.
- The removal of debris from a small-size IEG is very difficult in micro-EDM. Moreover, the accuracy of microfeatures generated is degraded while ejecting the debris through the narrow gap walls. However, in conventional EDM, the debris removal is not a problem as the IEG is large.
- High-precision features can be generated if the microtool is vibrated and the amount of tool wear is compensated in conventional EDM. However, in micro-EDM, microfeatures machining is highly precise and accurate if the process parameters are controlled significantly.
- The amount of electrode wear per discharge in micro-EDM is high and the accuracy of microfeatures degrades. However, for conventional EDM, the electrode wear is low.

4.5 System components of micro-EDM

A typical micro-EDM setup consists of a number of major system components, namely, the central control unit, the position control unit, the servo control unit, the machining chamber, and the dielectric circulating unit. The central control unit supervises the operator and the machine for conducting various operations in the setup. It also controls the operation between two units of the machine. The movement of the microtool towards the workpiece for machining is monitored and controlled by the servo control unit. In the machining chamber, the workpiece is clamped tightly and held for machining. The working tank is filled with dielectric fluid at that level so that the machining zone is immerged in the dielectric to avoid fire hazards. The motion of the tool in the Z axis and the movement of workpiece in the X and Y axes are controlled by the position control unit. The dielectric circulating unit provides the required amount of fresh dielectric to the machining chamber and filtration of used dielectric to remove debris from it. The schematic of the basic units of the micro-EDM system is presented in Fig. 4.3. In the following section, details of major components of the micro-EDM setup are discussed.

4.5.1 Pulse generator

To produce the pulse trains for micro-EDM, various types of pulse generators are used. Very short pulse-on-time of a several dozen nanoseconds can be produced by the RC pulse generator. One of the drawbacks of the RC pulse generator is that it is incapable of producing high discharge frequency pulses as it requires time to charge the capacitor. Moreover, it has difficulties for controlling the pulse interval. Therefore, the workpiece is subjected to thermal damage as long as the dielectric property is not recovered after the discharge. This leads current to flow through the same plasma channel without recharging the capacitor (Masuzawa, Sata, & Kinoshita, 1971).

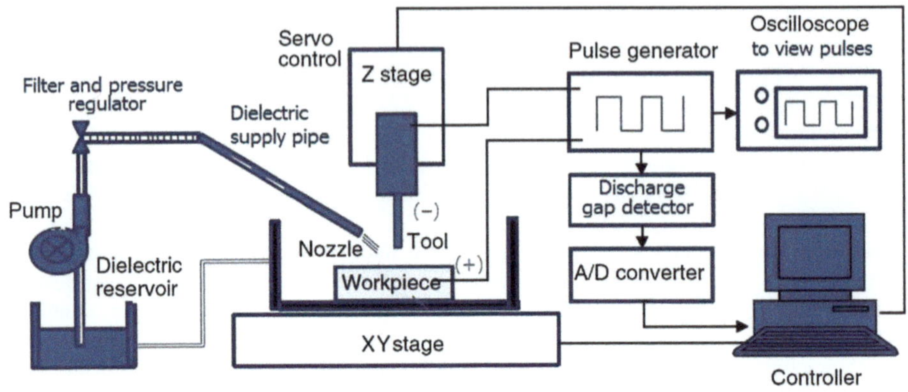

Figure 4.3 Schematic diagram of a typical micro-EDM system showing different subsystems.

The pulse discharge in a transistor-type pulse generator is achieved by making the switching component ON and OFF and therefore, there is no need to charge any capacitor. Thus, compared with the RC pulse generator, the transistor-type pulse generator provides a higher discharge frequency, which improves the material erosion rate. With the transistor-type pulse generator, by detecting the discharge condition in the IEG, the discharge process can be controlled easily. The main drawback of this type of pulse generator is the time required to transmit the pulse by the switching component and the pulse control circuit components. Therefore, the nanosecond pulses for micro-EDM cannot be generated by the transistor-type pulse generator.

To reduce the delay time of the transistor-type pulse generator, the transistor-type isopulse generator was developed and successfully employed for micro-EDM operation (Han, Yamada, Kawakami, & Kunieda, 2004). For rough and semi-finish machining, a field effect transistor is used to cut off the discharge current. Instead of observing the gap voltage, the pulse current is observed for detecting the discharge. As the current sensor provides an output less than 5 V and acts as an input to the pulse control circuit, it eliminates the voltage attenuation circuit. In this way, the delay time is shortened to a significant amount and ultimately, about 80 ns pulse duration is achieved.

As in the RC-type pulse generator, stray capacitance determines the minimum discharge energy per pulse; therefore, it acts as the limitation for generating microfeatures on the workpiece (Masuzawa & Fujino, 1980). A crater diameter of less than 2 μm cannot be achieved because of the difficulty in eliminating the stray capacitance (Han, Yamada, Kawakami, & Kunieda, 2003). To avoid these problems, a capacity coupling-based pulse generator was developed. With this generator, the effect of the stray capacitance can be eliminated as electric feeding is done without touching the microtool electrode. Thus, the discharge crater dimension of the nanometer domain is realized.

4.5.2 Servo control unit

The servo-controlled electrode feed mechanism is the most important part of the micro-EDM process as it ensures a stable and efficient machining condition by monitoring the gap condition between the microtool and the workpiece. The high-efficiency and high-stability machining process in micro-EDM is ensured by precise detection of the discharge states and constant servo feed control (Zhang, Jia, Liu, & Li, 2012). High-dimensional microfeatured is generated by implementing a proper gap control system (Rajurkar et al., 2006). In micro-EDM, the IEG or discharge gap is in the micron range, consequently requiring a specially-designed servo controller to maintain this gap. The microcontroller sends a digital signal for driving the servo motor attached to the servo feed system.

4.5.3 Dielectric circulating unit

One of the essential components on the micro-EDM system is the dielectric circulating system. The unit has various components such as a dielectric fluid, a pump to deliver the dielectric fluid to the machining chamber, a reservoir for the dielectric, microfilters to separate the machining debris from the used dielectric, pipes and nozzles to supply the fluid to the machining zone, and a flushing pressure control unit to maintain the flushing pressure.

The dielectric fluid serves as a spark conductor, concentrating the spark energy to an extremely narrow region. When the voltage is applied across the tool electrode and the workpiece, a potential difference is established across the two electrodes because of the electrical resistivity of the dielectric fluid. As the potential difference between the electrodes reaches the breakdown strength of the dielectric fluid, discharge occurs and conduction of current takes place. Once the total spark energy is discharged, the dielectric fluid regains its dielectric strength due to the supply of fresh dielectric fluid in the narrow gap. The maximum potential difference that a unit thickness of a dielectric medium can withstand is known as the dielectric strength.

The main functions of the dielectric fluid are as follows:

- To provide insulation in the IEG between the microtool and the workpiece.
- To make ionization possible, i.e., building the discharging channel.
- To carry the conductive particles into the ionized channel forming a bridge over which sparks jump and current flows.
- To flush away the eroded particles (debris) from the machining zone produced during machining and from the discharge gap, i.e., the work–tool gap retaining only a small amount of conductive particles.
- To build up a new electric field in the discharge channel with the use of conductive particles.
- To cool the section of the tool electrode and the workpiece, which are heated by the discharge.
- To extinguish the sparks after the discharge is completed.

4.6 Micro-EDM process parameters

In micro-EDM, the performance of the machining operation is directly associated with a number of process parameters. These process parameters are mainly divided into three categories: (1) electrical process parameters, (2) nonelectrical process parameters, and (3) gap control and motion parameters. In the following sections, the important process parameters involved in the micro-EDM operation are discussed.

4.6.1 Electrical process parameters

4.6.1.1 Discharge energy

In micro-EDM, the most significant process parameter is discharge energy. This parameter is a collection of other operating parameters, which are related to the energy of the discharge created between the electrodes at the IEG. For a different type of pulse generator, the calculation for discharge energy is different. The MRR is directly related to discharge energy during the micro-EDM operation. On the contrary, the TWR also increases, which deteriorate the surface finish and accuracy of microfeatures generated.

4.6.1.2 Gap and discharge voltage

Gap voltage is the voltage in the gap between two electrodes. The total energy of the spark is determined by the applied voltage. Depending on the setting of the voltage, the IEG is set by the servo control. A larger value of the IEG improves the flushing of debris from the machining zone and makes the next discharge stable, ultimately improving the MRR. However, the surface finish deteriorated because of the large size of the crater dimension at high-voltage conditions. The voltage of the IEG at which discharge occurs between the microtool and the workpiece is known as discharge voltage. The discharge voltage mainly depends on the breakdown strength of the dielectric and the IEG.

4.6.1.3 Peak current

The average current is the average of amperage in a spark gap measured over a complete cycle. This is read on the ammeter during the process. The theoretical average current can be measured by multiplying the duty cycle and the peak current, i.e., the maximum current available for each pulse from the power supply. The amount of energy/power used for discharge is mainly determined by the peak current. A higher value of the peak current signifies a better machining efficiency in terms of the MRR. At the same time, the surface finish deteriorates and the TWR increases.

4.6.1.4 Pulse duration

Pulse duration or pulse-on-time is the time interval within which the applied current is flowing through the IEG of the two electrodes. During this period, breakdown of dielectric and removal of material from the workpiece surface take place. Large values for pulse duration mean higher MRR. Broader and deeper craters are achieved at a longer pulse duration setting and consequently, a rough-machined surface is attained. On the contrary, smaller craters obtained at low pulse durations provide a smoother surface finish.

4.6.1.5 Duty factor

This is an important parameter in the micro-EDM process. Duty factor represents the percentile value of the ratio of pulse duration to total cycle time. Duty ratio is calculated using the following equation:

$$\text{Duty factor (DF)\%} = \frac{\text{Pulse duration}}{\text{Total cycle time}} \times 100 \tag{4.1}$$

If the duty factor is high, the flushing time is very low, which might lead to a short-circuit condition, and a small duty factor indicates a high pulse off time and low machining rate.

4.6.1.6 Pulse frequency

Pulse frequency is the measure of the number of cycles per unit of time, i.e., in 1 s.

A high surface finish is achieved at a high-frequency setting. A high pulse frequency value results in a lower value of pulse duration. A smaller pulse duration and a high pulse frequency result in the generation of small craters and less thermal damage of the machined surface.

4.6.1.7 Polarity

Polarity refers to the electrical conditions determining the direction of the current flow relative to the electrode. The polarity condition of the electrodes is of two types, (1) straight polarity and (2) reverse polarity. Straight polarity is that condition when the microtool is connected to the cathode (−), whereas reverse polarity is that condition in which the tool electrode is connected to the anode (+) and the workpiece to the cathode (−). To achieve high MRR from the workpiece, the tool electrode is used as the cathode and the workpiece as the anode. Depending on the application, some electrode/work material combinations provide better results when the polarity is changed. Generally for graphite electrodes, a positive polarity gives a better wear condition, whereas a negative polarity gives better machining speed.

4.6.2 Nonelectrical process parameters

4.6.2.1 Tool electrodes

The tool electrode is important to achieve effective and efficient machining conditions. The thermal properties of the tool electrode material play a significant role during micro-EDM as it is a thermal process. Materials of higher melting, boiling points and heat conductivity are used to fabricate the microtool for the micro-EDM process (Jahan, Wong, & Rahman, 2009). The criteria for selecting the tool materials are as follows:

- Erosion characteristics, i.e., MRR and wear ratio;
- Machining possibilities, i.e., ease of shaping the electrode;
- Market condition, i.e., cost and availability;
- Conductivity;
- Thermal diffusivity;
- Hardness and toughness of the electrode material;
- Strength;
- Density; and
- Melting and boiling temperatures

Materials such as copper, brass, tungsten, tungsten copper alloy, steel, zinc-based alloy, and copper graphite are particularly suitable for fabricating microtool electrodes.

4.6.2.2 Workpiece materials

One of the criteria to determine the machinability of workpiece materials in micro-EDM is electrical conductivity. Machinability also depends highly on specific heat, thermal conductivity, and the melting and vaporization points of the workpiece material (Mahardika, Tsujimoto, & Mitsui, 2008).

4.6.2.3 Dielectric fluids

As in the dielectric fluid, the micro-EDM process takes place; therefore, several properties of dielectric fluid, such as viscosity, dielectric strength, cooling capability, and chemical compositions play significant roles for the efficient and stable discharge during machining. For a safe machining operation and a stable sparking condition, the dielectric strength and the flash point temperature of the dielectric fluid should be higher. Furthermore, low viscosity and specific gravity are two desirable properties of the dielectric fluid. These properties significantly affect the machining efficiency and consequently improve the MRR, lower the TWR, and enhance the surface finish of the machined features.

4.6.3 Gap control and motion parameters

4.6.3.1 Servo feed

To properly maintain the discharge gap width and avoid arcing and short-circuiting between the microtool and the workpiece, the servo feed control system plays a vital role during the micro-EDM process. As soon as the value of the average gap voltage approaches more than the preset threshold voltage of the pulse generator, the feed rate of the servo increases, compensating the discharge gap between the electrodes and vice versa (Kunieda et al., 2005).

4.6.3.2 Electrode rotation

Microtool rotation during the micro-EDM process significantly affects the machining performance. Utilizing the tool rotating method, the flushing of debris is improved and the overall surface finish and accuracy of microfeatures are enhanced. During micro-EDM, employing tool rotation, the tangential force of the microtool, provides enhanced effectiveness for stable discharge by smoothly ejecting the debris from a narrow IEG and improves the overall machining rate (Yan, Huang, Chow, & Tsai, 1999). A higher rotating speed for microtool electrodes also reduces the relative TWR.

4.6.3.3 Tool geometry and shape

In micro-EDM, the tool geometry solely depends on the microfeature to be generated. The common tool geometries being used are square, rectangular, cylindrical, and circular. The shape of the microtool electrode has significant effects on the electrode wear ratio (EWR) during micro-EDM. Depending on the tool shape and geometry, the flushing efficiency varies. Researchers reported that vibration-assisted micro-EDM using a helical microtool improves the machining rate and reduces the taper and discharge gap when drilling a deep microhole (Hung, Lin, Yan, Liu, & Ho, 2006).

4.6.3.4 Workpiece and tool vibration

The vibration of the tool electrode and the workpiece during micro-EDM is one of the efficient strategies for a considerable improvement in micro-EDM performance. During vibration of the microtool or workpiece, the forward and backward motion of the tool or workpiece changes the discharge gap and consequently, the dielectric fluid pressure in the IEG also changes constantly. When the microtool advances toward the workpiece, the dielectric fluid is forced out from the machining zone. When the microtool moves away from the machine zone, the fresh dielectric is then taken by the discharge gap, thus increasing the overall flushing efficiency. With the vibration of the microtool and the workpiece, debris removal is enhanced as the vibration continuously changes the pressure in the narrow gap (Jahan, Rahman, & Wong, 2012).

4.6.3.5 Types of dielectric flushing

Dielectric flushing has an important role in removing the debris from the machining zone and consequently, it enables the stable discharge condition by supplying fresh dielectric fluid in the gap. In general, there are mainly two types of flushing: pressure flushing and suction flushing. Depending on the type of flushing, the amount of flushing pressure is provided. In micro-EDM operation, jet flushing is more frequently used to effectively generate high-aspect-ratio microfeatures such as microholes. In other cases, side flushing is commonly used. If the jet flushing is provided from one direction, the debris may accumulate in the downstream, which creates an irregular gap width; as a result, the accuracy of microfeatures deteriorates (Levy & Ferroni, 1975). To avoid this situation, jet flushing from both sides and sweeping-type flushing are sometimes recommended.

4.6.3.6 Flushing pressure

To maintain a stable and effective machining condition, it is very much essential to flush out the eroded particle from the IEG and to cool the electrode and the workpiece so that localized and concentrated discharge is avoided (Kunieda et al., 2005). Higher flushing pressure is preferable for effective debris removal, stable machining, and high-aspect-ratio microfeature generation. However, as stiffness of the microtool is low, high flushing pressure may deteriorate dimensional accuracy because of microtool vibration or deflection.

4.7 Performance criteria in micro-EDM

4.7.1 Material removal rate

The MRR is the amount of material erosion from the workpiece per unit of time. It expresses the speed of the machining of the workpiece. A high machining rate is always desirable because it is directly related to productivity. At the same time, with high productivity aspect, the desired dimensional accuracy and surface quality are significant features in micro-EDM. The MRR in micro-EDM is calculated by the volume of the material removed or the difference in the weight of the workpiece before and after the machining operation, as given in Eq. (4.2). Process parameters (electrical and nonelectrical) have significant effects on MRR. Higher MRR can be achieved at high discharge voltage, peak current, pulse duration, and duty cycle. However, other desirable process performances such as TWR, surface finish, and dimensional accuracy are also important aspects and taken into account in micro-EDM.

$$\text{Material removal rate (MRR)} = \frac{\text{Weight of workpiece before machining} - \text{Weight of workpiece after machining}}{\text{Machining time}} \quad (4.2)$$

4.7.2 Electrode wear ratio

EWR is the ratio of the amount of material removed from the microtool electrode to the amount of material removed from the workpiece by volume. Sometimes, TWR is also calculated by dividing the difference in weight of the microtool before and after machining with the machining time, as seen in Eq. (4.3). High TWR causes inaccurate machined features and poorer surface finish substantially. During micro-EDM drilling operations, tool wear results in shortened tool length; therefore, the total amount of tool feed provided is larger than the workpiece thickness. Process parameters have considerable influences on TWR. Low TWR and high surface quality are achieved at lower peak current and pulse duration. TWR also depends on several other factors such as melting point, thermal conductivity, and density of the tool material (Jahan et al., 2009). Depending on the tool geometry and complicacy, TWR also varies. Proper optimization of the process parameters can substantially control the wear rate and improve the accuracy of the machined features.

$$\text{Tool wear rate (TWR)} = \frac{\text{Weight of micro-tool before machining} - \text{Weight of micro-tool after machining}}{\text{Machining time}} \tag{4.3}$$

4.7.3 Surface roughness

The surface roughness of the machined features mainly depends on the crater size (diameter and depth) formed by each discharge. In addition, if the dielectric circulation in the discharge gap is not efficient, then some molten material from the tool and the workpiece resolidifies on the microfeature surface, making the surface rough. Thus, an effective flushing technique has a significant effect on the surface finish of the features. Crater dimensions also largely depend on the pulse energy of discharge and other process parameters such as peak current, pulse frequency, and pulse duration (Jahan, Rahman, & Wong, 2011). The required amount of flushing pressure can decrease the roughness of the machined surface. Moreover, the properties of the material of the microtool and the workpiece have considerable effects on the surface finish.

4.7.4 Overcut

Overcut (OC) is an unnecessary dimensional inaccuracy in micro-EDM, especially during the microhole drilling process. OC is the diametral difference of the entrance microhole to the microtool tip diameter when a cylindrical tool electrode is used, as shown in Eq. (4.4). However, for other geometry of the microtool, it is the excess breadth or width of the machined cavity than the tool tip dimension. Whereas the debris from the machining zone is ejected out by the flushing pressure, the material removal takes place from the sidewall of the microfeature surface because of secondary sparking phenomena, resulting in OC. A number of process parameters such as peak current, pulse duration, and flushing pressure significantly affect the amount of OC.

$$\text{Overcut (OC)} = \frac{\text{Diameter of entry hole} - \text{Diameter of tool electrode}}{2} \quad (4.4)$$

4.7.5 Diametral variance at entry and exit holes

This performance criterion is related to through microhole drilling in micro-EDM. The diametral variance at the entry and exit holes is measured by differentiating the microhole diameter at the entry and exit sides from end-to-end hole on the workpiece. During microdrilling operation of a high-aspect-ratio hole, if the secondary sparking occurs for a long time, then a large diametral difference is obtained. The diametral variance at the entry and exit (DVEE) holes is affected by a number of process parameters such as peak current, pulse duration, duty cycle, and flushing pressure.

4.7.6 Circularity

Circularity is one of the major performance criteria during microhole drilling in micro-EDM. Circularity is defined as the degree of roundness of a circular hole. The circularity of the microhole is calculated based on Eq. (4.5).

$$\text{Circularity} = 4\pi \frac{\text{Area of through micro-hole}}{[\text{Perimeter of through micro-hole}(P)]^2} \quad (4.5)$$

At a higher flushing pressure, the high-aspect-ratio microtool loses its stiffness, and vibration occurs during microhole drilling. As a result, the degree of roundness of the microhole, i.e., circularity, deteriorates.

4.7.7 Machining time

Machining time is the duration of machining to generate a particular microfeature on the workpiece surface in micro-EDM. Machining time is directly related to MRR. However, several process parameters such as peak current, pulse frequency, pulse duration, and proper flushing greatly influence the machining time criterion.

4.8 Titanium alloys as advanced engineering materials

The high strength, low weight, and outstanding corrosion resistance possessed by titanium and its alloys have led to a wide and diversified range of successful applications. These applications demand high levels of reliable performance in surgery and medicine, such as bone and joint replacement, dental implants, cardiovascular devices, and surgical instruments. Other engineering applications are in aerospace, automotive, chemical plant, power generation, oil and gas extraction, sports, and other major industries.

Titanium alloy was developed in the early 1950s for defense and aeronautic applications because of its very high strength-to-weight ratio. The unique combination of high strength, low weight, and excellent corrosion resistance of titanium alloy has made it suitable for a wide variety of industrial applications. Commercially pure titanium is used primarily for its corrosion resistance. Further titanium alloy can withstand pitting, crevice and cavitations, corrosion, erosion, and stress corrosion cracking in salt water, marine atmospheres, and a broad range of acids, alkalis, and industrial chemicals. Its uses are not only confined and concentrated in aerospace engine and airframe components but extends to major nonaerospace applications in marine offshore, power generation industries, and biomedical applications as well. Ti-6Al-4V is the most widely used titanium alloy, accounting for more than half of all titanium tonnage worldwide.

4.9 Literature review of micro-EDM of Ti-6Al-4V

Several research studies have been conducted on the development of micro-EDM, employing innovative strategies as well as exploring the effects of process parameters for improving the overall performances of micro-EDM during machining of Ti-6Al-4V. Researchers have also extensively employed several statistical techniques for optimizing the output criteria. Pradhan, Masanta, Sarkar, and Bhattacharyya (2009a) carried out an experiment on the micro-EDM process of Ti-6Al-4V by a brass electrode with a diameter 500 μm. A Taguchi methodology-based statistical approach is considered to construct the experimental settings. Material removal rate, tool wear rate, overcut and taper were chosen as observed performance criteria, whereas peak current, pulse-on-time, flushing pressure, and duty ratio were considered as process parameters. Machining performances are affected mostly by the peak current and pulse-on-time. Using response surface methodology (RSM), Pradhan and Bhattacharyya (2009b) carried out an experiment on microhole drilling in Ti-6Al-4V with a brass electrode. Process performance parameters such as MRR, TWR, and OC were measured after micro-EDM. RSM, and artificial neural network (ANN) with back-propagation-algorithm-based mathematical models that have been developed to correlate the performance criteria with process parameters. Furthermore, single and multiobjective optimization of the machining characteristics of micro-EDM during the microhole machining operation on Ti-6Al-4V was also conducted. Ali, Rahman, and Aris (2011) investigated the influence of mixing silicon carbide (SiC) powder in dielectric on MRR during machining of Ti-6Al-4V in micro-EDM. The investigation revealed that relative to conventional dielectric, SiC powder mixed in dielectric enhances the optimization of MRR. Optimum MRR of 7.31 μg/min is achieved at a powder concentration of 24.75 g/L and discharge energy of 56.77 μJ during micro-EDM of Ti-6Al-4V. Meena and Azad (2012) investigated for optimizing the MRR, TWR, and OC during micro-EDM of Ti-6Al-4V alloy. Gray relational analysis (GRA) and analysis of variance were performed to achieve the optimal parametric setting, i.e., voltage, pulse frequency, current, and pulse width. Porwal, Yadava, and Ramkumar (2014) carried out an experimental investigation for machining a Ti-6Al-4V thin sheet during hole sinking electrical discharge micromachining (HS-EDMM). An integrated model (ANN-GRA-PCA) of a single hidden layer BPNN was developed to predict MRR, TWR and the taper of the microhole. GRA coupled with PCA hybrid optimization strategy was utilized for optimizing the process criteria, and an optimal combination of the process parameters was reported as 140 V gap voltage and 100 nF capacitance. Tiwary, Pradhan, and Bhattacharyya (2014) investigated the influence of process parameters such as pulse-on-time, peak current, gap voltage, and flushing pressure on MRR, TWR, OC, and taper of microhole during the micro-EDM process of Ti-6Al-4V by using RSM. Experiments were conducted on a 1 mm thick titanium alloy material using a brass electrode with a diameter of 300 μm. To determine the optimum process parameters, such as pulse-on-time, peak current, gap voltage, and pulse frequency, combined RSM and fuzzy-TOPSIS method was used. Kuriachen and Mathew (2014) described the effect of gap voltage, capacitance, rotational speed of electrode, and feed rate on the MRR of microelectro discharge milling of Ti-6Al-4V. A quadratic regression model was developed based on the RSM−Box Behnken experimental design. An increase in MRR showed a direct relationship with capacitance and the rotational speed of the electrode. Plazaa et al. (2014) studied the influence of micro-EDM parameters on MRR, TWR, machining time, and quality of microhole during machining of Ti-6Al-4V by using helical microtool electrodes. The influences of the helix angle and the flute depth of the helical microtool on process performances were also examined. Moses and Jahan (2015) carried out an experimental investigation by machining blind and through microholes and microslots on brass and Ti-6Al-4V materials. The qualities of microfeatures such as dimensional accuracy, surface finish, and profile accuracy were measured. In addition, several arrays of microfeatures were machined, such as single through microhole, single blind hole, letter *H* blind, three blind slots, and three through slots. Tiwary, Pradhan, and Bhattacharyya (2015) investigated the influence of various process parameters such as pulse-on-time, peak current, gap voltage, and flushing pressure on MRR, TWR, OC, and taper of microhole during machining of Ti-6Al-4V. RSM was utilized to develop a mathematical relationship between the input process parameters and the responses. Multiobjective optimization was achieved to attain the optimal parametric setting, i.e., pulse-on-time of 1 μs, peak current of 2.5 A, gap voltage of 50 V, and flushing pressure of 0.20 kgf/cm^2. The optimal values of responses achieved were MRR of 0.0777 mg/min, TWR of 0.0088 mg/min, OC of 0.0765 mm and taper of 0.0013. Kuriachen and Mathew (2015a) carried out an investigation to machine Ti-6Al-4V with tungsten carbide electrode employing SiC microparticle-suspended dielectric during micro-EDM milling. The effects of various process parameters such as voltage, capacitance, and powder concentration on MRR and TWR were studied. The recommended process parametric setting was powder concentration of 5 g/L, capacitance of 0.1 μF, and voltage of 115 V to achieve high material removal and low TWR. Kuriachen, Varghese, Somashekhar, Panda, and Mathew (2015b) developed a heat transfer-based predictive thermal model to simulate the single-spark microelectric discharge machining process. Finite volume method was utilized to solve the model. Crater geometry and temperature distribution in the workpiece at various process parametric settings were predicted using the Gaussian distribution of heat flux, percentage distribution of energy among the workpiece, tool electrode, and dielectric. Kuriachen and Mathew (2015c) attempted to develop a mathematical model that can predict the radius of the single-spark during micro-EDM of Ti-6Al-4V. A three-level full factorial experimental design was utilized in the experiments. Capacitance and voltage were considered as process parameters. The results revealed that the spark radius increases proportionally within capacitance except at higher energy levels where a double sparking phenomenon was observed.

The literature review on micro-EDM of Ti-6Al-4V shows that most of the studies investigate the effects of process parameters on several process performances—MRR, TWR, OC, and surface roughness, among others—and the optimization of the process by utilizing statistical tools to improve the micro-EDM performance. However, many issues remained unsolved in micro-EDM, such as improving MRR, methodology for compensating the microtool wear, improving the stability of discharge, and improving the accuracy of microfeatures implementing innovative ideas. To solve these important issues and to improve overall micro-EDM efficiency, several new micro-EDM machining strategies have been developed, and researchers have utilized these innovative strategies during micro-EDM of Ti-6Al-4V. The following sections discuss the experimental investigation and analysis of the micro-EDM process of Ti-6Al-4V by extensively utilizing some innovative machining strategies. These strategies include micro-ED machining at reverse polarity, rotating the microtool electrode, employing nonhydrocarbon based dielectric fluid, and utilizing powder-mixed dielectric fluids.

4.10 Investigation of micro-EDM process employing innovative machining strategies

The micro-EDM process has some inherent problems associated with it, which demand careful attention and exhaustive research studies to achieve the desired and effective criteria yield while machining microcomponents. Manufacturers and users of the micro-EDM set-up always look forward to achieving high productivity with increased accuracy and surface integrity. Therefore, a systematic approach and categorical research for most suitable process parametric settings become invariably prerequisite to achieve effective machining performance of micro-EDM for real-time utilization. The effectiveness and efficiency of the machining performance of micro-EDM is affected by several process parameters.

Considering all the problems associated with the micro-EDM process, new machining techniques that can address the problems of EDM micromachining urgently need to be developed to achieve precise and accurate microcomponents. In the first part of the research, a novel technique of machining with a changing polarity in a systematically designed time domain is explored. This novel technique of changing polarity in an exponential time domain is employed to generate straight-through microholes in micro-EDM on a Ti-6Al-4V workpiece. In the second part, the investigation is conducted with a rotating microtool for machining straight-through microhole on a Ti-6Al-4V workpiece, and the effect of microtool rotation is evaluated. In the third part, micro-EDM is carried out by employing kerosene and deionized water as dielectric fluids, and the machining performance criteria were compared by varying the significant process parameters during through microhole machining. In the fourth part of the experiment, micro-EDM process of Ti-6Al-4V is conducted by employing powder-mixed dielectrics, and the effects of various process parameters are investigated. Comparative study and analysis of microhole drilling are performed using pure and powder-mixed dielectrics.

4.10.1 Changing the polarity of electrodes

4.10.1.1 Experimental method and conditions

The experimental condition has been designed such that the polarity during micro-EDM machining has been changed in an exponential time domain to improve the machining condition, debris removal, and machining efficiency to achieve a higher geometrical accuracy for the microhole. In a constant polarity machining condition, the job is positive and the tool electrode is negative, i.e., normal polarity. However, in changing the polarity machining condition, the polarity of the job and the tool electrode have been changed in an exponential time domain. The machining begins with normal polarity for the first 10 min, with polarity changing for the next 3 s and again switching back to normal polarity, and the machining continues for another 9 min before the second change. The amount of carbon deposits and machining debris increases as microhole depth increases and its removal becomes difficult. Therefore, with an increase in the microhole depth, the rate of change of polarity must change to facilitate the removal of the deposited carbon and debris from the machining zone. Thus, the duration of machining with normal polarity is reduced after every change, and this process continues until the through microhole is produced on the workpiece. However, machining with reverse polarity is kept constant at 3 s in each change.

The time chart has been prepared after conducting several trial experiments to find out the time required to machine the through microhole in a 1 mm thick Ti-6Al-4V alloy sheet with a 300 μm diameter brass tube electrode. The dielectric fluid used is kerosene. Past studies and experimental investigations on micro-EDM indicate that peak current and pulse-on-time are the most influential parameters. Therefore, these dominating parameters have been selected as process parameters in the present micro-EDM experimentation. To study the effects of pulse-on-time (T_{on}) and peak current (I_p), the experimental planning has been carried out first by only varying the peak current from 0.5 to 2 A while keeping pulse-on-time (T_{on}), duty factor (t), and flushing pressure (P_r) constant at 10 μs, 95%, and 0.5 kgf/cm^2, respectively, and second, by only varying the pulse-on-time (T_{on}) from 1 to 20 μs while keeping peak current (I_p), duty factor (t), and flushing pressure (P_r) constant at 1 A, 95%, and 0.5 kgf/cm^2, respectively.

Each experiment at that particular parametric setting was conducted three times, and the average of the three were considered for calculating the MRR, TWR, OC, and DVEE of the microhole. Here, MRR and TWR are calculated based on the weight difference measured in a precision weighing machine of Mettler Toledo, Switzerland with the least count of 0.01 mg. The experiments were conducted on ZNC R50 EDM (Manufacturer: Electronica Machine Tools Pvt. Ltd., Pune, India).

4.10.1.2 Experimental results and analysis

The variations of MRR with peak current and pulse-on-time with constant and changing polarity while keeping all other process parameters constant, i.e., pulse-on-time at 10 μs in case of a varying peak current, and peak current at 1 A in case of a varying pulse-on-time, duty factor at 95%, and flushing pressure at 0.5 kgf/cm^2, as shown in Figs. 4.4A, B, respectively. Fig. 4.4A shows that MRR increases monotonically in both cases, with the increase in peak current from 0.5 to 1.5 A, but for changing polarity, it decreases as peak current increases from 1.5 to 2 A. The magnitude of MRR in both cases is almost equal in the considered peak current range. The low MRR at a smaller peak current could be due to lower discharge energy when machining in both constant and changing polarity. However, MRR at a changing polarity is low compared with that at a constant polarity, as shown Fig. 4.4A, because of the change in the position of maximum liberation of heat energy due to sparking. The increase in MRR with increasing peak current is attributed to a larger discharge energy.

Fig. 4.4B shows the variation of MRR with pulse-on-time. This figure indicates that the MRR variation is almost opposite in nature for changing and constant polarities at lower pulse-on-time values between 1 and 5 μs. MRR decreases from 1 to 2 μs for constant polarity, but in the same pulse-on-time range, MRR increases for changing polarity. In the range from 2 to 5 μs, MRR increases and decreases sharply for constant and changing polarity, respectively. However, from 5 to 10 μs, MRR increases gradually in both cases. Furthermore, in the pulse-on-time range of 10–20 μs, MRR increases slowly for constant polarity but decreases for changing polarity. It can also be observed that maximum MRR is obtained at the lower range of pulse-on-time, i.e., 1–5 μs. In the case of the changing polarity approach to machining, owing to change in polarity, the position of maximum liberation of heat energy due to sparking is changed; at the same time, the number of sparking per cycle is increased for lower pulse-on-time. Thus, the total heat energy generated in the discharge phenomenon is increased during the parametric setting, and a high MRR is obtained.

Figs. 4.5A, B show the variation of TWR with peak current and pulse-on-time, respectively. As expected, Fig. 4.5A shows that TWR increases proportionally with the increase in peak current from 0.5 to 2 A at a fixed pulse-on-time of 10 μs in both constant and changing polarity. The increase in TWR with the increase in peak current is due to the increase in discharge energy. The increase in magnitude of discharge energy rapidly deteriorates the tool geometry as the thermal energy is concentrated in a very small area (the size of the electrode in this case). However,

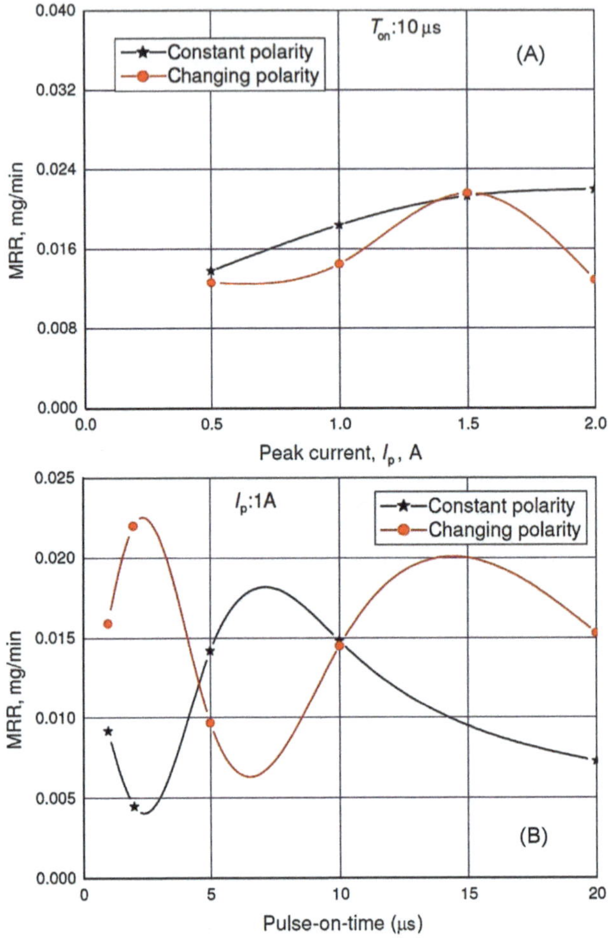

Figure 4.4 Variation of MRR with (A) peak current and (B) pulse-on-time (Pradhan & Bhattacharyya, 2008).

Fig. 4.5A reveals that TWR is greater with changing polarity; this is exactly opposite to the expectation outlined above, where tool wear was thought to be less in this case. This occurrence might be attributable to the removal of carbon deposits from the tool during normal polarity.

However, when pulse-on-time was varied while peak current, duty factor, and flushing pressure remained constant, as shown in Fig. 4.5B, TWR decreased significantly with changing polarity relative to constant polarity. This indicates that pulse-on-time is the more critical factor in tool wear than peak current. Therefore, for minimum TWR, it is better to use changing polarity with smaller pulse-on-time (1−10 µs) and low peak current of 1 A to yield a higher MRR and a low TWR. This observation reveals that shorter pulse-on-time, which means a higher frequency of sparking, and low peak current are suitable for micro-EDM.

Figs. 4.6A, B show the variation of OC with peak current and pulse-on-time, respectively. It is observed from Fig. 4.6A that OC decreases sharply from 0.5 to 1 A for both constant and changing polarity conditions and thereafter increases monotonically with an increasing peak current in both machining conditions. This observation clearly indicates that the optimum peak current setting is 1 A at the present process parameter range. If the peak current is low, the discharge energy is also low, which means longer machining time, exposing the sidewall of the hole to secondary sparking and thus resulting in larger OC. However, the increase in OC at a higher peak current is due to higher discharge energy and larger debris concentration in the gap because of higher MRR. Fig. 4.6B also indicates that OC increases steeply from 1 to 5 μs, decreases sharply from 5 to 10 μs, and again increases at

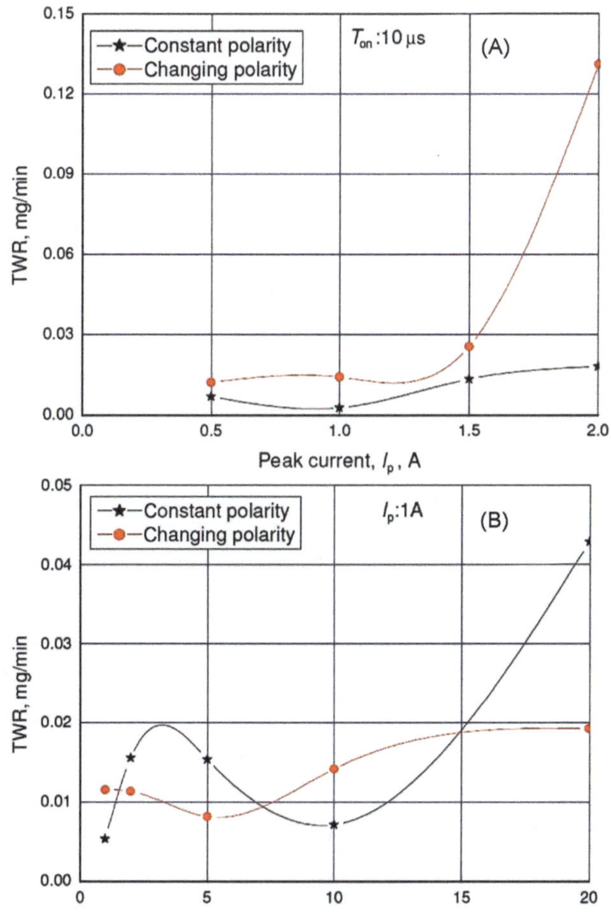

Figure 4.5 Variation of TWR with (A) peak current and (B) pulse-on-time (Pradhan & Bhattacharyya, 2008).

10–20 μs for constant polarity. Thus, OC fluctuates with the change in pulse-on-time for constant polarity. However, for changing polarity, OC is found to decrease gradually for pulse-on-time from 1 to 20 μs, suggesting that the changing polarity technique achieves a lower OC in the machining of microholes, hence increasing the geometrical accuracy of the machined microhole in Ti-6Al-4V. Further, throughout the range of pulse-on-time from 1 to 20 μs, the magnitude of OC is far lower with changing polarity than with constant polarity, which is an indicator in itself that changing polarity yields low OC and results in the improvement of micro-hole geometry. This effect is attributable to the fact that the carbon particles, which are by-products of micro-EDM, are deposited on the surface of the tool electrode,

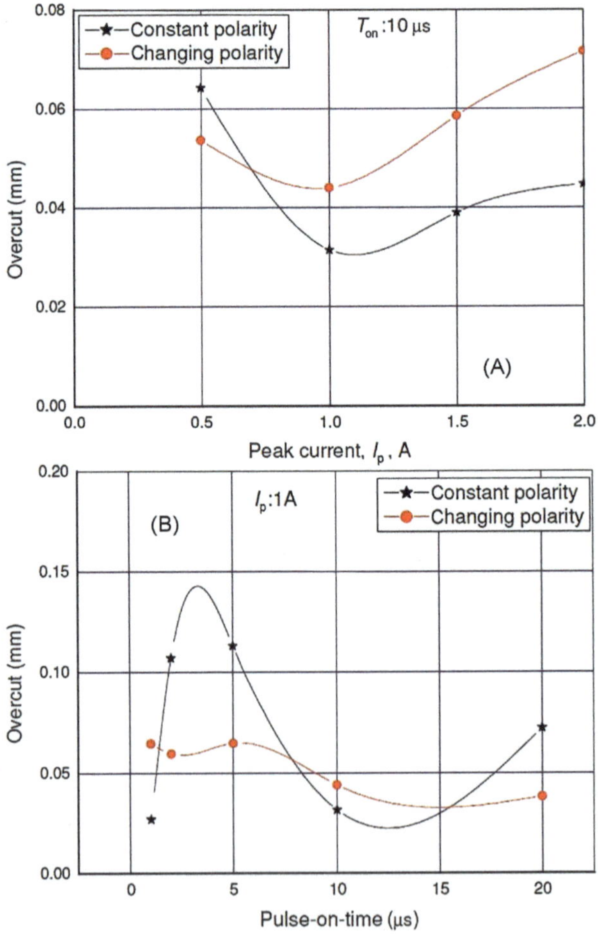

Figure 4.6 Variation of OC with (A) peak current and (B) pulse-on-time (Pradhan & Bhattacharyya, 2008).

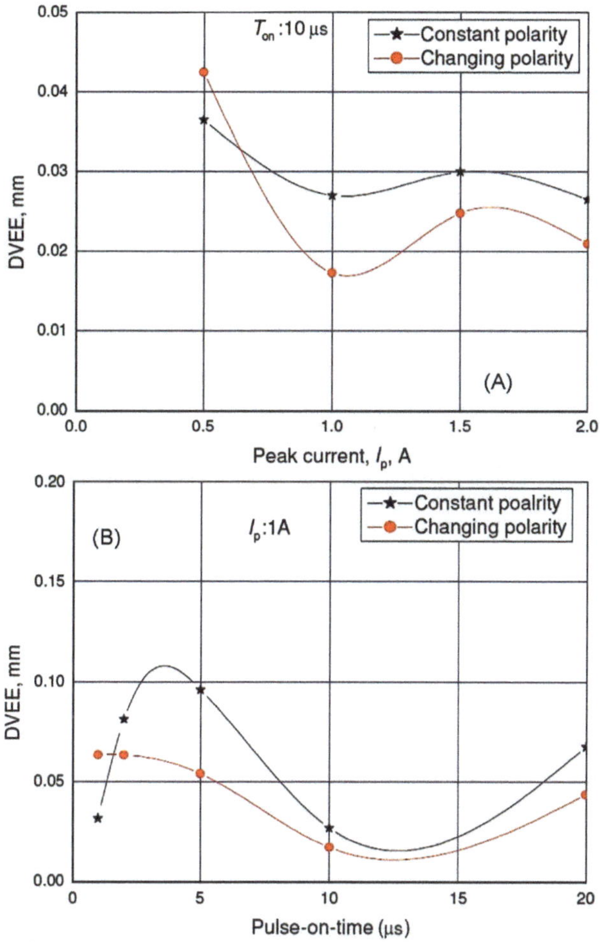

Figure 4.7 Variation of DVEE with (A) peak current and (B) pulse-on-time (Pradhan & Bhattacharyya, 2008).

helps prevent secondary sparking. Thus, only the end face or bottom face is exposed for sparking, reducing OC and resulting in straight-through microhole generation, thus improving the accuracy of microhole machining.

It is observed from Figs. 4.7A, B that DVEE decreases sharply with the increase in peak current and pulse-on-time. The lowest DVEE is obtained at 1 A and 10 µs, as seen in the figures for both constant and changing polarity machining conditions. DVEE is also observed to be less with changing polarity than with constant polarity throughout the peak current and pulse-on-time ranges considered in the experiments. With the novel technique of changing the polarity, a straight microhole can be achieved. As the depth of the microhole increases, the sparking points shift radially inward. When the polarity is changed, the pointed tip of the microtool wears

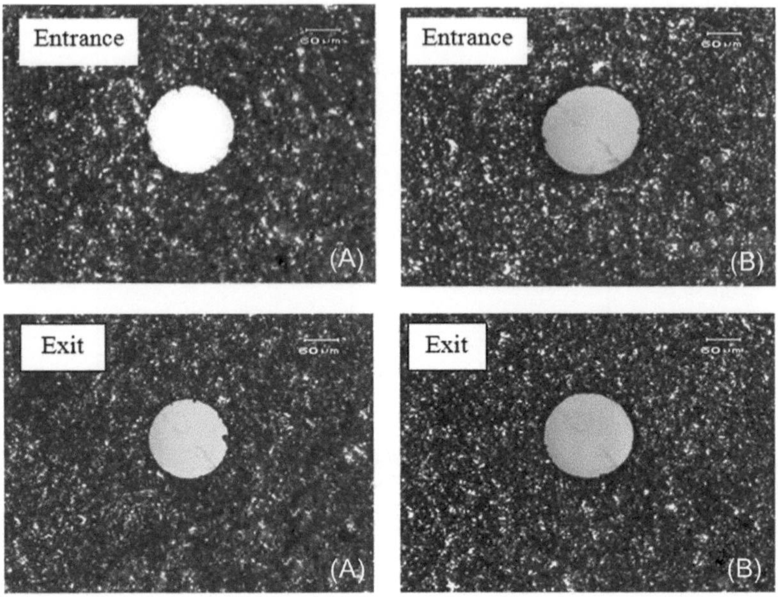

Figure 4.8 Optical micrographs of microholes with (A) constant and (B) changing polarity machined at 1 A/10 μs/95% duty factor /0.5 kgf/cm^2 (Pradhan & Bhattacharyya, 2008).

off uniformly, making the tool end broader and helping machine a straight microhole, thereby decreasing DVEE.

Optical micrographs of the microholes (entry and exit diameter) machined with constant and changing polarity at 1 A/10 μs/95% duty factor/0.5 kgf/cm^2, and 1 A/ 20 μs /95% duty factor/0.5 kgf/cm^2 are shown in Figs. 4.8A, B and 4.9A, B, respectively. The DVEE at a parametric combination of 1 A/10 μs/95% duty factor/ 0.5 kgf/cm^2 for the constant polarity machining condition is 0.0270 mm, and for the changing polarity condition, the value is 0.0173 mm. The comparison of these data and the micrographs clearly indicate that DVEE and the geometrical shape are better when using the changing polarity condition for the micromachining of microholes by the micro-EDM. The SEM micrographs of machined microholes at the parametric settings of 1 A/10 μs/95% duty factor/0.5 kgf/cm^2 and 1 A/20 μs/95% duty factor/0.5 kgf/cm^2 with constant and changing polarity techniques are shown in Figs. 4.10A, B and 4.11A, B, respectively. These figures show that the thickness of recast layer formed on the microhole surface by using the polarity changing technique is less that that using the constant polarity technique, corroborating the fact that the surface quality of the machined microhole has improved with the new machining technique. This observation clearly indicates that with the changing polarity technique, a better machining condition is achieved during microhole machining by the micro-EDM process.

Thus, taking all aspects into account as discussed above and considering all major factors involved in micro-EDM machining, changing the polarity conditions

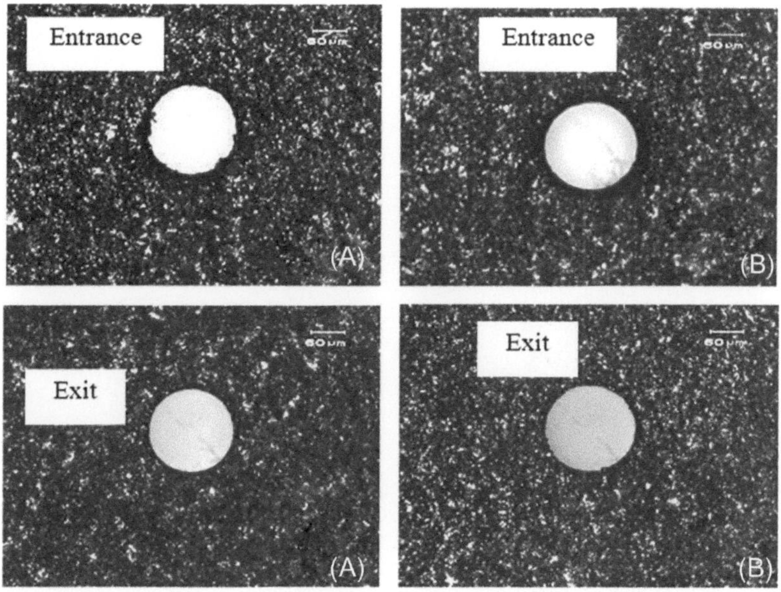

Figure 4.9 Optical micrographs of microholes with (A) constant and (B) changing polarity machined at 1 A/20 μs/95% duty factor /0.5 kgf/cm^2.

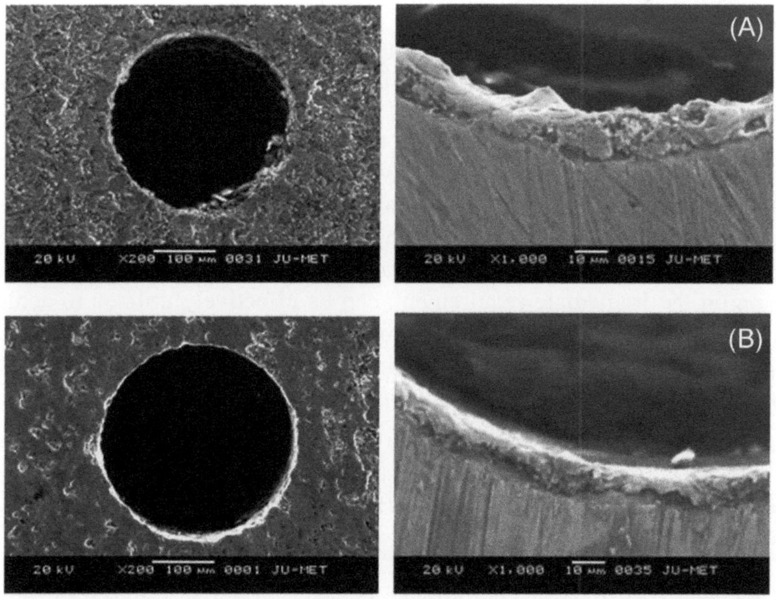

Figure 4.10 SEM micrographs of microholes with (A) constant and (B) changing polarity machined at 1 A/10 μs/95% duty factor /0.5 kgf/cm^2.

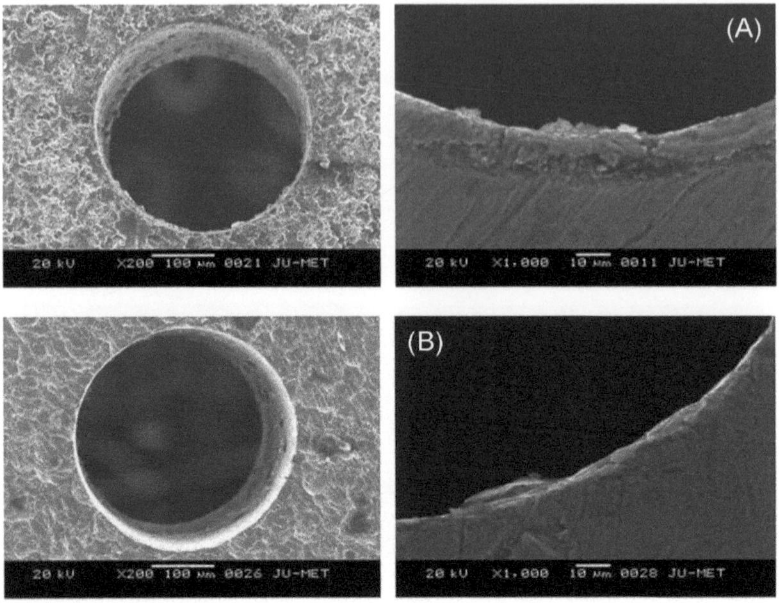

Figure 4.11 SEM micrographs of microholes with (A) constant and (B) changing polarity machined at 1 A/20 μs/95% duty factor /0.5 kgf/cm^2.

and evaluating their effects extensively in this experimental investigation for determining the most effective parametric combinations, and exploring all possible aspects of choosing them, the following optimal parametric combinations have been selected. To achieve higher productivity, the optimal parametric combination with the changing polarity machining condition is 1 A/2 μs/0.5 kgf/cm^2/95%. The optimal parametric combination with the changing polarity machining technique is 1 A/5 μs/0.5 kgf/cm^2/95% for the least TWR and 1 A/20 μs/0.5 kgf/cm^2/95% for the least OC. To achieve a higher dimensional accuracy, i.e., the least DVEE, the optimal parametric combination with the changing polarity machining technique is 1 A/10 μs/0.5 kgf/cm^2/95%. One of these various optimal parametric combinations, depending on the immediate requirement, can be effectively utilized to achieve the same for the best possible machining conditions by using the polarity changing technique in an exponential time domain in order to improve the specific machining criteria in EDM during micromachining operations.

4.10.2 Rotating the microtool electrode

4.10.2.1 Experimental method and conditions

In this research investigation, the effects of peak current (I_p), pulse-on-time (T_{on}), and rotational speed of the microtool electrode are explored during microhole machining in micro-EDM. The experiments were conducted using the same ZNC

R50 EDM machine to evaluate the effects of the rotation of electrodes with respect to MRR, TWR, OC, and DVEE on a Ti-6Al-4V workpiece of 1 mm thickness with a brass electrode 300 μm in diameter. During machining, kerosene is used as the dielectric fluid. To rotate the tool electrode, a tool rotational attachment has was used in which the rotational speed can be varied between 1−300 rpm at a resolution of 2 rpm. The ranges of peak current (I_p) and pulse-on-time (T_{on}) selected were 0.5−2 A and 1−20 μs, respectively. The experiments were conducted in two stages: (1) varying only peak current from 0.5 to 2 A keeping pulse-on-time, duty factor, and flushing pressure as constant at 10 μs, 95%, 0.5 kgf/cm^2, respectively, and (2) varying only pulse-on-time from 1 to 20 μs, with peak current, duty factor, and flushing pressure constant at 1 A, 95%, and 0.5 kgf/cm^2, respectively, with stationary and rotating electrodes with a rotational speed of 150 rpm in each case. The peak current was fixed at 1 A because for micromachining, very low current density is not sufficient to melt and vaporize the workpiece, and very high current density leads to higher TWR and larger thermal damage of the workpiece surface. On the other hand, pulse-on-time is fixed at 10 μs because short duration helps reduce tool wear, thus being more beneficial for micromachining. A moderate rotational speed of the electrode was selected, i.e., 150 rpm after conducting trials as this speed facilitates the removal of debris from the machining zone, thereby keeping tool wear at a minimum.

4.10.2.2 Experimental results and analysis

By using the microtool electrode rotating facility with a developed tool holder, the experiments were conducted to evaluate the effects of micro-EDM parameters on process criteria, namely, MRR, TWR, OC, and diametral variation at entry and exit (DVEE) of the machined microholes.

The variations of MRR with peak current (I_p) and pulse-on-time (T_{on}) with stationary and rotating electrode are shown in Figs. 4.12A, B, respectively. The graph indicates that MRR increases with an increase in peak current as was expected for both stationary and rotating electrodes. MRR increases because higher peak current leads to higher discharge energy. The same figure shows that with a rotating electrode, a higher MRR is obtained. The higher MRR with a rotating electrode may be attributed to better removal of sludge and carbonized particles from the machining zone because of the centrifugal force of rotation. This improved sludge removal due to the rotating effect of the electrode helps expose the actual machining surface, which then improves the overall machining condition, leading to a higher MRR. Thus, for maximum MRR from within the considered range of parametric setting, the best parametric combination in the present case study is 2 A/10 μs/0.5 kgf/cm^2/ 95% /150 rpm.

Figs. 4.13A, B, respectively, show the variations of TWR with peak current (I_p) and pulse-on-time (T_{on}) with stationary and rotating electrode. It can be observed from the graph that TWR increases as the peak current increases from 0.5 to 1.5 A. However, TWR is observed to decrease from 1.5 to 2 A. The increase in TWR with the increase in peak current can be attributed to the increase in discharge energy and the rotational effect of the electrode. A further increase in peak current increases machining efficiency and decreases tool wear because the tool electrode is subjected to high-energy electric field for a shorter duration. Thus, this figure clearly indicates that for microhole machining with low TWR, a smaller peak current is suitable. Owing to a rotational effect, the magnitude of TWR further decreases, which is evident from the figure under consideration. This may be attributed to enhanced removal of debris caused by the tangential force of rotation. Thus, for least TWR, the best parametric combination from within the considered range of parametric settings is 1 A/10 μs/0.5 kgf/cm^2/95%/150 rpm.

Figs. 4.14A, 4.14B respectively show the variations of OC with peak current (I_p) and pulse-on-time (T_{on}) with stationary and rotating electrodes. It is observed from

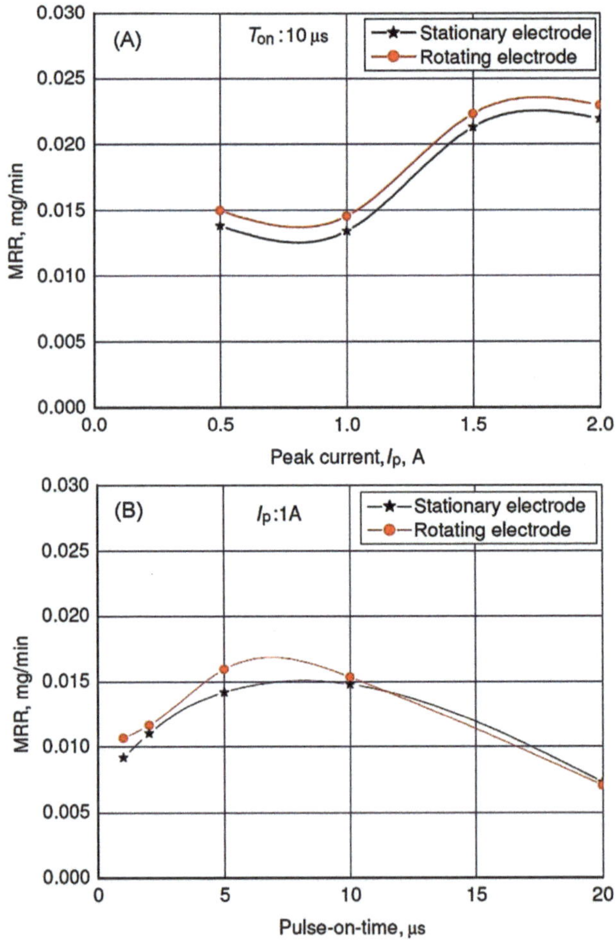

Figure 4.12 Variation of MRR with (A) peak current and (B) pulse-on-time (Pradhan, Sarkar, Kibria, & Bhattacharyya, 2009c).

the figure that OC decreases with an increase in peak current ranging from 0.5 to 1 A, which may be attributed to the increase in discharge energy with the increase in peak current and enabling faster machining, thereby reducing the effective machining time. The reduction in machining time means less exposure of the tool electrode to discharge energy, which is responsible for tool wear. However, with peak current beyond 1 A, OC increases monotonically. An increase in peak current above 1 A leads to larger discharge energy, causing a larger MRR and resulting in a larger OC. Thus, the most suitable parametric combination for the least OC from within the considered range of parametric settings is 1 A/10 μs/95% duty factor/ 0.5 kgf/cm^2/150 rpm.

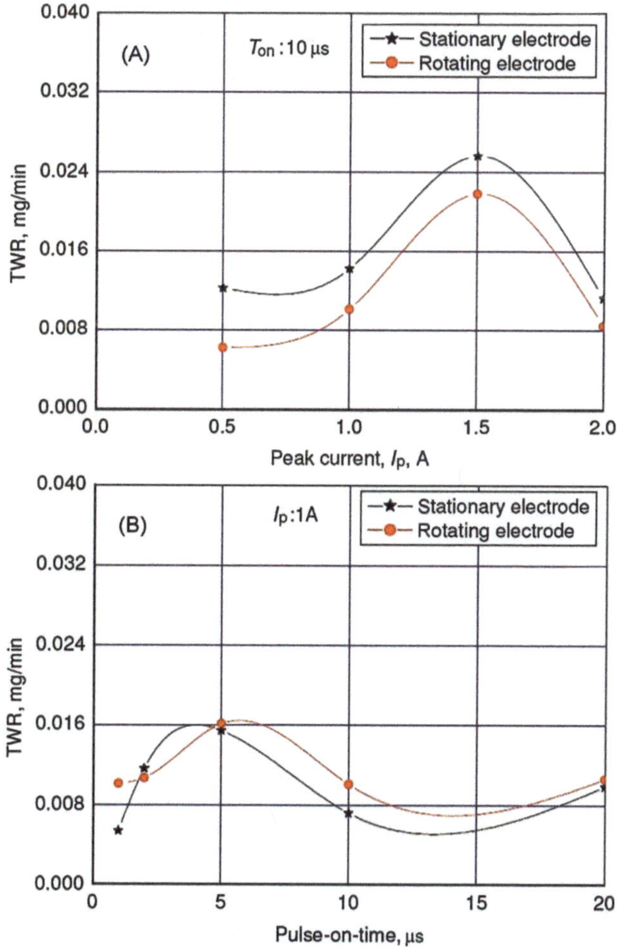

Figure 4.13 Variation of TWR with (A) peak current and (B) pulse-on-time (Pradhan et al., 2009c).

The variations of DVEE with peak current (I_p) and pulse-on-time (T_{on}) with stationary and rotating electrodes are shown in Figs. 4.15A, B, respectively. The same graph shows that with the increase in peak current, DVEE decreases. Furthermore, the magnitude of DVEE with a rotating electrode is less than that with a stationary electrode throughout the considered range of peak current. As the depth of the microhole increases, the sparking point shifts radially inward. When the peak current is increased, the thermal energy density increases at the pointed tip of the microtools. This high density discharge energy rapidly melts and vaporizes the sharp microtool tips, subsequently broadening the microtool end, which finally helps achieve a straight-through microhole by decreasing DVEE. Further, the rotational speed of the tool helps remove the sludge and debris efficiently from the

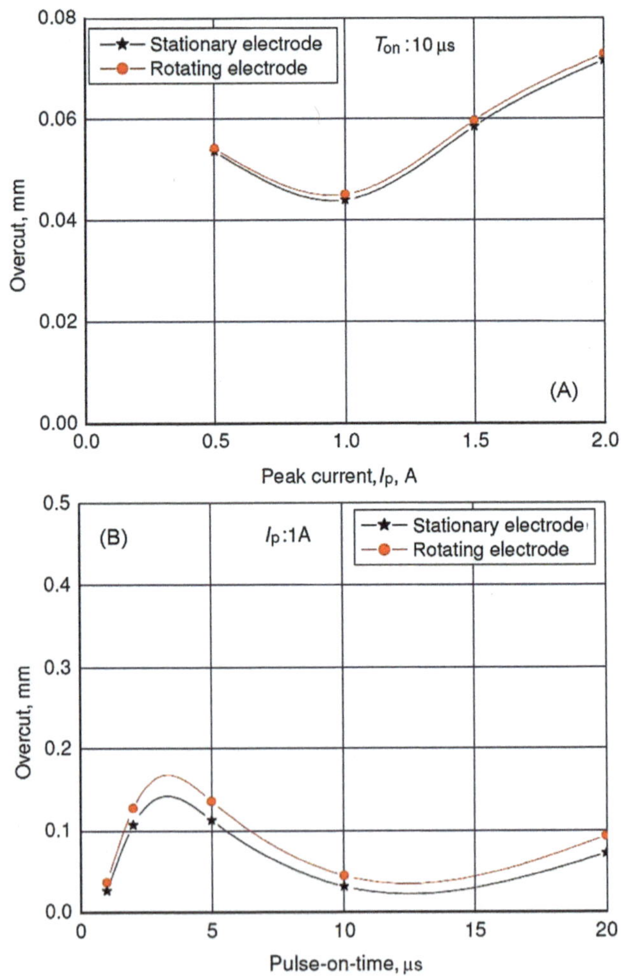

Figure 4.14 Variation of OC with (A) peak current and (B) pulse-on-time.

machining zone and reduces the chances of secondary discharge sparking, thereby providing stable machining conditions for achieving microhole with less DVEE. The lowest DVEE achieved is with the parametric combination of 2 A/10 μs/95%/ 0.5 kgf/cm^2/150 rpm.

The improvement in microhole geometry with the increase in peak current coupled with rotation of the tool electrode may be attributed to better flushing of debris due to rotation of electrode, uniform tool wear, and evenly distribution of discharge energy.

Fig. 4.16A shows the optical micrographs of the microhole diameters at the entry and exit machined with the parametric setting of 1 A/5 μs/95% duty factor/0.5 kgf/cm^2, i.e., with the stationary microtool. Fig. 4.16B shows the optical micrographs of the microhole diameters at entry and exit machined with the

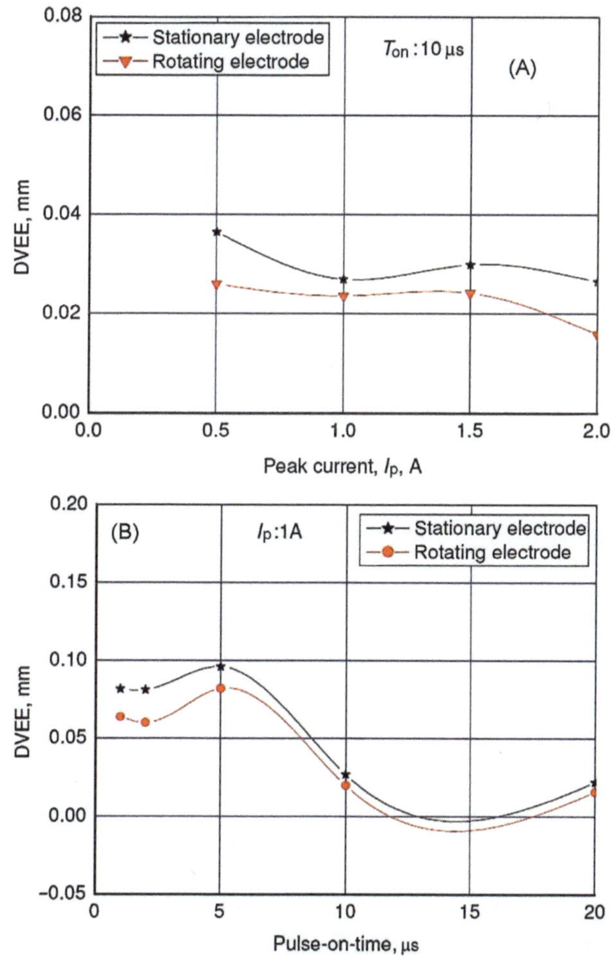

Figure 4.15 Variation of DVEE with (A) peak current and (B) pulse-on-time.

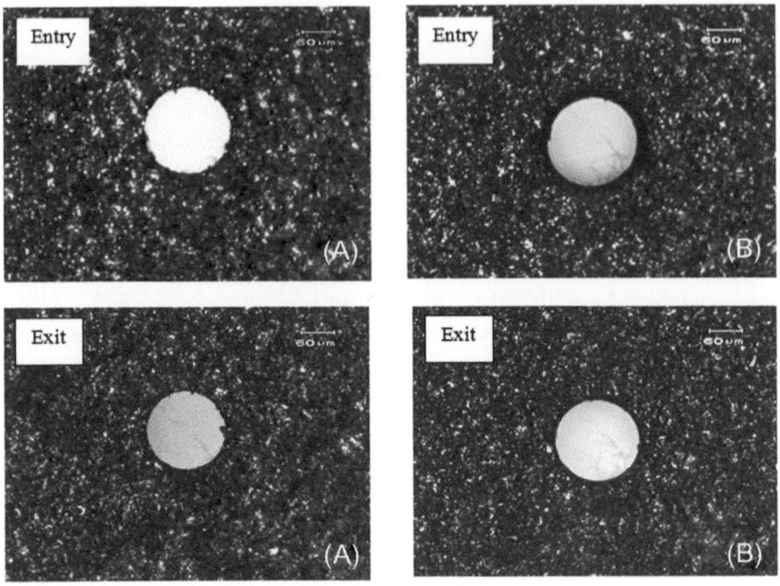

Figure 4.16 Optical micrographs of microholes with (A) stationary and (B) rotating tool electrode machined at 1 A/5 μs/95% duty factor /0.5 kgf/cm^2.

parametric setting of 1 A/5 μs/95% duty factor/0.5 kgf/cm^2/150 rpm, i.e., with the rotating microtool. It can be observed from these two micrographs that with the rotating tool electrode, straight-through microhole can be fabricated.

4.10.3 Comparative study of using kerosene and deionized water dielectrics

4.10.3.1 Experimental method and conditions

The use of various dielectrics (hydrocarbon oil, deionized water, EDM oil, etc.) has important effects for improvement of machining performance. Different dielectrics have different properties in terms of dielectric strength, degree of recovery capability, degree of fluidity, and chemical compositions. Therefore, it is very important to carry out extensive research for employing various type of dielectric fluid during micro-EDM of Ti-6Al-4V. In this research, a comparative study was performed for using pure kerosene and deionized water on micro-EDM performance criteria, i.e., MRR, TWR, OC, and DVEE. The same EDM system mentioned earlier is used for this experimental investigation. When deionized water is employed for machining of microhole, a separate dielectric circulating system was used as the micro-EDM system uses kerosene-based dielectric. The circulating system consists of a pump, a reservoir, piping, a pressure-regulator, and a filter. Ti-6Al-4V plates of size 13 × 15 × 1 mm are used to machine through microholes. Tungsten microtool electrodes of diameter 300 μm were used in this experiment. Peak current (I_p)

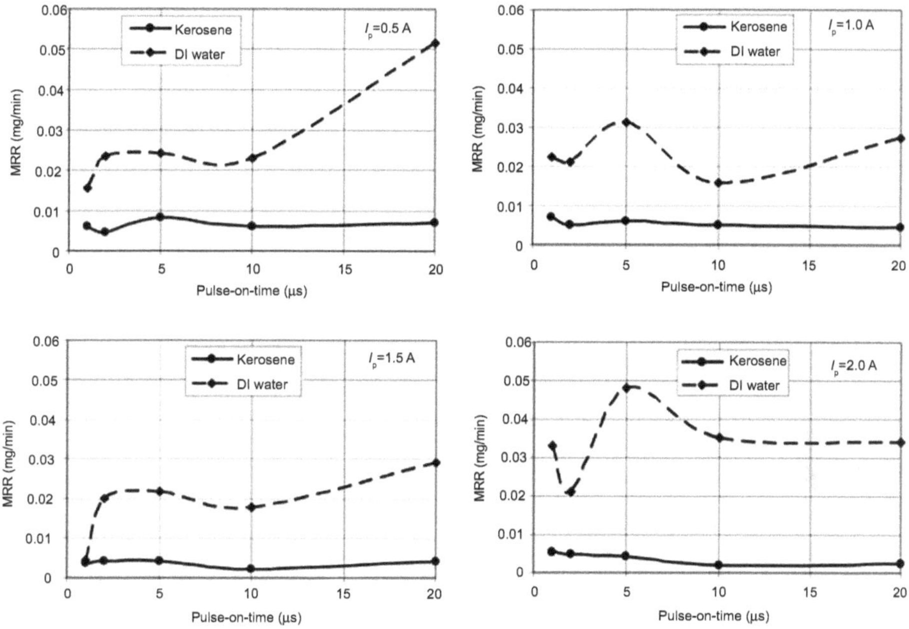

Figure 4.17 Influence of different dielectrics on MRR with varying pulse-on-time.

(0.5, 1, 1.5, 2 A) and pulse-on-time (T_{on}) (1, 2, 5, 10, 20 µs) were varied in this study, with other process parameters such as flushing pressure (0.5 kgf/cm^2) duty factor (95%) remaining constant. Performance criteria such as MRR, TWR, OC, and DVEE were measured after each experiment. Comparative investigation and analysis was performed to study the influence of dielectric liquid on machining performances.

4.10.3.2 Experimental results and analysis

Fig. 4.17 shows the comparative plot of MRR using kerosene and deionized water with varying peak current and pulse-on-time. The figure shows that the MRR is employing more deionized water than kerosene at all values of pulse duration. Kerosene is a chemical compound of carbon and hydrogen and that during machining in micro-EDM, it decomposes at a discharge temperature and produces a titanium carbide (TiC) layer on the machined surface. On the other hand, at high temperature discharge, deionized water decomposes and produces a titanium oxide (TiO_2) layer on the machined surface. The melting point of TiC is higher (3150°C) than that of TiO_2 (1750°C). Thus, large discharge energy is required for improving the MRR using kerosene. In addition, the size of the craters produced on the machined surface that employs kerosene dielectric is less that using deionized water, thereby improving the MRR.

Fig. 4.18 show the comparative plot of TWR using two different dielectrics varying peak current and pulse-on-time. The figure reveals that higher TWR is obtained

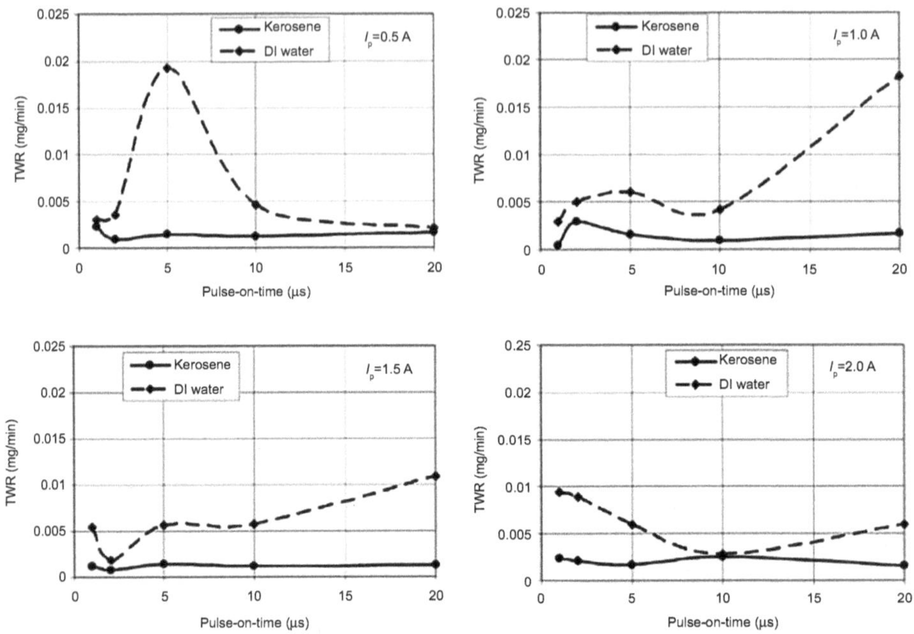

Figure 4.18 Influence of different dielectrics on TWR with varying pulse-on-time.

using deionized water than machining with kerosene. Kerosene decomposes in high discharge energy and produces carbon particles. These particles adhere to the microtool electrode surface and further restricts rapid wear of the tool. On the other hand, deionized water does not produce any carbon particle; consequently, TWR is higher enough by using deionized water.

Fig. 4.19 show the comparative plot of OC with varying peak current and pulse-on-time using kerosene and deionized water. The plot shows that the OC of the machined microholes is larger when using deionized water at pulse durations of 1 and 2 µs. However, at higher pulse durations, the OC is larger when using kerosene. Deionized water releases oxygen during discharge and influences machining stability and increases the possibility of a crater formation. When these debris particles try to eject out through a small gap of tool surface and microhole walls, secondary sparking occurs, resulting in higher OC. However, at a higher pulse duration, machining stability and efficiency increase because of more pulses per second, resulting in a higher OC with kerosene. High peak current results in a higher OC using deionized water compared with kerosene.

Fig. 4.20 shows the comparative results of the diametral variance at entry and exit using kerosene and deionized water with varying peak current and pulse-on-time parameters. The figure shows that the DVEE of the microholes increases at low discharge duration when deionized water is used. However, a further increase in pulse duration results in a decrease in DVEE when deionized water is used. The least DVEE is obtained at a pulse duration of 5 µs and peak current of 1.5 A.

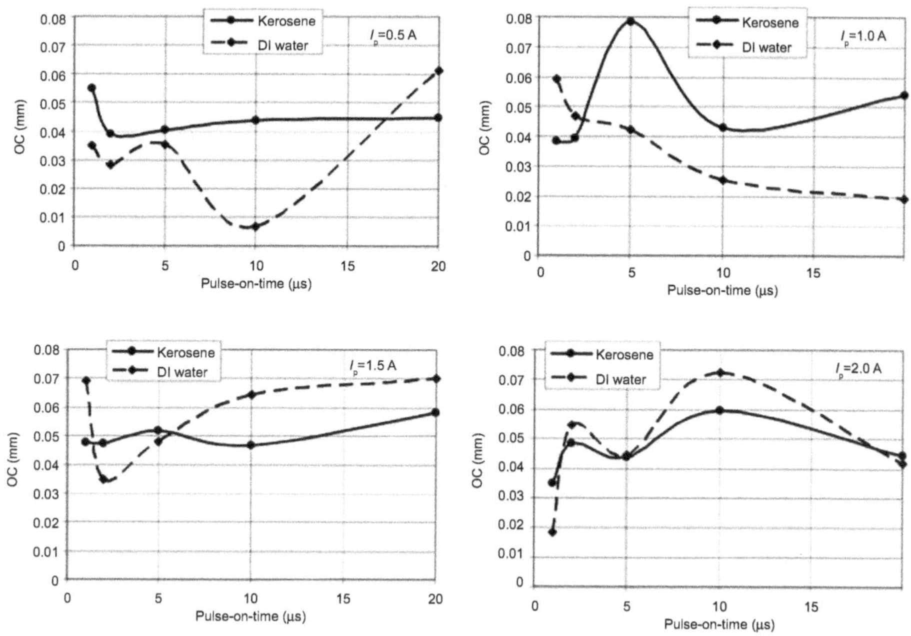

Figure 4.19 Influence of different dielectrics on OC with varying pulse-on-time.

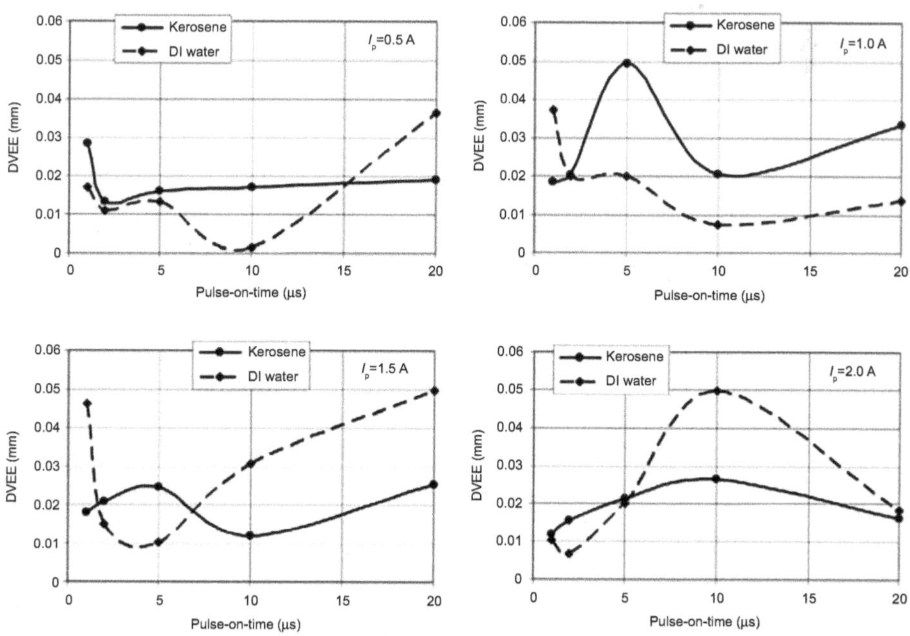

Figure 4.20 Influence of different dielectrics on DVEE with varying pulse-on-time.

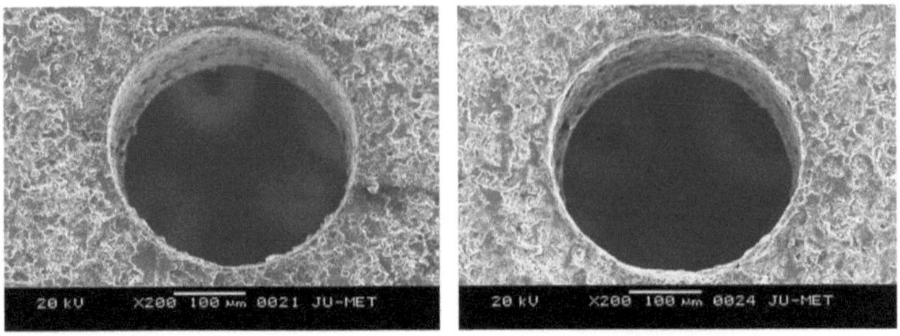

Kerosene (1 A/5 μs) De-ionized water (2 A/10 μs)

Figure 4.21 SEM micrographs of machined microholes generated employing kerosene and deionized water (Kibria, Sarkar, Pradhan, & Bhattachayya, 2010).

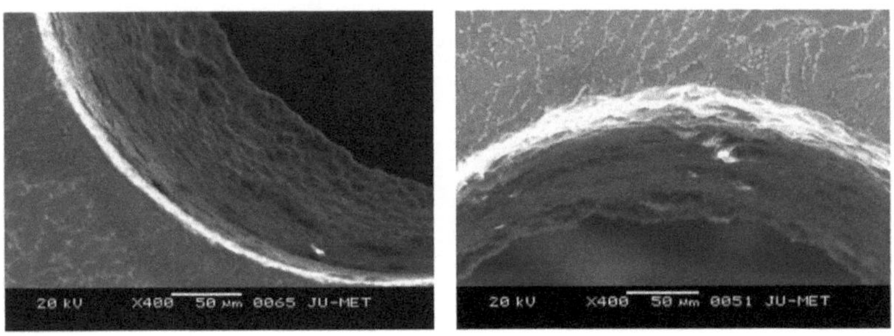

Kerosene Deionized water

Figure 4.22 SEM micrographs of inner surface topography of microholes machined at 2 A/10 μs (Kibria et al., 2010).

On the other hand, by employing deionized water, the DVEE is reduced at peak currents of 0.5 and 1 A. As the peak current increases, the diameter variance also increases. Thus, a straight-through microhole is difficult to achieve. SEM micrographs of machined microholes using pure dielectrics are shown in Fig. 4.21. These figures confirm that kerosene dielectric produces microholes with higher accuracy than does deionized water. Fig. 4.22 shows the inner surface topography of the machined microholes for both dielectrics at 2 A/10 μs parametric combination. The figure shows a smoother inner microhole surface when pure deionized water rather than pure kerosene is used. In Fig. 4.23, the SEM micrographs of the microhole edges are shown to examine the white/recast layer formation using kerosene and deionized water. The thickness of the white layer is much lower when deionized water rather than kerosene is used. Moreover, with an increase in the pulse-on-time, the thickness of the white layer increases. As pulse duration increases, the effective machining time also increases. Therefore, more debris is generated, and this debris adheres to the microhole surface and is resolidified as deionized water, which has a higher cooling rate than kerosene.

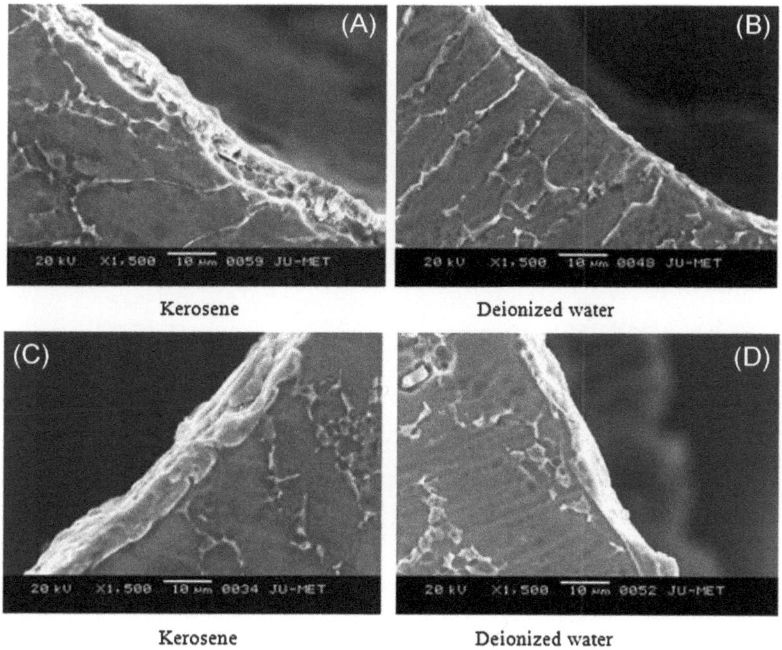

Figure 4.23 SEM micrographs of white layer formed on microhole edge (A) and (B) machined at 0.5 A/1 μs and (C) and (D) machined at 1.5 A/10 μs (Kibria et al., 2010).

4.10.4 Comparative study of mixing a boron carbide additive in dielectrics

4.10.4.1 Experimental method and conditions

In micro-EDM, debris in the IEG facilitates the ignition process and further increases the gap size and overall flushing conditions (Luo, 1997). The absence of debris particles in the gap can result in arcing between the electrodes and further leads to a lack of precise feeding mechanism. However, excess debris leads to uneven discharge and short-circuiting. Some debris particles in the machining gap provide increased discharge transitivity, gap size, breakdown strength, and deionization (Jeswani, 1981). Past studies were performed on EDM by employing powder-mixed dielectrics. However, no research reportedly uses powder-mixed dielectric during micro-EDM of Ti-6Al-4V. In the present study, boron carbide (B_4C) powder is mixed (size 8–10 μm and of concentration 4 g/L) in both dielectrics to investigate the different micro-EDM response criteria. B_4C powder exhibits high chemical resistance and hardness, excellent wear, and abrasion resistance, among others. These properties may provide effective and efficient discharge conditions at the machining zone and enhanced machining performances. Ti-6Al-4V plates measuring 13 × 15 × 1 mm were used as workpieces, and a cylindrical tungsten microtool with a diameter 300 μm was used for microhole machining. MRR, TWR, OC, and diametral variance at entry and exit holes were measured after each experiment.

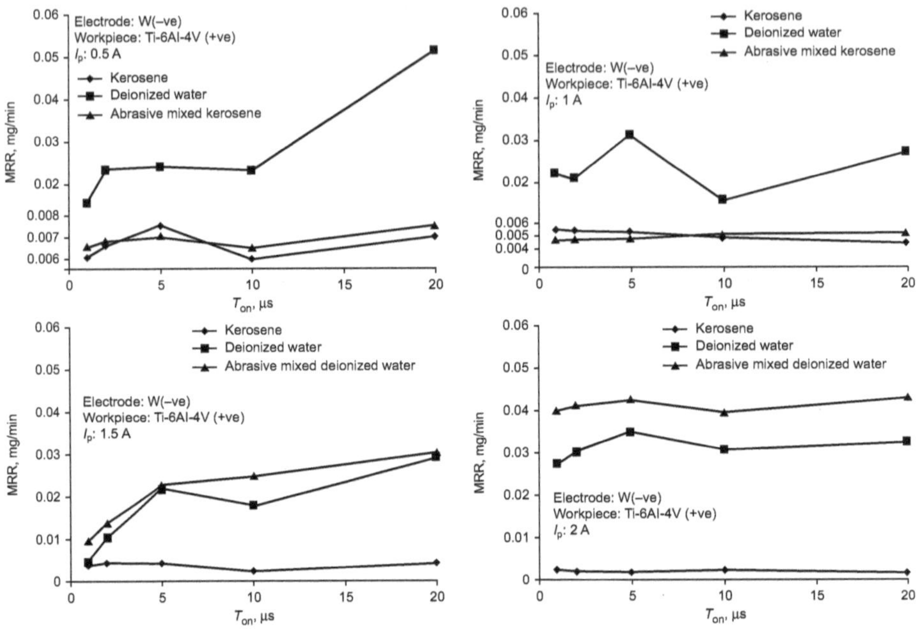

Figure 4.24 Variation of MRR with pulse duration at various peak current using different dielectrics (Kibria et al., 2010).

4.10.4.2 Experimental results and analysis

Fig. 4.24 shows comparative plots of MRR using pure kerosene, pure deionized water, and boron carbide (B_4C) mixed dielectrics at different pulses-on-time and peak currents. MRR is high when deionized water rather than kerosene is used for all values of pulse duration and peak current. Again, in the case of B_4C powder additives in kerosene, MRR increases with an increase in pulse duration at peak currents of 1.5 and 2 A. Moreover, MRR is larger using powder-mixed dielectrics compared with machining with pure dielectrics at higher values of pulse duration. Due to longer effective machining time per pulse, the increase of MRR with pulse duration is revealed using B_4C mixed kerosene. The presence of boron carbide additive in kerosene further helps in the uniform distribution of discharge energy and better conduction of discharge current thereby enabling better machining condition. In case of B_4C powder mixed in deionized water, MRR is more than pure deionized water at peak currents of 1.5 and 2 A. Therefore, the addition of carbide powder particles in dielectrics prevails better machining efficiency because to uniform distribution of discharge energy in the machining zone.

In Fig. 4.25, comparative results of TWR with pulse duration at various peak currents are shown. At peak currents of 0.5 and 1 A, TWR is less when B_4C mixed kerosene rather than pure kerosene is used. Boron carbide abrasive mixed kerosene results in less tool wear. TWR is lower at a peak current of 1 A than at 0.5 A when additive mixed kerosene is used because higher discharge energy generates more

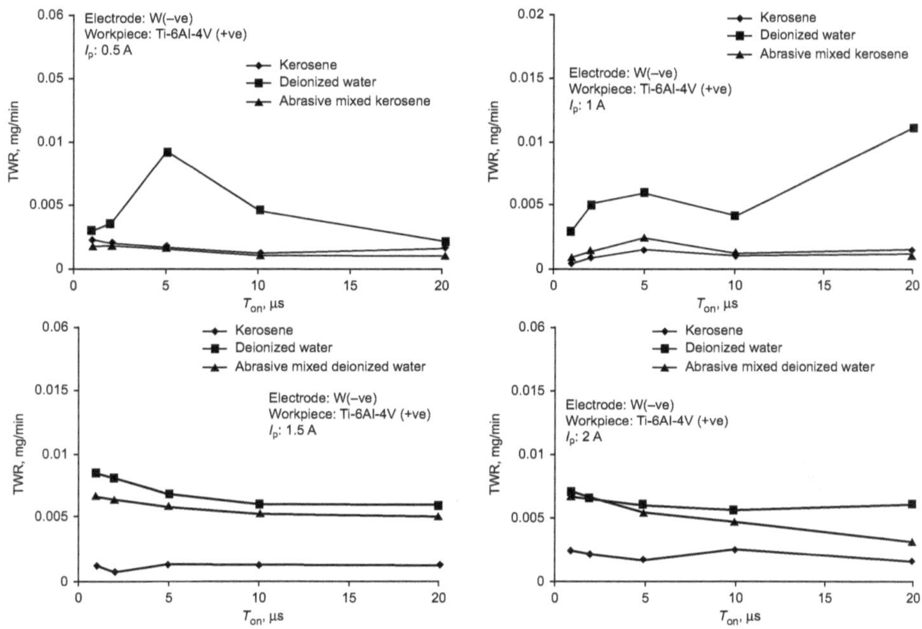

Figure 4.25 Variation of TWR with pulse duration at various peak current using different dielectrics (Kibria et al., 2010).

carbon to adhere onto the microtool surface, thereby preventing secondary sparking. Powder mixed deionized water results in lower TWR at 2 A than at 1.5 A because more carbon particles are deposited on the microtool.

The comparative plots of the OC of microholes are shown in Fig. 4.26 at various pulse durations and different peak currents. The OC of the machined microholes is less when deionized water is used at peak currents of 0.5 and 1 A. However, at peak currents of 1.5 and 2 A, the OC is greater when deionized water rather than pure kerosene is used. In addition, OC is found to be large when powder-mixed dielectrics rather than pure dielectrics are used. The reason is that the suspended additive particles remove the molten layer from the machining zone and further reduce the formation of a thick white layer. OC decreases with an increase in pulse duration by using B_4C suspended kerosene. However, a larger OC is found in B_4C mixed deionized water.

In Fig. 4.27, comparative plots of DVEE are shown for varying peak currents and pulse-on-time parameters. The DVEE of the microholes employing deionized water is lower than that of the microholes employing kerosene at peak currents of 0.5 and 1 A. However, at peak currents of 1.5 and 2 A, DVEE is larger when deionized water is used. Boron carbide powder-mixed kerosene results in large DVEE compared with pure kerosene at low peak currents of 0.5 and 1 A. As machining progresses, the additive boron carbide particles create more carbon adhesion on the work surface, further resulting in reduced material removal at the exit side of the

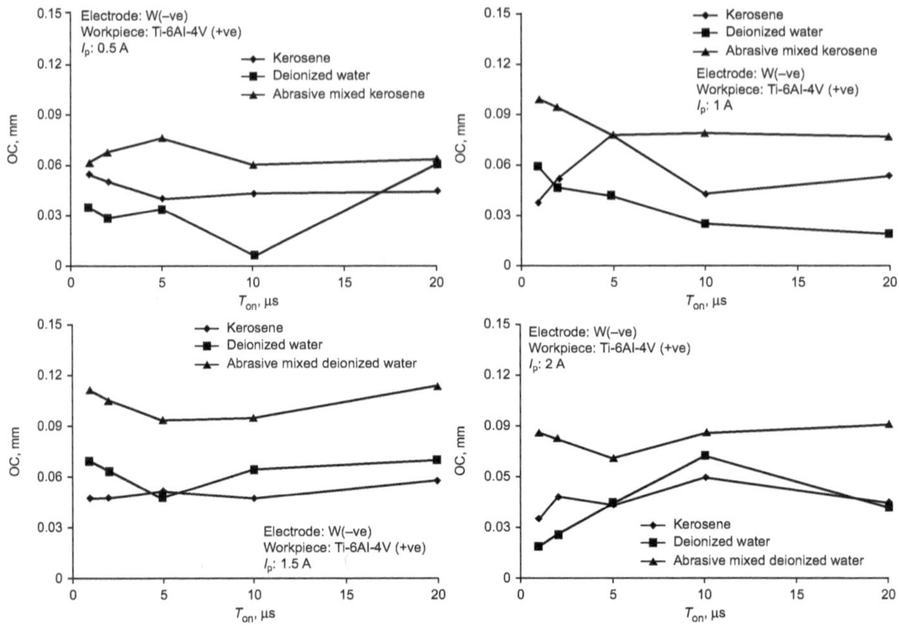

Figure 4.26 Variation of OC with pulse duration at various peak current using different dielectrics (Kibria et al., 2010).

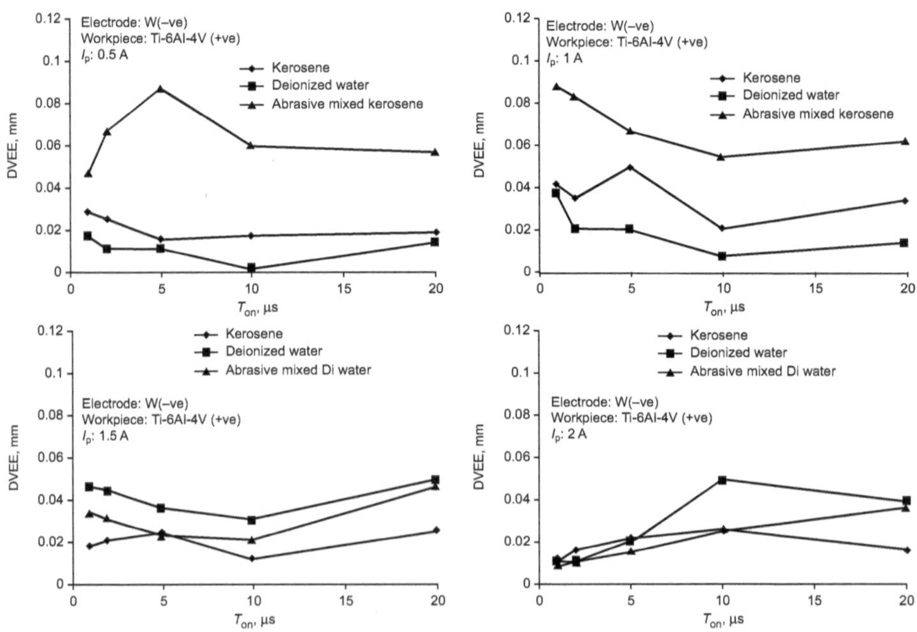

Figure 4.27 Variation of DVEE with pulse duration at various peak current using different dielectrics (Kibria et al., 2010).

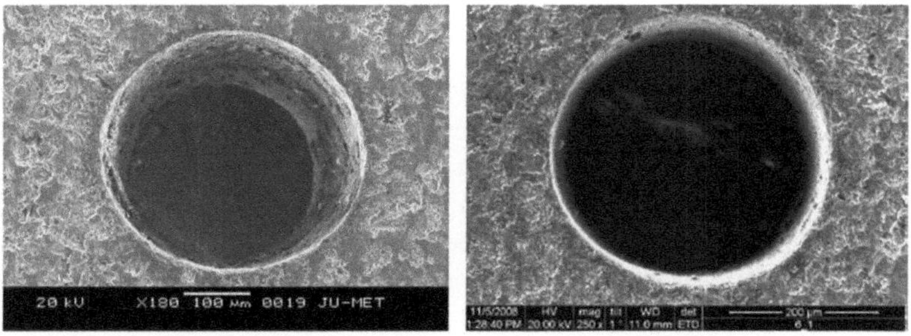

Kerosene (1 A/5 μs) Deionized water (2 A/10 μs)

Figure 4.28 SEM micrographs of machined microholes using powder mixed dielectrics (Kibria et al., 2010).

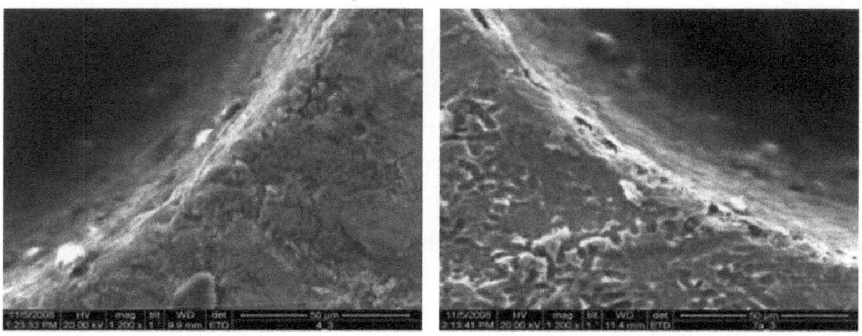

Kerosene (0.5 A/1 μs) Deionized water (1.5 A/10 μs)

Figure 4.29 SEM micrographs of white layer of machined microhole's edges using powder mixed dielectrics (Kibria et al., 2010).

microhole and increased variance at entry and exit diameters. However, at higher peak currents of 1.5 and 2 A, using powder particles help in the uniform distribution of discharge energy, which, in turn, leads to a higher dimensional accuracy of the microholes compared with that obtained when pure deionized water is used.

Fig. 4.28 shows the SEM micrographs of the inner surface of a machined microhole using powder-mixed kerosene and deionized water at parametric combinations of 1 A/5 μs and 2 A/10 μs of peak current and pulse-on-time. The micrographs reveal that with more discharge energy, an inaccurate microhole is generated using powder mixed deionized water, resulting from more secondary sparking phenomena. However, a smooth inner surface is produced using powder-mixed deionized water rather than powder-mixed kerosene. In Fig. 4.29, the SEM micrographs of a microhole's edge are viewed to examine the recast/white layer formation during microhole machining. It is revealed from these figures that the recast layer formed on the edges is much less than that produced using pure dielectrics. The reason is that the additive particles help remove the molten debris and restrict the formation of a thick white layer on the machined microhole edges.

Conclusions

The present chapter describes the overview of the micro-EDM process starting with the EDM principle and then briefly describes the micro-EDM process, its system components, significant process parameters, and the micro-EDM process performances. A brief literature review of the micro-EDM process of micromachining of Ti-6Al-4V materials is presented. Experimentation and analysis of microhole machining in micro-EDM are carried out, implementing four machining strategies, such as (1) changing the polarity of tool electrode and workpiece, (2) rotating the microtool electrode, (3) employing kerosene and deionized water as dielectrics as well as a comparative study of the process criteria, and (4) employing boron carbide additive in kerosene and deionized water and use it as dielectrics. These four sets of experimental investigation and analysis reveal that these novel machining strategies during microhole machining in micro-EDM in Ti-6Al-4V has significant influences on the performance criteria, such as MRR, TWR, OC, and diametral variance of the entry and exit of the microhole. A straight-through microhole can be generated by properly controlling the significant process parameters such as peak current, pulse-on-time, and flushing pressure. The micro-EDM process has great versatility and therefore has huge potential for fabricating various microstructures as well as microsystems and devices in a wide range of hard-to-machine materials. Micro-EDM is a slow process; thus, the novel machining strategies mentioned above must be implemented to improve the process performances, thereby improving the accuracy of the microstructure, surface finish, and efficiency. Furthermore, innovative hybrid micromachining processes can also be developed for micro-EDM to machine Ti-6Al-4V alloy in order to increase the yield of micromanufacturing.

Acknowledgements

The authors acknowledge the financial support and assistance provided by CAS Ph-IV programme of Production Engineering Department of Jadavpur University, Kolkata under University Grants Commission (UGC), New Delhi, India.

References

Ali, M. Y., Rahman, N. A. B. A., & Aris, E. B. M. (2011). Powder mixed micro electro discharge milling of titanium alloy: investigation of material removal rate. *Advanced Materials Research, 383–390*, 1759–1763.

D'Urso, G., & Merla, C. (2014). Workpiece and electrode influence on micro-EDM drilling performance. *Precision Engineering*, *38*(4), 903–914.

Gentili, E., Tabaglio, L., & Aggogeri, F. (2005). Review on micromachining techniques. *Courses Lectures, CISM International Centre for Mechanical Science*, *486*, 387–396.

Han, F., Yamada, Y., Kawakami, T., & Kunieda, M. (2003). Investigations on feasibility of sub micrometer order manufacturing using micro-EDM. In Proceedings of American Society of Precision Engineering (ASPE) *a*nnual *m*eeting, Vol. 30 (pp. 551–554).

Han, F., Yamada, Y., Kawakami, T., & Kunieda, M. (2004). Improvement of machining characteristics of micro-EDM using transistor type isopulse generator and servo feed control. *Precision Engineering*, *28*, 378–385.

Ho, K. H., & Newman, S. T. (2003). State of the art electrical discharge machining (EDM). *International Journal of Machine Tools and Manufacture*, *43*, 1287–1300.

Hung, J. C., Lin, J. K., Yan, B. H., Liu, H. S., & Ho, P. H. (2006). Using a helical micro-tool in micro-EDM combined with ultrasonic vibration for micro-hole machining. *Journal of Micromechanics and Microengineering*, *16*, 2705–2713.

Jahan, M. P., Rahman, M., & Wong, Y. S. (2011). Study on the nano-powder mixed sinking and milling micro-EDM of WC–Co. *International Journal of Advanced Manufacturing Technology*, *53*, 167–180.

Jahan, M. P., Rahman, M., & Wong, Y. S. (2012). Evaluation of the effectiveness of low frequency workpiece vibration in deep-hole micro-EDM drilling of tungsten carbide. *Journal of Manufacturing Processes*, *14*(3), 343–359.

Jahan, M. P., Wong, Y. S., & Rahman, M. (2009). A study on the fine-finish die-sinking micro-EDM of tungsten carbide using different electrode materials. *Journal of Materials Processing Technology*, *209*, 3956–3967.

Jeswani, M. L. (1981). Effect of the addition of graphite powder to kerosene used as the dielectric fluid in electrical discharge machining. *Wear*, *70*, 133–139.

Katz, Z., & Tibbles, C. J. (2005). Analysis of micro-scale EDM process. *International Journal of Advanced Manufacturing Technology*, *25*, 923–928.

Kibria, G., Sarkar, B. R., Pradhan, B. B., & Bhattachayya, B. (2010). Comparative study of different dielectrics for micro-EDM performance during microhole machining of Ti-6Al-4V alloy. *International Journal of Advanced Manufacturing Technology*, *48*(5–8), 557–570.

König, W., & Klocke, F. (1997). *Fertigungsverfahren-3: Abtragen Und Generieren* (Vol. 3) Berlin: Springer.

Kunieda, M., Lauwers, B., Rajurkar, K. P., & Schumacher, B. M. (2005). Advancing EDM through fundamental insight into the process. *CIRP Annals-Manufacturing Technology*, *54*(2), 64–87.

Kurafuji, H., & Masuzawa, T. (1968). Micro-EDM of cemented carbide alloys. *Journal of the Japan Society of Electrical-Machining Engineers*, *2*(3), 1–16.

Kuriachen, B., & Mathew, J. (2014). Modeling of material removal mechanism in micro electric discharge milling of Ti-6Al-4V. *Applied Mechanics and Materials*, *592–594*, 516–520.

Kuriachen, B., & Mathew, J. (2015a). Effect of powder mixed dielectric on material removal and surface modification in micro electric discharge machining of Ti-6Al-4V. *Materials and Manufacturing Processes*. Available from http://dx.doi.org/10.1080/10426914.2015.1004705.

Kuriachen, B., & Mathew, J. (2015b). Spark radius modeling of resistance-capacitance pulse discharge in micro-electric discharge machining of Ti-6Al-4V: an experimental study. *International Journal of Advanced Manufacturing Technology*. Available from http://dx.doi.org/10.1007/s00170-015-7999-9.

Kuriachen, B., Varghese, A., Somashekhar, K. P., Panda, S., & Mathew, J. (2015c). Three-dimensional numerical simulation of microelectric discharge machining of Ti-6Al-4V. *International Journal of Advanced Manufacturing Technology*, *79*(1), 147–160.

Levy, G.N., & Ferroni, B. (1975). Planetary spark erosion-applications and optimization. In Proceedings of the 16th MTDR conference (pp. 291–297).

Lim, H. S., Wong, Y. S., Rahman, M., & Edwin Lee, M. K. (2003). A study on the machining of high aspect ratio micro-structures using micro-EDM. *Journal of Materials Processing Technology*, *140*, 318–325.

Luo, Y. F. (1997). The dependence of interspace discharge transitivity upon the gap debris in precision electro-discharge machining. *Journal of Materials Processing Technology*, *68*, 127–131.

Mahardika, M., Tsujimoto, T., & Mitsui, K. (2008). A new approach on the determination of ease of machining by EDM processes. *International Journal of Machine Tools and Manufacture*, *48*, 746–760.

Masuzawa, T. (2000). State of the art micromachining. *Ann CIRP*, *49*(2), 473–488.

Masuzawa, T., & Fujino M. (1980). Micro pulse for EDM. In Proceedings of the *j*apan *s*ociety for *p*recision *e*ngineering *a*utumn *c*onference (pp. 140–142). Japan (in Japanese).

Masuzawa, T., Sata, T., & Kinoshita, N. (1971). The occurring mechanism of the continuous arc in micro-energy EDM by RC circuit. *Journal of Japan Society of Electrical-Machining Engineers*, *5*(9), 35–52.

Meena, V. K., & Azad, M. S. (2012). Grey relational analysis of micro-EDM machining of Ti-6Al-4V alloy. *Materials and Manufacturing Processes*, *27*(9), 973–977.

Moses, M. D., & Jahan, M. P. (2015). Micro-EDM machinability of difficult-to-cut Ti-6Al-4V against soft brass. *International Journal of Advanced Manufacturing Technology*, *81* (5), 1345–1361.

Plazaa, S., Sancheza, J. A., Pereza, E., Gila, R., Izquierdob, B., Ortegaa, N., et al. (2014). Experimental study on micro EDM-drilling of Ti6Al4V using helical electrode. *Precision Engineering*, *38*(4), 821–827.

Porwal, R. K., Yadava, V., & Ramkumar, J. (2014). Modelling and multi-response optimization of hole sinking electrical discharge micromachining of titanium alloy thin sheet. *Journal of Mechanical Science and Technology*, *28*(2), 653–661.

Pradhan, B. B., & Bhattacharyya (2008). Improvement in microhole machining accuracy by polarity changing technique for microelectrode discharge machining on Ti-6Al-4V. *Proceedings of the Institution of Mechanical Engineers Part B Journal of Engineering Manufacture*, *222*, 163–173.

Pradhan, B. B., & Bhattacharyya, B. (2009a). Modelling of micro-electro discharge machining during machining of titanium alloy Ti-6Al-4V using response surface methodology and artificial neural network algorithm. *Proceedings of the Institution of Mechanical Engineers Part B Journal Engineering Manufacture*, *223*(6), 683–693.

Pradhan, B. B., Masanta, M., Sarkar, B. R., & Bhattacharyya, B. (2009b). Investigation of electro-discharge micro-machining of titanium super alloy. *International Journal of Advanced Manufacturing Technology*, *41*, 1094–1106.

Pradhan, B. B., Sarkar, B. R., Kibria, G., & Bhattacharyya, B. (2009c). EDM with rotational electrode for machining micro holes in Ti-6Al-4V. *Journal of Institution of Engineers (India)*, *89*, 3–8.

Rajurkar, K. P., Levy, G., Malshe, A., Sundaram, M. M., McGeough, J., Hu, X., et al. (2006). Micro and nano machining by electro-physical and chemical processes. *Ann CIRP*, *55*(2), 643–666.

Tiwary, A. P., Pradhan, B. B., & Bhattacharyya, B. (2014). Application of multi-criteria decision making methods for selection of micro-EDM process parameters. *Advances in Manufacturing*, *2*(3), 251–258.

Tiwary, A. P., Pradhan, B. B., & Bhattacharyya, B. (2015). Study on the influence of micro-EDM process parameters during machining of Ti–6Al–4V superalloy. *International Journal of Advanced Manufacturing Technology*, *76*(1–4), 151–160.

Venkatesh, V. C., & Izman, S. (2007). *Precision engineering*. New Delhi: Tata McGraw Hill.

Yan, B. H., Huang, F. Y., Chow, H. M., & Tsai, J. Y. (1999). Micro-hole machining of carbide by electrical discharge machining. *Journal of Materials Processing Technology*, *87*, 139–145.

Yu, Z. Y., Rajurkar, K. P., & Shen, H. (2002). High aspect ratio and complex shaped blind micro holes by micro EDM. *CIRP Annals-Manufacturing Technology*, *51*(1), 359–362.

Zhang, L., Jia, Z., Liu, W., & Li, A. (2012). A two-stage servo feed controller of micro-EDM based on interval type-2 fuzzy logic. *International Journal of Advanced Manufacturing Technology*, *59*(5–8), 633–645.

Microelectrochemical machining: principle and capabilities

5

Z. Liu, C. Gao and J. Zhao
Nanjing University of Aeronautics and Astronautics, Nanjing, Jiangsu, China

5.1 Fundamentals of microelectrochemical machining

5.1.1 Principle of electrochemical machining

Electrochemical machining (ECM) is a nonconventional manufacturing process used as an alternative to conventional mechanical machining processes for difficult-to-cut metals. The advantages of ECM includes tool electrode wear-free, residual stress-free, absence of heat-affected zone, and recast layer. The ECM process was first patented by Gusseff in 1929. Significant advances during the 1950s and the 1960s developed ECM into a major technology in the aircraft and aerospace industries for shaping, finishing, deburring, and milling operations of large parts (Bhattacharyya, Munda, & Malapati, 2004; McGeough, 1974; Rajurkar, Zhu, McGeough, Kozak, & De Silva, 1999). At present, the ECM process plays an important role in the manufacture of a variety of parts, ranging from machining large, complex-shaped metallic pieces to fabricating microscaled parts that are a few micrometers in size. Many industries, such as aerospace, automotive, and electronics, apply the ECM process in the mass production of turbine blades, die and mold, shaving heads, artillery projectiles, and surgical implants (Kozak, Rajurkar, & Makkar, 2004; Rajurkar et al., 1999).

During the ECM process, both the anode (workpiece) and the cathode (tool electrode) are submerged in a constantly renewed electrolytic solution, and a voltage is applied. A distance between the anode workpiece and the cathode tool—i.e., the machining gap—must be maintained so that current passes through the gap and a chemical reaction occurs. Such a situation is sketched in Fig. 5.1. As a result, the anodic material is corroded, and the workpiece is shaped according to the geometrical features of the cathode (McGeough, 1974). The metal removal from the anode material produces an increase in the gap distance, which causes a decrease in the current; therefore, either the anodic workpiece or the cathodic tool is moved by a mechanical feed drive system to maintain the desired gap distance.

Since current passes across the machining gap by the ions of the electrolyte during ECM, a solution of high conductivity is used as the electrolyte, e.g., a 5–25 wt% aqueous solution of sodium chloride (NaCl). Certain acid solutions, such as HCl and H_2SO_4, have also been used as electrolytes, specifically for some situations where electrolyte cannot be easily renewed in the gap region. Generally, the most practical

Microfabrication and Precision Engineering. DOI: http://dx.doi.org/10.1016/B978-0-85709-485-8.00005-X

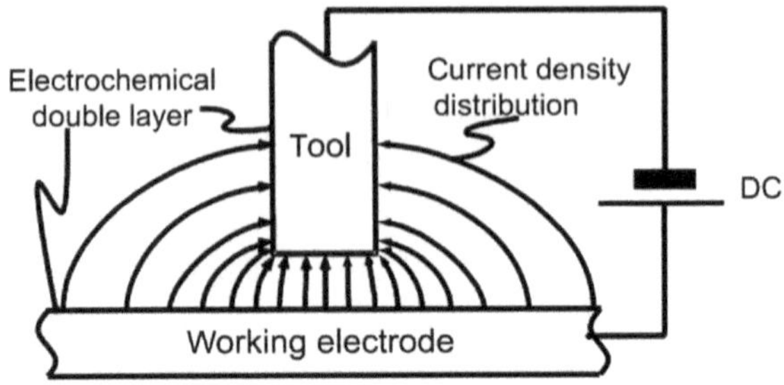

Figure 5.1 Schematic of electrochemical machining (Kock, Kirchner, & Schuster, 2003).

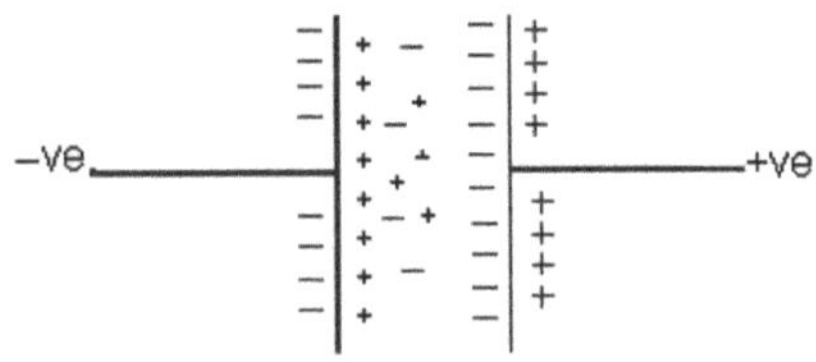

Figure 5.2 Schematic of the principle of the electrochemical double layer (Rajurkar et al., 2006).

electrolytes are solutions of neutral salts— e.g., NaCl, KCl, and $NaNO_3$. With ECM of the steel workpiece in a NaCl solution as an example, the chemical reaction that occurs at the anode (workpiece) is

$$Fe \rightarrow Fe^{2+} + 2e^-$$

$$Fe^{2+} + 2OH^- \rightarrow Fe(OH)_2\downarrow \tag{5.1}$$

$$4Fe(OH)_2 + 2H_2O + 2O_2 \rightarrow 4Fe(OH)_3\downarrow$$

and at the cathode (tool electrode) is

$$2H^+ + 2e^- \rightarrow H_2\uparrow \tag{5.2}$$

Metals such as tungsten and molybdenum can be machined only in alkaline electrolyte because highly protective films are formed on W or Mo anodes in neutral solutions.

Theoretically, it is the potential drop across the double layer (DL), which is a common term to describe the electrode/electrolyte interface (shown in Fig. 5.2), drives the electrochemical reactions in ECM (Park, Kim, & Chu, 2006). The DL influences the electron transfer rate across the interface of the electrodes, thereby affecting the electrochemical reaction rate of an ECM cell. The DL structure and its capability of transferring ions depend on parameters such as the electrode material, electrode porosity, type of electrolyte, extent of specific adsorption of ions and molecules, and temperature (McGeough, 1974).

5.1.2 Microelectrochemical machining using ultra-short pulsed current

ECM has received much attention in the fabrication of microparts since the 1980s (Bhattacharyya & Munda, 2003; Datta, Shenoy, & Rominkiw, 1996). However, it was not until end of the last century that the micro-ECM research accessed a prominent breakthrough by introducing ultra-short pulses current into the process, which leads to a strong confinement of electrochemical reactions down to nanometer precision to achieve high resolution (Kirchner, Cagnon, Schuster, & Ertl, 2001; Rajurkar et al., 2006; Schuster, Kirchner, Allongue, & Ertl, 2000). Pulsed current enables recovery of the gap conditions during pulse-off times, giving improved dissolution efficiency. It also shows a more flexible control than traditional ECM by varying the pulse duration and pulse-duty factor to optimize the removal rate and the surface finish.

The DL can be commonly characterized as a capacitance during ECM using an ultra-short pulsed voltage (Kirchner et al., 2001; Park et al., 2006; Rajurkar et al., 2006; Schuster et al., 2000). The DL is periodically charged and discharged on both electrodes during pulse duration. The charging time constant (τ) is a product of the DL capacitance (c) and the resistance of the electrolyte along the current path (ρ), i.e., $\tau = \rho c$. This resistance locally varies depending on the local separation of the electrode surfaces. Upon proper pulse duration, the DL areas where the electrodes are in close proximity are strongly charged by the voltage pulses, whereas, at farther distances the charging becomes progressively weaker. The rates of electrochemical reactions are exponentially dependent on the potential drop in the DL. Thus, chemical reactions are strongly confined to the polarized electrode regions in close proximity. In other words, if the duration of applied pulses is longer than the time constant (τ), the DL becomes charged high enough for dissolution. However, in other regions where the time constant (τ) is larger than the pulse duration, the DL is not charged sufficiently for dissolution. Fig. 5.3 shows a comparable results machined by ECM using pulses voltage with different pulse durations.

The abovementioned explanation can very well describe the localization principle of ultra-short pulsed micro-ECM; however, some researchers argued that

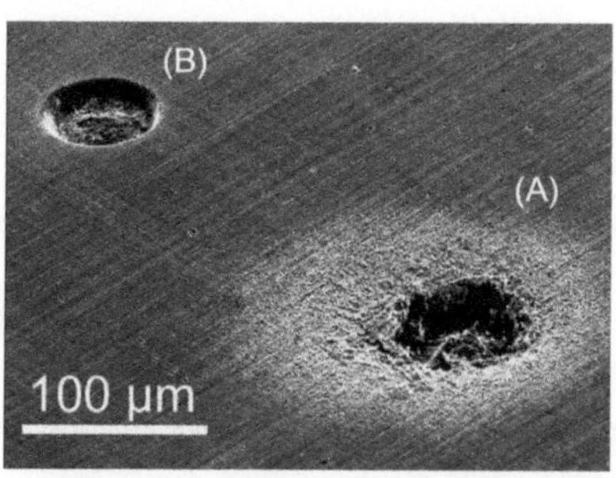

Figure 5.3 SEM image of two holes corroded into a Cu sheet by using pulsed voltage: (S) 5 ms pulse duration, (B) 100 ns pulse duration (Kock et al., 2003).

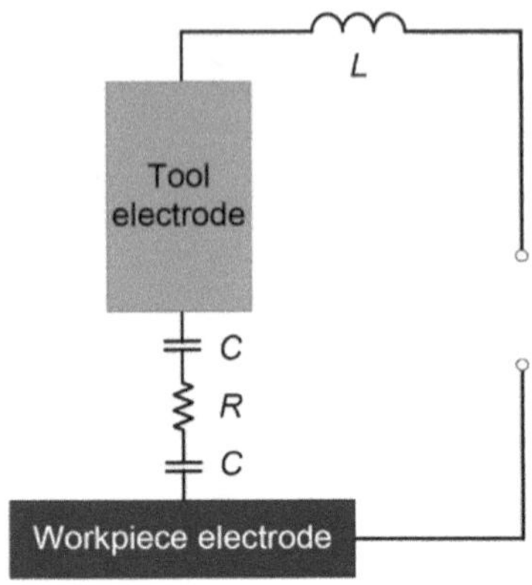

Figure 5.4 Micro-ECM cell model considering inductance (Park et al., 2006).

inductance effects cannot be disregarded in an actual micro-ECM system because of very high pulse frequency (Park et al., 2006). They developed an equivalent electrochemical cell model considering inductance (*L*), as illustrated in Fig. 5.4. Resistance (*R*) from electrolyte resistivity is determined by the current flow path between two electrodes. Capacitance (*C*) is the product of the specific DL capacitance and the whole area immersed in an electrolyte. The model separated all current flow through an electrochemical cell into a faradaic current, i.e., a reaction current and a transient current, i.e., nonfaradaic current (Bard & Faulkner, 2001). Faradaic current represents the material dissolution rate and nonfaradaic current only charges and discharges the DL capacitance. When ultra-short voltage pulses are applied, nonfaradaic current dominates transiently, and dissolution is affected by the cell impedance, which is the sum of resistance and capacitive reactance.

5.1.3 Miniaturization of cathode tool

The ECM is a manufacturing process in which a certain rate of anodic dissolution is maintained by feeding a preshaped tool towards the workpiece with a constant feed rate. Consequently, the workpiece surface will be shaped according to the geometrical features of the tool electrode, which is commonly macrosized. Principally, a micro-ECM cell also needs a preshaped microsized cathode to produce microstructures in the process. However, the introduction of ultra-short pulsed current in the micro-ECM process can significantly improve the metal dissolution localization to submicrometer regions; therefore, relatively simple-shaped micro tools—e.g., a cylindrical rod—can be applied to fabricate complex structures. In most cases, the size of the tool electrode determines the size of structures machined in micro-ECM (Brousseau, Dimov, & Pham, 2010; Liu, Zeng, & Zhang, 2014; Mathew & Sundaram, 2012).

Many technologies can be used for fabricating micrometer-scale tools—tools for micromechanical machining, electro-discharge machining (EDM), electrochemical etching, electrochemical manufacturing, and so on. Micromechanical machining is a process employing a microcutter to mechanically remove the material from the

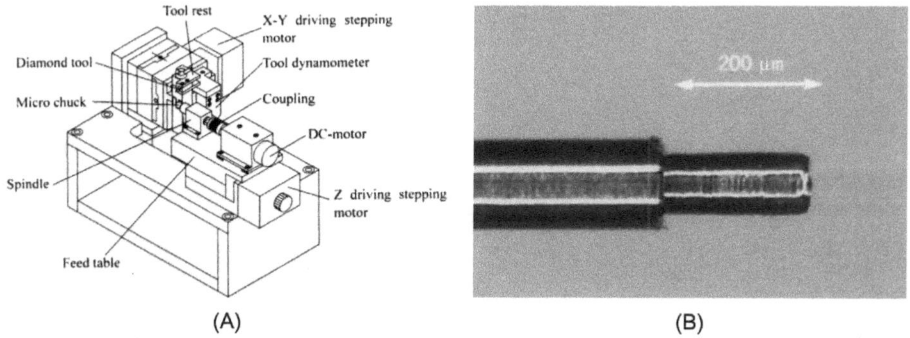

Figure 5.5 Microtool machined by a microlathe: (A) schematic of the microlathe, (B) the machined microtool (Lu & Yoneyama, 1999).

workpiece. Basically, the hardness of the cutter needs to be higher than that of the workpiece. EDM is an approach using repeated thermal discharge to erode metals by pulsed voltage applied between two electrodes. The process can be used to fabricate microtools with the advantages of noncontact between electrodes and high feasibility regardless of hardness and stiffness of the workpiece. Electrochemical etching, which is similar to ECM, is a method that relies on an acidic or alkaline solution to dissolve unwanted metals from the workpiece.

The literature reported some applications of fabricating a micrometallic tool by using variant approaches. As shown in Fig. 5.5, Lu & Yoneyama (1999) used a microdiamond single-point cutter to turn a workpiece into one with a diameter of 10 μm and a rotation speed of up to 15,000 rpm in a microlathe system. On the basis of the EDM technology, Masuzawa, Fujino, Kobayashi, Suzuki, and Kinoshita (1985) proposed the wire electro-discharge grinding (WEDG) method to prepare very thin rods sized to a few tens of micrometers in diameter. Ali and Ong (2006) applied the WEDG process to fabricate microtools 20 μm in diameter in tungsten carbide, as demonstrated in Fig. 5.6. Lim, Wong, Rahman, and Lee (2003) fabricated microelectrodes of 100 μm in diameter with high aspect ratios by using micro-EDM, which is classified into three types of sacrificial electrode: stational sacrifical block, rotating sacrifical disk, and guided running wire (Fig. 5.7).

Electrochemical etching is originally considered a fast and affordable method for fabricating metallic tips for scanning tunneling microscopy or atom probe field ion microscopy. The standard method involves dipping a tungsten wire into an electrolyte; etching then begins at the air/electrolyte reaction interface and continues into the immersed metallic wire (Duong & Kim, 2015; Fan, Hourng, & Wang, 2010), as shown in Fig. 5.8. Wang and Zhu (2009) introduced a two-step composite processing technology, combining EDM and electrochemical etching, to fabricate microelectrode array. Rectangular columns are machined by wire electrical discharge machining (WEDM) at the first stage; afterwards, electrochemical etching is used to erode the microelectrode array into cylindrical columns (shown in Fig. 5.9).

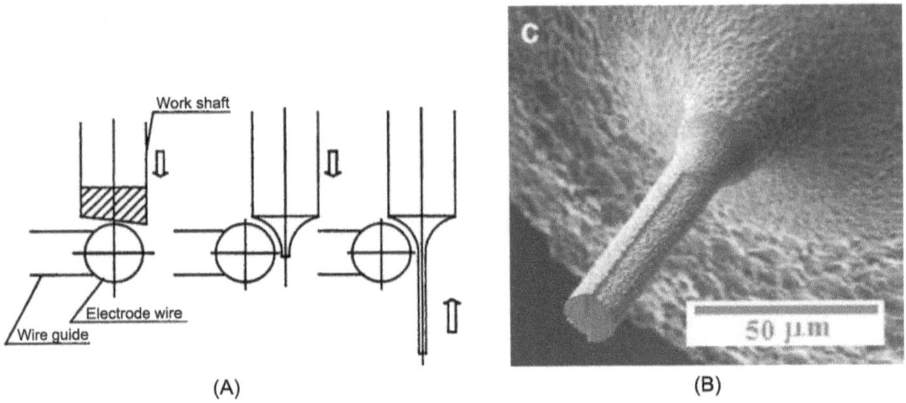

Figure 5.6 Microtool machined by WEDG: (A) schematic of WEDG, (B) the machined microtool (Ali and Ong, 2006).

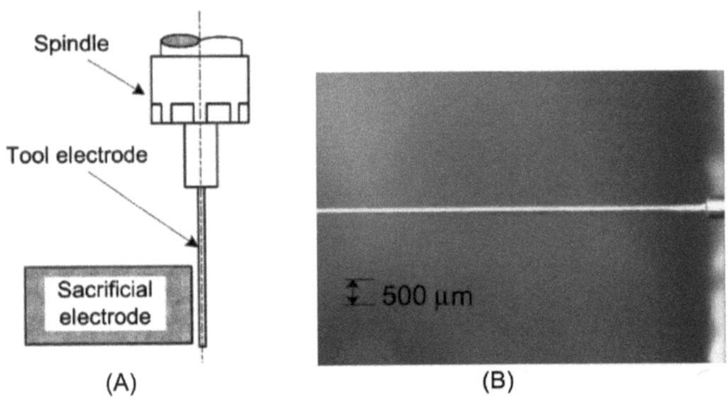

Figure 5.7 Microtool machined by micro-EDM: (A) schematic of microtool fabrication, (B) the machined microtool (Lim et al., 2003).

During the micro-ECM process, since electrolysis takes place as long as the current flows through the tool, using an uninsulated tool results in producing a taper shape at the machined side wall, as demonstrated in Fig. 5.10A. Coating an insulation layer at the sides of the tool aids the micro-ECM process to restrict current flows to an uninsulated area and consequently reduce stray corrosion effectively (Liu, Liu, Qiu, & Qu, 2009; Park et al., 2006). Therefore, in order to reduce the taper, it is necessary to insulate the sides of the tool, as shown in Fig. 5.10B. As a result, the machining gap is uniform regardless of the machining depth, and the taper shape of structures can be markedly prevented.

Compared with the cylindrical electrode, a disk-type electrode is also capable of reducing taper on the side wall profile during the process, as demonstrated in Fig. 5.11 (Kim, Ryu, Choi, & Chu, 2005). For a cylindrical electrode (Fig. 5.11A), the upper side of a wall is dissolved more than the bottom side during downward

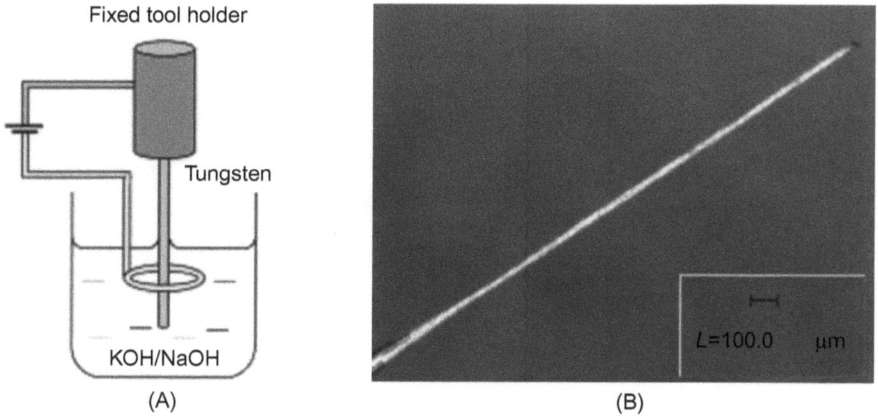

Figure 5.8 Electrochemical etching for microtool fabrication: (A) schematic of the principle (Duong & Kim, 2015), (B) a microtool fabricated by electrichemical etching (Pan, Xu, & Zhou, 2015).

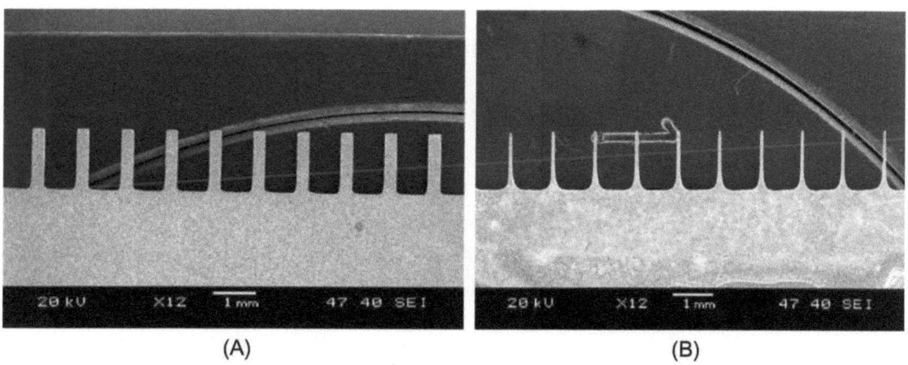

Figure 5.9 Microelectrode array prepared by a two-step process: (A) in the first step, a rectangular column is machined by WEDM; (B) in the second step, the rectangular column is etched to cylindrical shapes (Wang & Zhu, 2009).

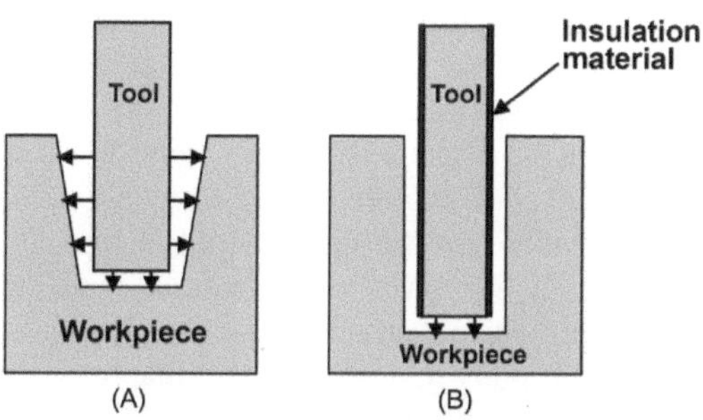

Figure 5.10 Schematic of side insulation on tool surface in micro-ECM: (A) uninsulated tool, (B) insulated tool.

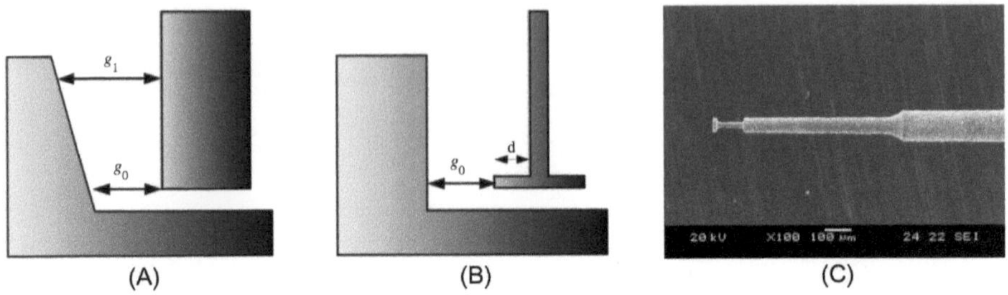

Figure 5.11 The effect of the electrode shape on the side wall profile: (A) a cylindrical electrode, (B) a disk-type electrode, (C) disk-type tool electrode made by micro-EDM (Kim et al., 2005).

feeding of the tool. However, by applying a disk-type electrode (Fig. 5.11B), dissolution can be restricted to the disk region with less dissolution to the side wall, especially in the case of using a passive electrolyte such as $NaNO_3$.

5.2 Variety of micro-ECM processes

5.2.1 Microelectrochemical drilling

Microelectrochemical drilling (ECD) is a process to machine microholes, applying a shaped tube or a cylindrical cathode tool feeding axially toward the workpiece. A number of microdrilling applications have been developed in aviation, space, automobile, electronics and computer, as well as medical and optics manufacturing in extremely hard metallic materials. Sen and Shan (2005) summarized electrochemical micro- and small drilling processes into shaped tube electrolytic machining (STEM), capillary drilling (CD), electro-stream drilling (ESD), and jet electrolytic drilling (JED), as illustrated in Fig. 5.12. They pointed out that most of the abovementioned processes use a weak acidic solution as the electrolyte, resulting in the dissolution of material in metal ions, which can be easily carried away from the machining region, and achieving smooth and deep holes of high aspect ratio. The achievable minimum hole size by the STEM, CD, ESD, and JED were ø0.5, ø0.2, ø0.1, and 0.125 mm, and the typical aspect ratios of that could be approximately 16:1 for these four measures.

Park et al. (2006) studied the effect of the tool electrode size on micro-ECM drilling using a 0.1 M sulfuric acid as electrolyte, a cylindrical microtool electrode insulated on the side surface, and an ultra-pulsed power source. As shown in Fig. 5.13, they found that an insulated tool electrode was capable of increasing machining depth and effectively preventing the taper at the side of the machined hole. Yang, Park, and Chu (2009) investigated the characteristics of the ECD process, applying ultra-short pulsed current, a semi-cylindrical tool with a rotation rate of 9.7 rpm, and an ultrasonic vibration on the tool. They concluded that ECD with a semi-cylindrical tool combined ultrasonic vibrations that provided increased flow

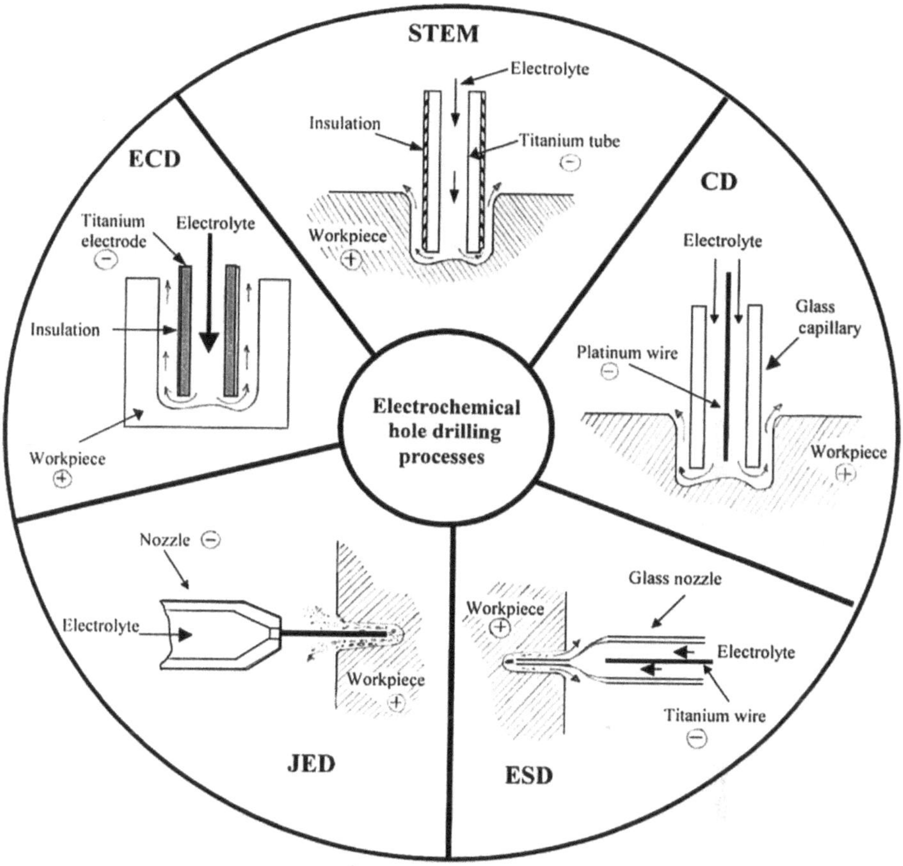

Figure 5.12 Variants of electrochemical drilling processes (Sen and Shan, 2005).

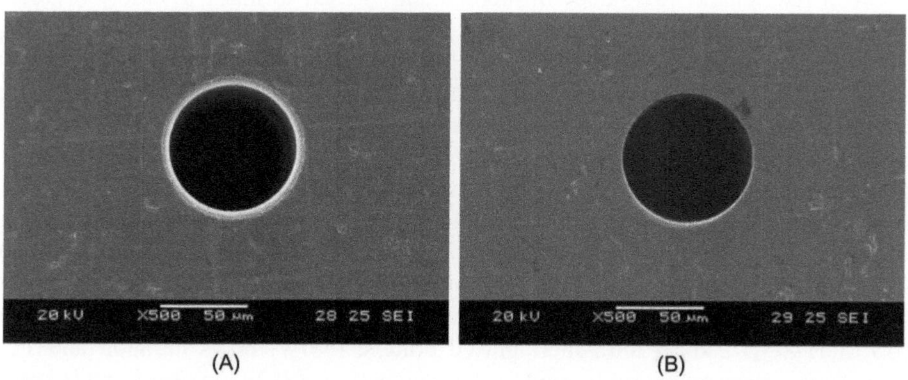

Figure 5.13 Microelectrochemical drilling with a cylindrical tool electrode in 304 SS: (A) entrance, (B) exit. (Park et al., 2006).

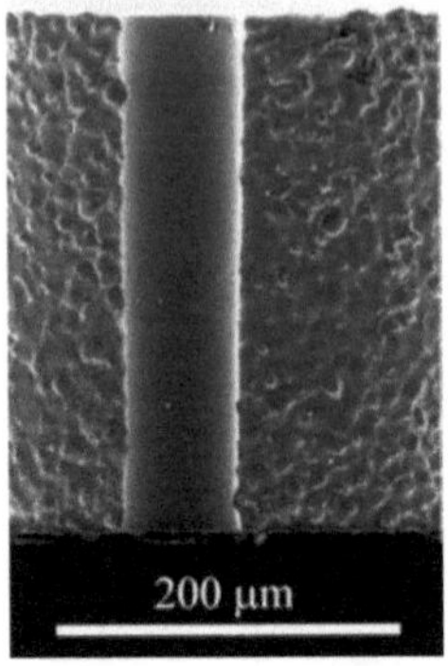

Figure 5.14 Side view of a machined hole produced by microelectrochemical drilling with a semi-cylindrical tool electrode and ultrasonic vibrations (Yang et al., 2009).

space for the electrolyte, which results in a higher aspect ratio compared with the process in the absence of vibrations (Fig. 5.14). Ultrasonic vibrations improved electrolyte diffusion and convection, as well as bubble elimination, and a considerable amount of machining time was saved. Liu, Qiu, Heng, and Qu (2009) presented a micro-ECD method of periodically retracting a tool electrode during the process. With the high-velocity retraction of the tool electrode, the machining region can be flushed using a fresh electrolyte, and corrosion sludge can be carried away periodically during the process. Consequently, the machining stability and localization can be improved.

5.2.2 Microelectrochemical milling

In the milling mode, a simple cylindrical tool typically has a diameter ranging between few tens to two hundred microns and is used to make a variety of structures. True three-dimensional contouring is possible by a proper tool position control (Rajurkar et al., 2006). During the ECM milling, the tool electrode moves along a path in a layer-by-layer fashion to achieve the desired geometry. On the one hand, the use of cylindrical tool avoids complex and time-consuming electrode design and fabrication; on the other hand, the up-to-date micromachining technologies cannot easily produce a complex preshaped tool electrode of a micrometer-scale without entailing efficiency and high costs. As a result, micro-ECM milling using a cylindrical tool is a particularly suitable approach for machining complex 3D microstructures. For example, Kim et al. (2005) fabricated various 3D microstructures through the electrochemical milling process. With a dilute electrolyte (0.1 M H_2SO_4) and an ultra-short pulsed voltage (6.0 V, 60 ns duration and 1 μs period), microstructures with good surface quality (Ra 0.28 μm) were made using layer-by-layer machining. Fig. 5.15A shows a microhemisphere on the top of a cylinder. Fig. 5.15B demonstrates a spiral machined into a Ni sheet with 3 ns pulses.

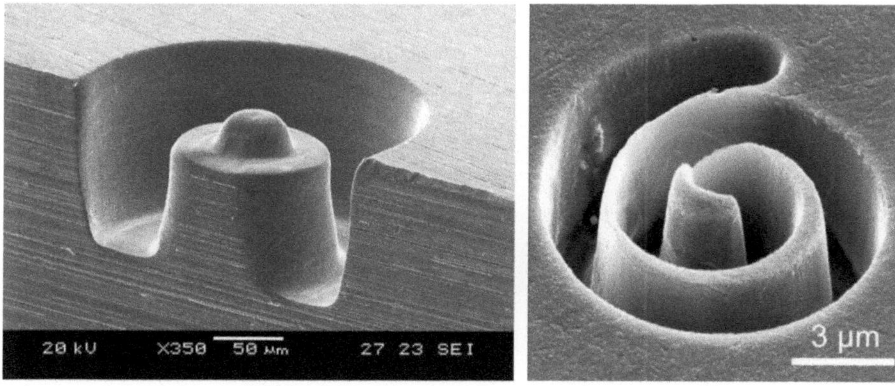

(A) (B)

Figure 5.15 Structures machined by micro-ECM milling: (A) a microhemisphere (Kim et al., 2005), (B) a spiral (Kock et al., 2003).

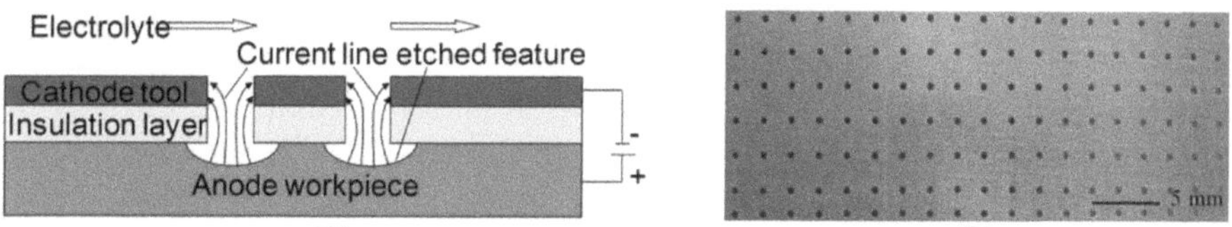

(A) (B)

Figure 5.16 A modified through-mask micro-ECM process: (A) schematic of the principle, (B) a machined hole array (Zhu et al., 2009).

5.2.3 Through-mask microelectrochemical machining

The through-mask ECM, capable of patterning thin films or foils by wet etching, generally involves selective material removal through photoresist masks. Compared with the common chemical etching process, the through-mask ECM method offers better control and flexibility for microfabrication. Other advantages include higher machining rate, use of nontoxic electrolyte, and possibility of machining a wide range of electrically conductive materials (Datta, 1997). Through-mask micro-ECM process is receiving considerable attention in the electronics and other high-tech industries because of high throughput for making mass microstructures. A standard through-mask micro-ECM begins by bonding a sheet of inert photoresist on the metal anode workpiece. Lithography, which includes procedures of spin coating, prebaking, exposure, development, and postbaking, are employed to pattern the photoresist. The photoresist is only a one-time mask and must be peeled off from the anode workpiece after machining (Datta & Landolt, 2000).

Zhu, Qu, Li, and Zeng (2009) demonstrated a modified through-mask micro-ECM process, as shown in Fig. 5.16, in which a mask with a patterned insulation plate coated with metal film as cathode is closely attached to a workpiece plate instead of bonding a photoresist layer on the workpiece in the standard through-mask

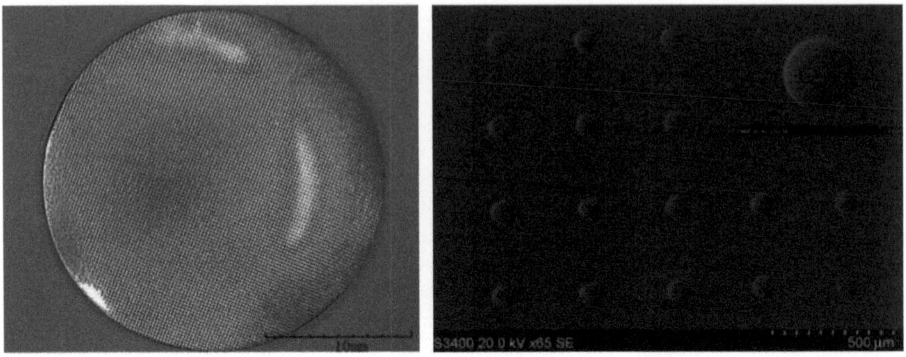

(A) (B)

Figure 5.17 Movable dry-film mask micro-ECM: (A) microdimple array on a flat surface prepared using the process, (B) a scanning electron microscopy image of the microdimple array (Qu et al., 2015).

micro-ECM. Compared with the standard through-mask micro-ECM, the modified process offers advantages, such as short lead time and low cost because the mask could be re-used. Fig. 5.17 shows a machining result achieved using a movable dry-film mask micro-ECM process (Qu, Zhang, & Chen, 2015). The process utilizes a microscale pattern transfer without photolithography of substrates to reduce the ratio of etched profile diameter and micropattern diameter. The electrochemical tool carrying the dry-film mask remains in close contact with the anodic workpiece surface during the machining process.

5.2.4 Microwire electrochemical machining

Microwire electrochemical machining (microwire ECM) is another promising machining method. Fig. 5.18A shows the schematic of the principle of the microwire ECM process in which a microwire is used as a tool electrode. In contrast to wire EDM, a thinner wire can be used as the wire will not be worn out, (Kim et al., 2005; Wang, Zhu, Zeng, & Liu, 2011). In the process, pulse voltage is applied between the workpiece and the wire electrode in an electrolyte cell. The workpiece is electrochemically dissolved and a narrow slit is produced as the cathode wire moves toward the anode workpiece. Fig. 5.18B illustrates a microstructure machined by wire ECM. Fig. 5.19 presents a method of fabricating a microtool using wire ECM in which a micrometallic rod is controlled to allow rotation and a wire electrode of tens of microns feeds to the rod to perform the cutting (Liu, Zeng, & Zhang, 2014).

5.2.5 Microelectrochemical jet machining

Electrochemical jet machining (ECJM), as described by Kozak, Rajurkar, and Balkrishna (1996), employs an electrolyte jet impinging on a metal workpiece to

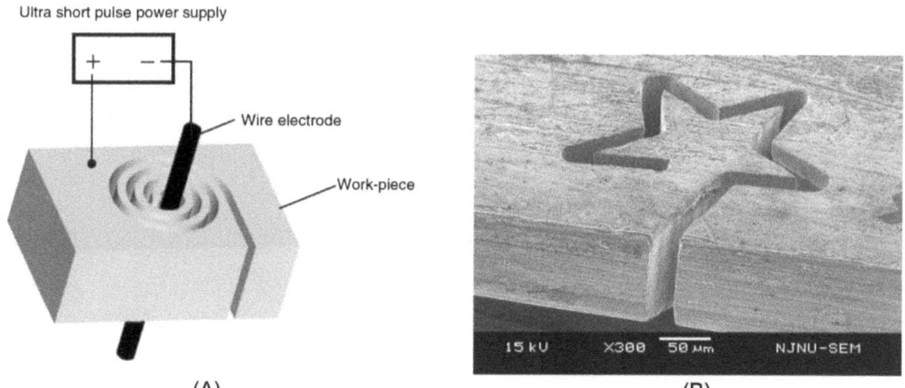

Figure 5.18 Microwire electrochemical cutting: (A) a schematic of the principle (Wang et al., 2011), (B) a microstructure machined by wire ECM (Zhu, Wang, & Qu, 2007).

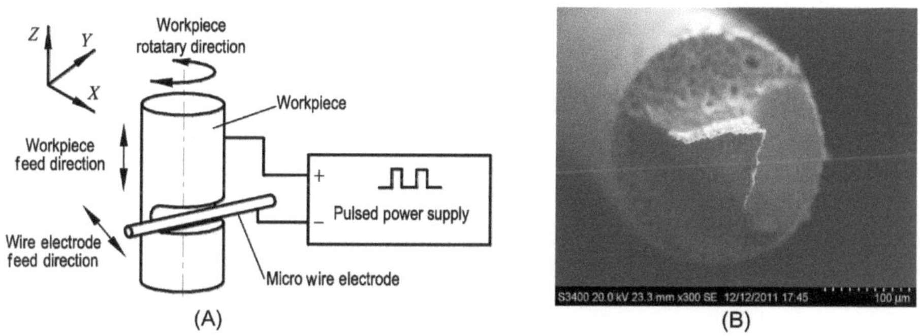

Figure 5.19 Fabrication of a microtool by wire ECM: (A) a schematic of the principle, (B) a machined microtool (Liu, Zeng & Zhang, 2014).

achieve anodic dissolution of the target material. The nozzle through which the electrolyte jet emerges forms the cathode while the workpiece acts as the anode, and a DC voltage is applied across the electrolyte jet and the workpiece (Liu, Nouraei, Papini, & Spelt, 2014), as demonstrated in Fig. 5.20A. ECJM can be used to drill narrow cooling holes in aircraft turbine blades and to fabricate patterns of microgrooves and slots in microelectronic parts without using masks to define the features. One advantage of ECJM is that it does not require the entry of the nozzle into the machined geometry because an adequate current flow and the machining rate can be achieved while maintaining a proper standoff distance between the nozzle cathode and the workpiece. Other advantages are the relatively low cost of the equipment and the absence of any requirement for a preformed cathode having the shape of the desired machined feature. Natsu, Ooshiro, and Kunieda (2008) used ECJM to fabricate microgrooves in stainless steel by using a 50 μm diameter scanning nozzle. Hackert-Oschaetzchen, Martin, and Meichsner (2013) presented

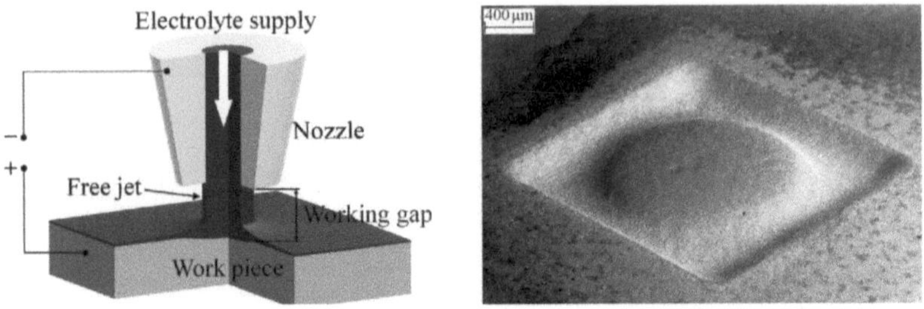

(A) (B)

Figure 5.20 Microelectrochemical jet machining: (A) schematic of the principle, (B) a convex surface machined by electrochemical jet machining (Hackert-Oschaetzchen et al., 2013).

microstructures, with one of them illustrated in Fig. 5.20B, machined by ECJM in tungsten carbide using electrolyte solutions—sodium nitrate, sodium chloride, and sodium hydroxide.

5.3 Hybrid processes associated with microelectrochemical machining

ECM has continuously received interest as an industrial machining process in the last two decades, especially for producing micro and small parts. One of the main advantages of ECM is that it does not cause any thermal stresses or heat-affected zones to the workpiece. This fact is the basis for some hybridizations of ECM with other processes such as EDM, striving to combine advantages of speed, and accuracy with a good surface finish by chemically dissolving any thermally affected layers (Silva, Pajak, Mcgeough, & Harrison, 2011). There are other hybrid processes that combine the micro-ECM to obtain a better localized removal of material.

5.3.1 Laser-induced electrochemical jet machining

Laser-induced electrochemical jet machining (LAECJM) is a hybrid process in which the laser beam is aligned coaxially with the jet of the electrolyte, creating a noncontact tool electrode. The process combines two different sources of energy simultaneously: energy of ions ECM and energy of photons (a laser beam). The main aim of combining a laser with a jet of electrolyte is to assist electrochemical dissolution from a specific workpiece surface area. Electrochemical dissolution is the main material removal mechanism supported by the parallel action of the low-power laser beam. The laser does not remove any material, but assists dissolution by its thermal activation. Thermal energy enhances the kinetics of electrochemical reactions providing faster dissolution. It also aids in breaking down the oxide layer

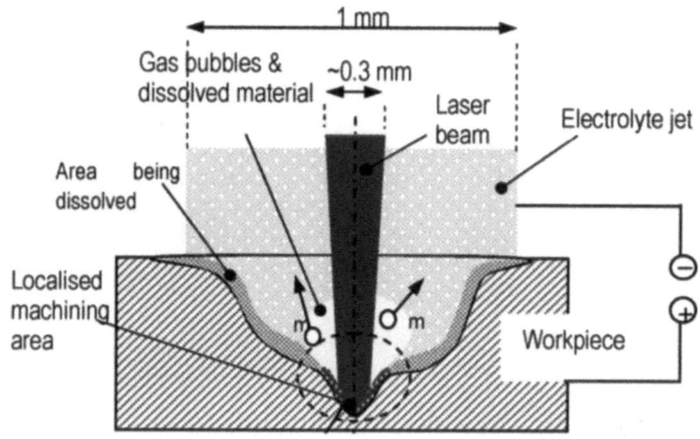

Figure 5.21 Schematic of laser-assisted electrochemical jet machining (Silva et al., 2011).

found on some materials in certain electrolytes that inhibit efficient dissolution (Silva et al., 2011; Pajak, Desilva, Harrison, & Mcgeough, 2006; Long, Xiong, Shi, & Liao, 2015). Fig. 5.21 illustrates the principles of hybrid LAECJM. For hole or slot machining, material removal is more profound in the axial rather than in the lateral direction for LAECJM.

5.3.2 Abrasive enhanced electrochemical jet machining

Abrasive enhanced electrochemical jet machining (AECJM), as shown in Fig. 5.22A, is a machining process which employs a premixed slurry jet (electrolyte mixed with abrasives) and a DC potential applied between the jet nozzle and target to remove material through anodic dissolution, erosion, and synergistic effects (Liu et al., 2014; Liu, Nouraei, Spelt, & Papini, 2015). It is a hybrid process that combines ECJM with abrasive slurry jet machining (ASJM), in which a premixed abrasive slurry is pumped at a relatively low pressure through an orifice to erode the target. The combined process of AECJM can achieve significantly higher metal removal rates than either ASJM or ECJM alone and produces smoother surfaces than ECJM alone. The major synergistic effects in ECJM are expected to depend on the target material and utilized process parameters. For example, in the case of the ESJM of ferrous metals using a NaCl solution, no oxide film will be developed on the target, and the corrosion simply enhances the rate of metal removal due to erosion. However, in the case of the ESJM of WC using an NaCl solution, an oxide film is expected to rapidly develop and be removed by impacting particles, thus exposing new material to the electrolyte, leading to a continuous anodic dissolution. Fig. 5.22B shows a microchannel machined by AECJM in tungsten carbide using sodium chloride as an electrolyte.

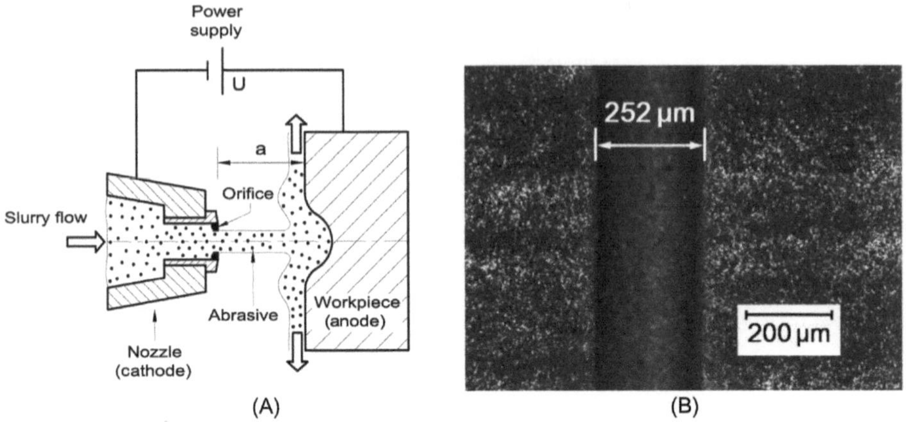

Figure 5.22 Abrasive enhanced electrochemical jet machining: (A) schematic of the process (Liu, Nouraei, Papini & Spelt, 2014), (B) a microchannel machined by the process in tungsten carbide (Liu et al., 2015).

5.3.3 Process combining EDM with ECM

In micro-EDM, the material is removed by vaporization and melting during each electric discharge. Therefore, the machined surface is made up of thermally damaged layers consisting of the white layer and the heat-affected zones (Nguyen, Rahman, & Wong, 2012). Microcracks and residual stresses are also observed in these layers. The surface machined by micro-EDM has high surface roughness due to discharged craters. On the other hand, micro-ECM has some valuable advantages, e.g., the machined surface is very smooth, stress-free with no burr and microcracks, and has no thermally affected layers. Hence, an appropriate combination of micro-EDM and micro-ECM could yield the advantages of these two processes while preventing their disadvantages (Zeng, Wang, Wang, Shan, & He, 2012). Fig. 5.23 shows the principle of a simultaneous micro-EDM and micro-ECM (SEDCM) process in low-resistivity deionized water (Nguyen et al., 2012). The primary aims of the process are to use electrochemical dissolution to lower the surface roughness and remove the thermally damaged zones created during EDM. Fig. 5.24 demonstrates the machining results of a microhole in the absence and presence of an electrochemical dissolution effect (Nguyen, Rahman, & Wong, 2013).

5.4 Conclusions

Technological advances made in microelectrochemical machining, which reflect the state of the art in academic and industrial achievements and applications, have been presented in this chapter. Micro-ECM has clearly emerged as an indispensable part of micromanufacturing technologies. The demonstrated processes offer the two- and three-dimensional machining capabilities of processing a wide variety of engineering

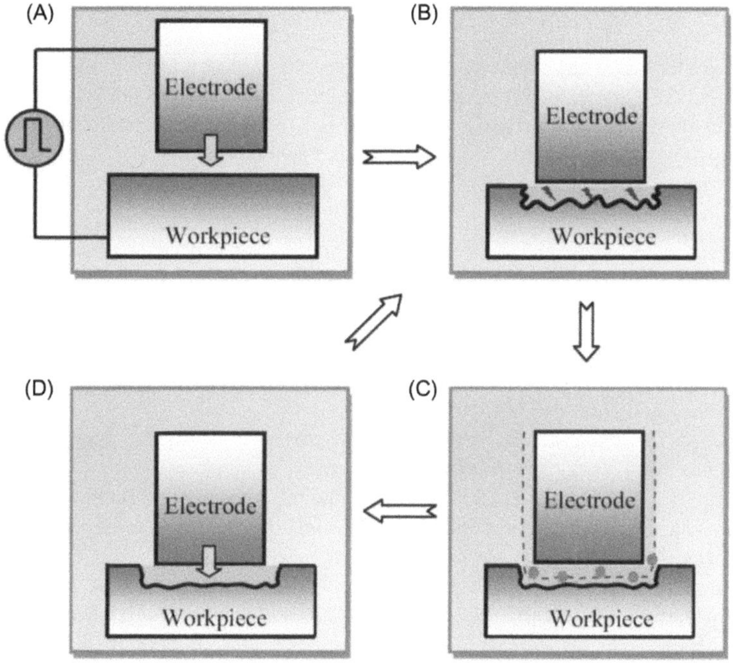

Figure 5.23 Principle of a simultaneous micro-EDM and micro-ECM (SEDCM): (A) process begins, (B) discharge, (C) dissolution, and (D) feeding (Nguyen et al., 2012).

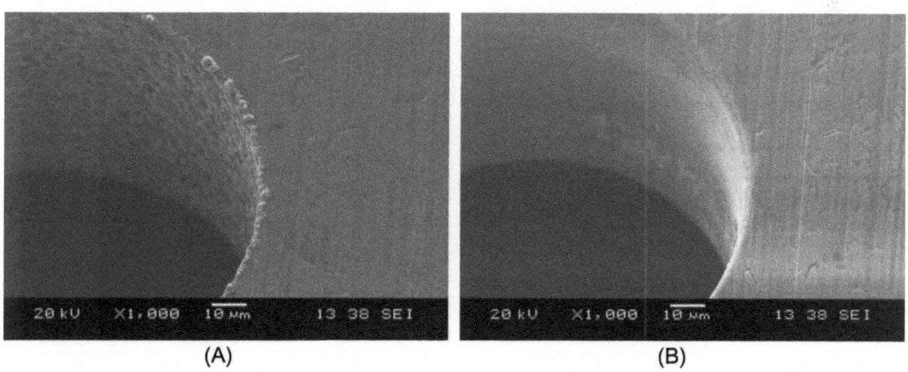

Figure 5.24 SEM images of microholes fabricated without (A) and with (B) the effect of material dissolution (Nguyen et al., 2013).

structures, including microholes, slots, grooves, tools, texturing, and complex surface in metals. Micro-ECM of complex dimensional features, development of machining and control systems, and combinations with other processes are in progress. Increasing requirements and applications of micro-ECM technologies will be foreseen in the future in industries of aerospace, automotive, electronics, and medical devices.

Acknowledgment

The authors acknowledge the support of Natural Science Foundation of Jiangsu Province of China (Grant No. SBK2015020754). Zhuang Liu gratefully acknowledges the Scientific Research Foundation for the Returned Overseas Chinese Scholars, State Education Ministry and Jiangsu Key Laboratory of Precision and Micro-Manufacturing Technology.

References

Ali, M. Y., & Ong, A. S. (2006). Fabricating micromilling tool using wire electrodischarge grinding and focused ion beam sputtering. *International Journal of Advanced Manufacturing Technology*, *31*, 501–508.

Bard, A. J., & Faulkner, L. R. (2001). *Electrochemical Methods: Fundamentals and Application*. New York, NY: John Wiley & Sons.

Bhattacharyya, B., & Munda, J. (2003). Experimental investigation into electrochemical micromachining (EMM) process. *International Journal of Materials Processing Technology*, *140*, 287–291.

Bhattacharyya, B., Munda, J., & Malapati, M. (2004). Advancement in electrochemical micro-machining. *International Journal of Machine Tools & Manufacture*, *44*, 1577–1589.

Brousseau, E. B., Dimov, S. S., & Pham, D. T. (2010). Some recent advances in multi-material micro- and nano-manufacturing. *International Journal of Advanced Manufacturing Technology*, *47*, 161–180.

Datta, A. M. (1997). Microfabrication by through-mask electrochemical micromachining. *Proceedings of SPIE-The International Society for Optical Engineering*, *3223*, 178–184.

Datta, M., & Landolt, D. (2000). Fundamental aspects and applications of electrochemical microfabrication. *Electrochimica Acta*, *45*, 2535–2558.

Datta, M., Shenoy, R. V., & Rominkiw, L. T. (1996). Recent advances in the study of electrochemical micromachining. *Transactions of the ASME*, *118*, 29–36.

Duong, T. H., & Kim, H. C. (2015). Electrochemical etching technique for tungsten electrodes with controllable profiles for micro-electrical discharge machining. *International Journal of Precision Engineering & Manufacturing*, *16*, 1053–1060.

Fan, Z. W., Hourng, L. W., & Wang, C. Y. (2010). Fabrication of tungsten microelectrodes using pulsed electrochemical machining. *Precision Engineering*, *34*, 489–496.

Hackert-Oschaetzchen, M., Martin, A., Meichsner, G., et al. (2013). Micro structuring of carbide metals applying jet electrochemical machining. *Precision Engineering*, *37*, 621–634.

Kim, B. H., Ryu, S. H., Choi, D. K., & Chu, C. N. (2005). Micro electrochemical milling. *Journal of Micromechanics & Microengineering*, *15*, 124–129.

Kirchner, V., Cagnon, L., Schuster, R., & Ertl, G. (2001). Electrochemical machining of stainless steel microelements with ultrashort voltage pulses. *Applied Physics Letters*, *79*, 1721–1723.

Kock, M., Kirchner, V., & Schuster, R. (2003). Electrochemical micromachining with ultrashort voltage pulses—a versatile method with lithographical precision. *Electrochimica Acta*, *48*, 3213–3219.

Kozak, J., Rajurkar, K. P., & Balkrishna, R. (1996). Study of electrochemical jet machining process. *Journal of Manufacturing Science and Engineering*, *118*, 490–498.

Kozak, J., Rajurkar, K. P., & Makkar, Y. (2004). Selected problems of micro-electrochemical machining. *Journal of Materials Processing Technology*, *149*, 426–431.

Lim, H. S., Wong, Y. S., Rahman, M., & Lee, M. K. E. (2003). A study on the machining of high-aspect ratio micro-structures using micro-EDM. *Journal of Materials Processing Technology*, *140*, 318–325.

Liu, Z., Liu, Y., Qiu, Z. J., & Qu, N. S. (2009). Effect of tool electrode insulation on electrochemical micro drilling accuracy. *Nanotechnology & Precision Engineering*, *7*, 355–360.

Liu, Z., Nouraei, H., Papini, M., & Spelt, J. K. (2014). Abrasive enhanced electrochemical slurry jet micro-machining: Comparative experiments and synergistic effects. *Journal of Materials Processing Technology*, *214*, 1886–1894.

Liu, Z., Nouraei, H., Spelt, J. K., & Papini, M. (2015). Electrochemical slurry jet micromachining of tungsten carbide with a sodium chloride solution. *Precision Engineering*, *40*, 189–198.

Liu, Z., Qiu, Z. J., Heng, C., & Qu, N. S. (2009). Electrochemical micro drilling of stainless steel with tool electrode jump motion. *Materials Science Forum*, *626–627*, 333–338.

Liu, Z., Zeng, Y., & Zhang, W. (2014). Fabrication of metal microtool applying wire electrochemical machining. *Advances in Mechanical Engineering*, *2014*, 382105–382105.

Long, Y. H., Xiong, L. C., Shi, T. L., & Liao, Z. Q. (2015). Experimental study on the processes of laser-enhanced electrochemical micromachining stainless steel. *Optik-International Journal for Light and Electron Optics*, *126*, 1826–1829.

Lu, Z., & Yoneyama, T. (1999). Micro cutting in the micro lathe turning system. *International Journal of Machine Tools & Manufacture*, *39*, 1171–1183.

Masuzawa, T., Fujino, M., Kobayashi, K., Suzuki, T., & Kinoshita, N. (1985). Wire electrodischarge grinding for micro-machining. *CIRP Annals-Manufacturing Technology*, *34*, 431–434.

Mathew, R., & Sundaram, M. M. (2012). Modeling and fabrication of micro tools by pulsed electrochemical machining. *Journal of Materials Processing Technology*, *212*, 1567–1572.

McGeough, J. A. (1974). *Principles of Electrochemical Machining*. London: Chapman and Hall.

Natsu, W., Ooshiro, S., & Kunieda, M. (2008). Research on generation of three-dimensional surface with micro-electrolyte jet machining. *CIRP Journal of Manufacturing Science and Technology*, *1*, 27–34.

Nguyen, M. D., Rahman, M., & Wong, Y. S. (2012). Simultaneous micro-EDM and micro-ECM in low-resistivity deionized water. *International Journal of Machine Tools & Manufacture*, *s54–55*, 55–65.

Nguyen, M. D., Rahman, M., & Wong, Y. S. (2013). Modeling of radial gap formed by material dissolution in simultaneous micro-EDM and micro-ECM drilling using deionized water. *International Journal of Machine Tools & Manufacture*, *66*, 95–101.

Pajak, P. T., Desilva, A. K. M., Harrison, D. K., & Mcgeough, J. A. (2006). Precision and efficiency of laser assisted jet electrochemical machining. *Precision Engineering*, *30*, 288–298.

Pan, Y., Xu, L. Z., & Zhou, H. X. (2015). Electrochemical etching of ultrafine tungsten wire poles. *Material Research Innovations*, *19*, S5.356–S5.361.

Park, B. J., Kim, B. H., & Chu, C. N. (2006). The effects of tool electrode size on characteristics of micro electrochemical machining. *CIRP Annals-Manufacturing Technology*, *55*, 197–200.

Qu, N. S., Zhang, X. F., Chen, X. L., et al. (2015). Modified microscale pattern transfer without photolithography of substrates. *Journal of Materials Processing Technology, 218*, 71–79.

Rajurkar, K. P., Levy, G., Malshe, A., Sundaram, M. M., Mcgeough, J., & Hu, X. (2006). Micro and nano machining by electro-physical and chemical processes. *CIRP Annals-Manufacturing Technology, 55*, 643–666.

Rajurkar, K. P., Zhu, D., McGeough, J. A., Kozak, J., & De Silva, A. (1999). New developments in electrochemical machining. *Annals of the CIRP, 48*, 567–579.

Schuster, R., Kirchner, V., Allongue, P., & Ertl, G. (2000). Electrochemical micromachining. *Science, 289*, 98–101.

Sen, M., & Shan, H. S. (2005). A review of electrochemical macro- to micro-hole drilling processes. *International Journal of Machine Tools & Manufacture, 45*, 137–152.

Silva, A. K. M., Pajak, P. T., Mcgeough, J. A., & Harrison, D. K. (2011). Thermal effects in laser assisted jet electrochemical machining. *CIRP Annals-Manufacturing Technology, 60*, 243–246.

Wang, M. H., & Zhu, D. (2009). Fabrication of multiple electrodes and their application for micro-holes array in ECM. *International Journal of Advanced Manufacturing Technology, 41*, 42–47.

Wang, S., Zhu, D., Zeng, Y., & Liu, Y. (2011). Micro wire electrode electrochemical cutting with low frequency and small amplitude tool vibration. *International Journal of Advanced Manufacturing Technology, 53*, 535–544.

Yang, I., Park, M. S., & Chu, C. N. (2009). Micro ECM with ultrasonic vibrations using a semi-cylindrical tool. *International Journal of Precision Engineering & Manufacturing, 10*, 5–10.

Zeng, Z., Wang, Y., Wang, Z., Shan, D., & He, X. (2012). A study of micro-EDM and micro-ECM combined milling for 3d metallic micro-structures. *Precision Engineering, 36*, 500–509.

Zhu, D., Qu, N. S., Li, H. S., Zeng, Y. B., et al. (2009). Electrochemical micromachining of microstructures of micro hole and dimple array. *CIRP Annals-Manufacturing Technology, 58*, 177–180.

Zhu, D., Wang, K., & Qu, N. S. (2007). Micro wire electrochemical cutting by using in situ fabricated wire electrode. *CIRP Annals-Manufacturing Technology, 56*, 241–244.

Microchannel fabrication via direct laser writing

6

S. Prakash and S. Kumar
Indian Institute of Technology Patna, Bihar, India

6.1 Introduction

Microelectro-mechanical systems (MEMS) have revolutionized the biological, chemical, control, and sensor-based application fields due to high throughput, low cost, low reagent requirement, smaller sizes, and smart structures. However, the story of such developments in miniaturization is not very new. In fact the earliest developments were started as early as in 1960 in the field of microelectronic circuitry. Soon, the improvisations over design were attempted in various other fields such as biological testing, chemical analysis, etc. In most of the later developed biological and chemical testing devices, certain type of fluid was made to flow in a tiny space sometimes referred as microchannels. Such developments led to garnish another important part of MEMS technology involving fluid flow and termed as microfluidics. The real success in microfluidics came in the late 1990s with the advent of inkjet printers where the fluid flowed through a tiny nozzle to create a spot of dimensions as small as 50 μm. Most popular microfluidic devices include gas chromatographs, electrophoresis devices, electroosmotic systems, separation devices, micromixers, micropumps, bioanalytical devices, microreactors, chemical devices, microheat exchangers, DNA amplifiers, cytometers, etc. MEMS devices are miniaturized embedded systems, having some micromachined parts enabling higher level functions and mostly used in control and sensor based applications.

All the MEMS and microfluidic devices consist of various microfeatures such as microcavities, microholes and surface structures. Microchannels are an integral part of most of such devices. The dimensional features and shape of microchannels vary according to their nature of applications. However, most of such dimensions lie in the range of 10–500 μm (Yuan & Das, 2007). The shape of such microchannels also depends upon the type of device or the purpose it is intended for. Rectangular, square, U-shape, V-shape, semicircular and Gaussian are the most used cross-sectional shapes for many of the applications. The circular cross-section microchannels are fabricated slightly below the top surface.

Broadly, MEMS and microfluidic devices can be categorized into three different kinds on the basis of their application field: (1) biological devices (2) chemical devices, and (3) electronics and mechanical engineering related devices (Prakash & Kumar, 2015). The physics behind all of such devices varies and molecular forces, capillary forces, surface effects, etc. become predominant in the understanding of

Microfabrication and Precision Engineering. DOI: http://dx.doi.org/10.1016/B978-0-85709-485-8.00006-1

such applications. In fact, such a study, though complex, may produce many fruitful results and lead to significant technological innovations.

Most of the biological devices in nature use some sort of microfluidic systems to grow such as plants, trees and living bodies using antigravity benefits of capillarity effects. The actual spread of natural biological microfluidic systems is such precise and complex that humans may not reach to reproduce it artificially to a satisfactory level. However, many of the human organ and related natural microfluidic systems can be simplified and replicated on a lab-on-chip (LOC) device for biological testing (Bhatia & Ingber, 2014). In recent times, the microfluidic devices have become very common for DNA analysis.

Chemical microfluidic devices boast many benefits such as lower diffusion distances, low Reynolds numbers, high interface-to-volume ratio (specific interface areas; solid/liquid or liquid/liquid) and low heat capacities for performance enhancement. Utilizing these advanced technologies, the new (LOC)-based chemical devices can perform many complex tasks that may not have been possible in a large normal scale (Tanaka et al., 2007). The first microfluidic based chemical device was developed way back in the 1970s in the form of gas chromatographs, which actually formed the basis of DNA and RNA analysis later (Terry, Jerman, & Angell, 1979). Tuckerman and Pease (1981) first introduced the forced cooling of integrated circuits for the purpose of compacting and miniaturizing these circuits. These microchannel-based microfluidic devices have flourished in recent times and have been a constant point of research from the heat transfer point of view.

Micromachining is basically an underlying foundation stone for MEMS and microfluidic applications and devices. It is a set of tools used to create miniaturized structures on desired substrate material by means of energy. Different micromachining processes utilize different forms of energy like thermal, mechanical, chemical, electrical etc. Further, although the micromachining process essentially refers to a subtractive process, microfabrication can also be performed with the help of additive processes. Various microfabrication processes include lithography, etching, injection molding, embossing, micromilling, and laser-based micromachining processes (Becker & Locascio, 2002). Beside these, there are many specially developed processes limited to specific applications and have not been used widely.

Many such microfabrication processes require clean-room facilities and multistep processing by trained personnel, and any modification in design consumes a lot of time and money. Most of these processes are material-specific with very low reproducibility. Lasers are an amazing tool for microfabrication process which is capable of overcoming most of these issues but with a high initial cost. Lasers mainly consist of high intensity focused and coherent electromagnetic waves which, when interacting with a material, ablate or remove the material by different mechanisms depending upon the type of lasers. Lasers are capable of machining almost all types of materials (with suitable wavelengths) without the need of clean-room facilities. Further, they are a type of contact-less machining, thereby producing no mechanical stress. The time consumption is often too small to be compared with any other process, and it also does not require any type of specially designed fixtures. Overall, it is the most versatile machining tool for microfabricating over a range of materials with utmost simplicity and dependable repeatability.

6.2 Important materials for MEMS and microfluidic devices

There are a wide range of materials used for MEMS and microfluidic applications, each having their own advantages and disadvantages. These materials can be divided into three different kinds: (1) metals and alloys, (2) semiconductors, composites, and specially developed materials, and (3) glass and polymer-based materials.

6.2.1 Metals and alloys

Metals and alloys are predominantly used in microchannel-based cooling applications. With the advancement of smart technologies, the need of rejecting heat at faster rates has tremendously increased. Heat transfer-based applications have gained significant momentum, which led to the increase in research and development studies in this area (Husain & Kim, 2009). Metals have unique ability to absorb and reject heat at much higher rate than polymers and glasses. They are also more thermally and mechanically stable substrates when compared to others. Conventional metallic substrates like copper, stainless steel, and aluminum are mostly used for applications involving temperature not more than 300–400°C. Paulraj, Paul, and Peterson (2012) and Prajapati, Pathak, and Khan (2015) have fabricated and studied stainless steel and copper based microchannel heat exchangers for low temperature applications like electronics cooling. Aluminum and indium phosphide (InP)-based heat sinks can dissipate heat at a very high rate than many other substrates. The heat transfer rate from these substrates may exceed to 1000 W/cm^2 (Philips, 1990). For high temperature cooling applications, metal aluminides have presented a cost effective solution as a suitable substrate material. These materials are generally used for applications involving temperature more than 650°C. Nickel aluminides can be used for temperatures as high as 1600°C (Kanlayasiri & Paul, 2004). Table 6.1 enlists the common metals used in MEMS and microfluidic applications.

6.2.2 Semiconductors, composites, and specially developed materials

Silicon-based microelectronic devices were developed for faster digital circuits (Pal & Sato, 2009). Silicon is also used as a substrate material for highly efficient fuel cells (Kamitani, Morishita, Kotaki, & Arscott, 2011). Silicon-based applications are also used in many separation-based devices and gas chromatographs (Mery et al., 2008). Quartz is used in many microfluidic devices due to its high optical transparency, electric insulation qualities, and chemical inertness to many reagents (Qin & Li, 2002). Similar to silicon, germanium is also used for some microfluidic devices; however, it is not as common as silicon due to high wafer

Table 6.1 Commonly used metals for MEMS and microfluidic applications

S. no.	Metal	Applications
1.	Stainless steel	Heat exchangers (Paulraj et al., 2012), stents (Meng, Liao, Zhou, & Zhang, 2009)
2.	Copper	Heat exchangers (Mei, Jiang, & Meng, 2008; Parida, Mei, Jiang, Meng, & Ekkad, 2010), microelectronics
3.	Aluminum	Heat exchangers (Parida et al., 2010), Electronics and electrical interconnects, optical
4.	Titanium	Biomedical, stents, sensors, and actuators (Aimi, Rao, MacDonald, Zuruzi, & Bothman, 2004; Klein et al., 2008)
5.	Silver	Electrochemistry (Rais-Zadeh, Kohl, & Ayazi, 2008)
6.	Tungsten	Sensors (Ali, Udrea, Milne, & Gardner, 2008) and microelectronics (Klein et al., 2008)
7.	Platinum	Electrochemistry and biomedical (Rubehn, Bosman, Oostenveld, Fries, & Stieglitz, 2009), microheating element (Courbat, Briand, & de Rooij, 2008; Erickson, Li, & Krull, 2003)
8.	Nickel aluminides	Heat exchanger and high temperature cooling applications (Kanlayasiri & Paul, 2004)
9.	ITO	Masks, sensors (Wisitsoraat, Patthanasetakul, Lomas, & Tuantranont, 2007)
10.	Gold	Electronic interconnects, optics, electrochemistry, membranes (Voskerician et al., 2003)
11.	Chromium	Adhesive layers (Wang, Lee, & Chiang, 2007, MEMS)
12.	Inconel	Microreactors, microfluidics, and MEMS bistable switch (Fleury et al., 2007)
13.	NiTi (Nickel Titanium)	Shape memory alloy, microrobotics (Vasilyev et al., 2015)
14.	Tantalum	RF MEMS switches (Persano et al., 2010)

cost (approximately 500 times) (Buriak, 2002). Gallium nitrite (GaN) based microchips are very frequently used in microelectronics because of their high band gap. They are commonly found in LEDs (light emitting diodes), display units, and imaging sensors (Jiang, Lin, Jin, & Li, 2002). InP (Indium Phosphide) is used in MEMS applications for optical waveguide devices (Pruessner et al., 2006). Schott Glaswerke developed an optically transparent glass ceramic also called Foturan, for microfluidic applications, which can sustain high thermal load (Fernandez-Pradas et al., 2009). Many new ceramic materials have also been developed for MEMS and microfluidic applications in recent times.

6.2.3 Glass and polymer-based materials

One of the primary requirements of many microfluidic devices is optical transparency. Glasses are not only optically transparent but mechanically and thermally

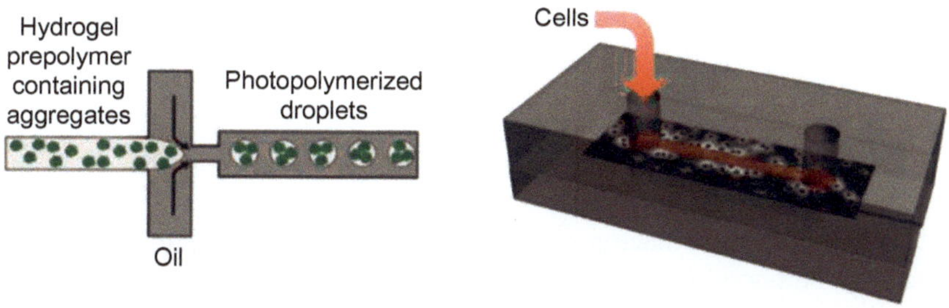

Figure 6.1 A PDMS-based lab-on-chip device for liver and kidney (Bhatia & Ingber, 2014).

stable materials. They are also inert to many organic solutions and reagents, making them a first choice for many of such applications (Malek, 2006). However, the device fabricating processes on glass substrates require a series of steps and are time-consuming. Lin, Lee, Lin, and Chang (2001) fabricated a glass based microfluidic system in 10 h with multiple steps.

Fabrication of microfluidic structures on substrates like glass, silicon, quartz, etc. is cumbersome, tough, and time-consuming. Also, often, the cost becomes so high that it cannot be fabricated into a large number for disposable use. With the application of transparent polymers as a substrate material, many of such problems can be overcome. The fabrication process is simpler for many of the polymer-based microfluidic devices and can be fabricated en masse easily (Castaño-Álvarez et al., 2008). Polydimethyl siloxane (PDMS) is a very frequently used polymer for microfluidic substrate because of its excellent optical transparency, higher thermal stability, and low cost. PDMS has been used for many biomedical microfluidic devices, electrophoresis devices, microreactors, and separation devices (Fujii, 2002). Fig. 6.1 show the PDMS-based liver and kidney test module fabricated by Bhatia and Ingber (2014). Polycarbonate as a substrate material offers lower cost, high impact resistance, good machining quality, and high glass transition temperature (145°C) (Ogonczyk, Wegrzyn, Jankowski, Dabrowski, & Garstecki, 2010). Polystyrene (PS) (Barker et al., 2000; Barker, Tarlov, Canavan, Hickman, & Locascio, 2000), polyethylene terephthalate glycol (PETG), and polyvinyl chloride (PVC) (Waddell, Locascio, & Kramer, 2002) have also been used by a few authors as microfluidic substrate materials. Cyclic olefin polymers like Topas (Castano-Alvarez, Fernandez-Abedul, & Costa-Garcia, 2005) and Zeonor (Mela et al., 2005) were also studied as a polymeric substrate material for electrophoresis applications.

Polymethyl methacrylate (PMMA) is one of the most prominent polymeric materials for microfluidic devices. Its unique ability to get a clean laser cut makes it a unanimous choice for many low temperature microfluidic device since its glass transition temperature is comparatively low (105–115°C). The optically transparent PMMA is the most suitable material for CO_2 lasers and excimer lasers (Snakenborg, Klank, & Kutter, 2004). Utilizing this material, total fabrication time can be minimized and mass production of disposable devices becomes feasible. Rapid prototyping of microfluidic devices also encourage the innovation for LOC applications since simulation does not provide the accurate results in many of such cases (De Marco et al., 2010).

6.3 Lasers for microfabrication

The conventional micromachining techniques like electrodischarge machining, wet etching, embossing, injection molding, and mechanical micromilling have inherent limitations of types of material, feature size, mechanical stress, fixtures, corrosion, and high machining time. Lasers are the most well suited tool for micromachining applications overcoming all these limitations. Laser machining can be used for MEMS devices, microelectronics and optoelectronic devices, and microfluidic systems on almost all type of substrate materials. The process is mainly subtractive (not always), contact-less, single step, and do not require special fixture or material holding devices. Lasers can machine optically transparent material, metals, ceramics, semiconductors, thin films, hard and soft materials, and various composites. Lasers used for micromachining purpose can be classified in two ways (Fig. 6.2): the timescale basis and wavelength scale basis. The timescale basis divides the lasers into three categories, i.e., (1) continuous wave (CW) lasers, (2) short pulse lasers, and (3) ultrashort pulse lasers. On the basis of wavelengths, lasers can mainly be divided into three kinds: (1) mid infra-red (MIR), (2) infrared (IR), and (3) ultraviolet (UV) lasers.

6.3.1 *Timescale based division*

When the laser beam interacts with any material, the laser pulsewidth, i.e., pulse-ON time plays a very critical role. Depending upon the energy density in pulsewidth duration, material interactions of laser beam differ considerably. The material response depends upon lattice heating time (t_l) and electron cooling time (t_e) apart

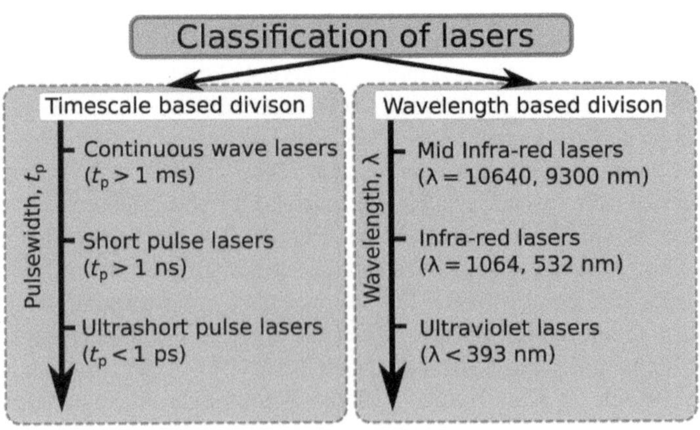

Figure 6.2 Classification of lasers used for micromachining.

from pulse-ON time. Time required to heat the lattice is much longer than electron cooling time because of higher heat capacity of the lattice. Electron cooling time is of the order of picosecond (10^{-12} s) while lattice heating time depends upon the material's heat capacity and is usually much longer than electron cooling time (Yao, Chen, & Zhang, 2004). According to different times of interaction, the physics of material ablation also changes. Based on these different types of material removal mechanisms, lasers can be subdivided into three categories.

6.3.1.1 Continuous wave laser

In CW lasers, the pulsewidth is generally greater than 1 ms. It is called CW because energy does not vary with time but remain almost constant over entire period of pulse-ON time. Laser material interaction for such lasers involves heating melting and vaporization cycle. These types of lasers are particularly useful in laser welding, bending, forming, cladding, and alloying. Most of the industrial CO_2 lasers work in continuous mode and classical heat transfer equations can be utilized to solve such heating problems. Often, CW lasers are used with gated super-pulsing instead of continuous mode with gated ON/OFF. Gated super-pulsing allows the larger power pulsed output at the start of each pulse for a brief amount of time. However, the maximum output power remains same as nominal CW mode with a reduced duty cycle (Fig. 6.3).

6.3.1.2 Short pulse lasers

In short pulse lasers, the pulsewidth is of the order of nanosecond. The pulse energy within the pulsewidth or pulse-ON time varies with time. Peak power in short pulse lasers may increase many thousand times than average power. Short pulse lasers generally include Q-switched Nd:YAG lasers, fiber lasers and Nd:YVO_4 lasers. Since the peak power is extremely high in short pulse lasers for a shorter duration, the heat-affected zone (HAZ) is normally lower than CW lasers. The micromachining process still involves heat as CW lasers but the process takes less time than CW lasers. Short pulse lasers have been found to be more suitable for micromachining applications on metals, semiconductors, and composites because of high cutting precision.

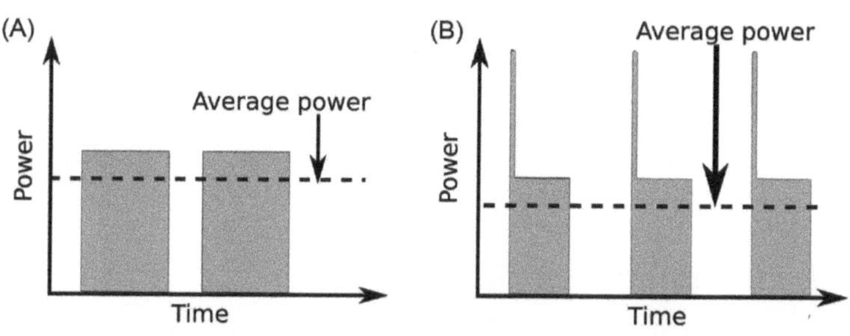

Figure 6.3 Continuous wave lasers (A) gated ON/OFF, (B) Gated super-pulsing.

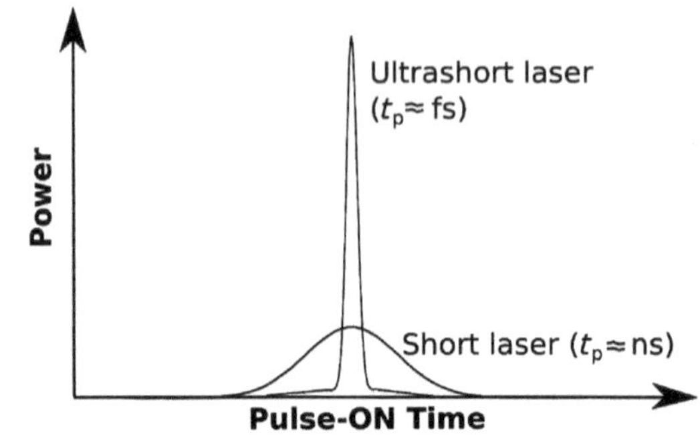

Figure 6.4 Short and ultrashort lasers.

6.3.1.3 Ultrashort pulse lasers

For ultrashort lasers, the material interaction time is of the order of femtosecond or picoseconds. Usually, when the pulsewidth of a laser beam is shorter than electron cooling time (10^{-12} s), the laser is termed as ultrashort laser. In this short time, the energy is transferred to the lattice via free electrons. Material removal takes place via cold ablation process unlike short and CW mode lasers. Due to such a short material interaction time, the peak powers of ultrashort lasers are much higher than short lasers (Fig. 6.4). The material ablation phenomenon is predominated by its intensity and wavelength effect is secondary for such cases. For most of the ultrashort lasers, material ablation takes place via a process called multiphoton absorption. Ultrashort lasers are expensive and require high quality beam optics. However, femtosecond lasers have been the most dependable solutions for fabricating microstructures as far as surface smoothness and repeatability of produced dimensions are concerned (Prakash & Kumar, 2015). Most of the applications of ultrashort lasers are still in the research domain because of excessive maintenance requirements. Ultrashort lasers can be focused on a tinier spot compared to short and CW lasers.

6.3.2 Wavelength based division

Another important classification of lasers is based on wavelengths. Wavelengths also affect the material removal process, hence plays a very important role. The frequency of the photons of a beam is directly proportional to photon energy and inversely proportional to beam wavelength; the nature of material interaction also changes with it. Considering the most used and easily available lasers for micromachining purpose, they can be divided into three categories, namely mid infrared, infrared, and ultraviolet lasers.

6.3.2.1 Mid infrared lasers (mid IR)

CO_2 laser falls in the category of mid infrared lasers. CO_2 laser is available at two different wavelengths, i.e., 10.6 and 9.3 μm. However, it is more common to use 10.6 μm wavelengths for CO_2 lasers because of availability of laser sources at this wavelength. The processing of polymers is better at 9.3 μm compared to 10.6 μm (Powell, 1998) because of better absorption characteristics. For microfabrication purpose, CO_2 lasers were found to be used for polymers, glasses, and organic materials only. They are not preferred for metals because of producing high HAZ and uncontrolled surfaces. Due to larger wavelength, the material removal process follows the thermal cycle, i.e., heating, melting, and vaporization in most of the target material.

6.3.2.2 Infrared lasers (IR lasers)

Infrared lasers mostly operate at 1.064 μm which is smaller than mid infrared lasers. Nd:YAG, fiber, diode, Nd:YVO_4, and semiconductor lasers mostly emit the beams in this regime. Frequency doubled green lasers (532 nm) also fall in this zone. Most of these lasers are Q-switched and operate in pulsed mode. The photon energy of such laser beams is larger than mid IR lasers. These lasers are especially suitable for metals and semiconductors due to better absorptivity and reflectivity characteristics. However, these lasers were not found to be compatible for most of the polymers. Due to excessive transmission, they are not used in many transparent materials including glass, acrylic, and PDMS. The photon energies for wavelength of infrared radiations of 1064 nm and 532 nm are 1.17 eV (1.9×10^{-19} J) and 2.33 eV (3.7×10^{-19} J), respectively.

6.3.2.3 Ultraviolet lasers (UV lasers)

Most common ultraviolet lasers include excimer lasers (351–157 nm) and frequency tripled Nd:YAG lasers (355 nm). Because of high photon energy and good absorption properties, these lasers are mostly used with polymers. The polymer ablation of UV lasers is relatively cleaner with less HAZ when compared to mid IR laser processing of polymers. However, since excimer lasers are tough to maintain and require typical optics, they are limited in actual use. The photon energies corresponding to 355 nm and 266 nm are 3.53 eV (5.6×10^{-19} J) and 4.66 eV (7.5×10^{-19} J) respectively, which is much larger than IR photon energy. Due to this much higher energy, UV lasers can break the chemical bonds of many materials including polymers and glasses that require 3–7 eV of energy. This results in cold ablation of the material.

6.4 Material removal mechanisms

All the lasers used for microfabrication purpose basically have a laser source, a set of reflecting mirrors and focusing lenses (Fig. 6.5). Laser beams generally follows the Gaussian energy distribution, however, top-hat distribution is also used in many excimer lasers. Timescale is an important factor to determine type of ablation of lasers. Based on the different timescales involved with CW, short, and ultrashort lasers, associated HAZ has been presented in Fig. 6.6. The largest HAZ appears in CW lasers appended by short pulse lasers and ultrashort pulse lasers respectively.

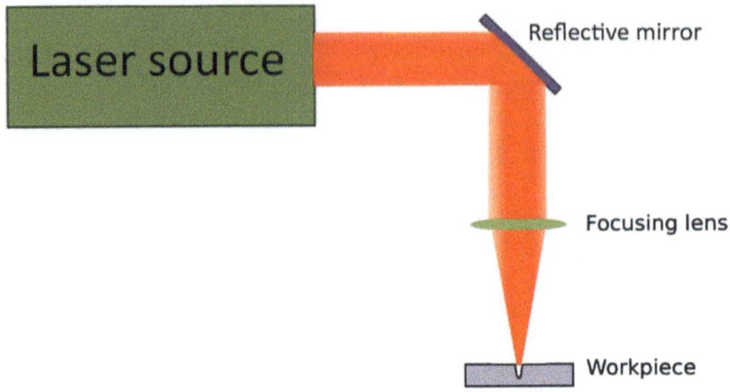

Figure 6.5 Schematic of basic laser machining set up.

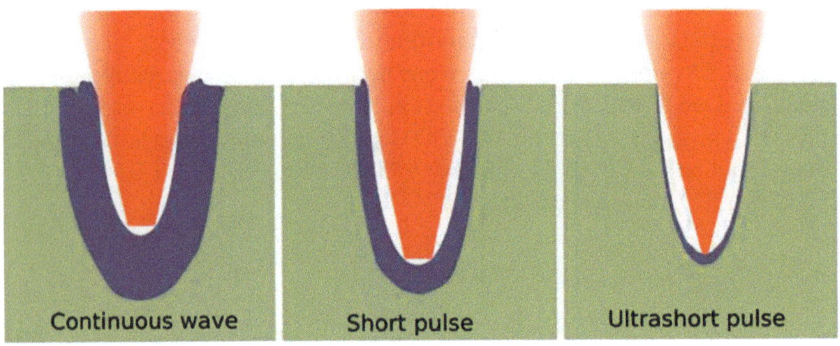

Figure 6.6 Material ablation at different timescales.

Based on the above description of lasers, it is evident that laser ablation can be divided into two distinct categories, i.e., thermal ablation and cold/photo ablation.

6.4.1 *Thermal ablation*

Whenever a laser beam strikes a surface, part of it is absorbed while the remaining part is reflected and transmitted following Eq. 6.1.

$$\alpha + \rho + \tau = 1 \tag{6.1}$$

where, α represents absorptivity, ρ reflectivity, and τ transmissivity of the material with respect to particular wavelength of the beam and surface smoothness characteristics. These values vary with material to material as well as with beam polarization. Inside the material, the light decays exponentially with depth following Beer-Lambert law (Eq. 6.2) (Brown & Arnold, 2010).

$$I = I_O e^{-\mu z} \tag{6.2}$$

Where, $I(z)$ is laser beam intensity at depth z, I_o is original incident intensity and μ is absorption coefficient. For good absorbers, the absorption coefficients vary in the range of $10^5-10^6\ cm^{-1}$ and attenuation length in the order of 10^{-5} and 10^{-6} cm (Duley, 1983).

In laser material interaction, the energy inside the material is transferred by free electrons in metals, lattice vibration in insulators, and via a combination of both in semiconductors. This absorbed energy converts into heat in a very short span of time (10^{-13} s for metals and $10^{-12}-10^{-6}$ s for nonmetals) (Dahotre & Harimkar, 2008). The confinement of heat into a localized zone depends upon many material properties such as diffusivity, conductivity, and heat capacity. Depending upon these properties, the heat is conducted into the material and convected or radiated from the surface. For low thermal conductivity materials, the heat remains accumulated into a localized zone and material melting or evaporation takes place as soon as the energy crosses the melting or evaporation threshold. Normal heat transfer equations can be solved for heating, melting, and vaporization phenomenon depending upon different types of heat sources, i.e., surface or volumetric. In this process, the material removal takes place either due to direct vaporization or due to ejection of melt layer from the cutting zone. Since the process involves heating as the primary requirement, ablation process is called thermal ablation. In case of short pulse lasers, which also remove the material by photothermal mechanism, the HAZ is smaller than CW lasers. This can be attributed to larger peak power of pulse laser compared to CW lasers. Due to this, evaporation takes place very quickly and less heat is conducted into the material. This also causes localized and focused heating with larger peak temperature (Banerjee, Ogale, Das, Mitra, & Subramanian, 2005).

6.4.2 Cold ablation/photochemical ablation/photo ablation

The mechanism of cold ablation is entirely different from photothermal ablation in the sense that the material removal phenomenon is independent of thermal cycle. In general, there are two distinct mechanisms of cold ablation. In the first kind, when the photon energy of laser beam exceeds the chemical binding energy of materials, the material ablation happens due to breaking of bonds. Due to high energy photons, faster electronic excitation takes place, which further causes ionization by breaking chemical bonds between atoms (Molian, Pecholt, & Gupta, 2009). This is the most usual case in UV laser ablation of polymers and glasses.

However, it was also observed that cold ablation takes place even when the photon energy is not sufficient to break the molecular bonds. This happens when the peak energy of the laser pulse is extremely high. This usually happens in ultrashort lasers like femtosecond and picosecond lasers. This type of material ablation phenomenon occurs due to multiphoton absorption. In multiphoton absorption, bond breaking takes place due to simultaneous absorption of two or more photons. If the peak intensity of the pulse reaches beyond $10^{13}\ W/cm^2$, the ablation process starts from vast pool of valence electrons stripped off atoms via Zener tunneling unlike multiphoton absorption (Joglekar, Liu, Meyhöfer, Mourou, & Hunt, 2004). For such a high peak power, the nonlinear absorption properties of material become predominant irrespective of material being opaque to associated wavelength. Heat diffusion time for any material is normally larger than pulse interaction time for such types of lasers. Since the material cannot respond fast enough, ionization takes place predominantly and hydrodynamic expansion of plasma takes away nearly all the input laser energy leaving nothing to conduct in the material to form heat. A few authors have advocated the phenomenon of Coulomb explosion in which a gentle ablation takes place at threshold energy resulting in a clean and HAZ-free surface (Sokolowski-Tinten et al., 1998). Coulomb explosion was found to be dominant in dielectric materials rather than metals and semiconductors (Cheng et al., 2013). Practically, the cold ablation phenomenon is always accompanied by little amount of photothermal ablation depending upon the material type (Dyer, 2003).

6.5 Laser microprocessing of materials

Although there are strictly only two methods of laser ablation, the actual ablation behavior depends upon the type of material. There are distinct types of lasers suitable for different types of materials. Wavelengths of specific lasers suit specific type of materials only. Therefore, it is essential to understand the behavior of laser beams with different kinds of material while keeping in view of the micromachining applications. In this section, few laser machining examples have been discussed on different types of substrates to aid in judging and determining different types of material responses. Further laser micromachining can be performed in open surroundings as well as in a wet environment. The cut quality of the material is also affected by the associated environment.

6.5.1 Direct laser micromachining in open surroundings

6.5.1.1 Metals and alloys

Most of the metals have high thermal conductivity. Hence the rate of heat diffusion into the material is also very high, resulting into comparatively larger conduction loss. Unless, the heat is not supplied beyond a certain energy threshold, the material ablation remains poor. This is the reason that mostly pulsed lasers are used in metal micromachining processes because of their higher peak energy compared to CW lasers. Kathuria (2005) discussed the laser micromachining of stainless steel in light of stent devices and pointed out that when compared to CO_2 lasers and IR lasers, UV range lasers were found to produce the most promising results. A third harmonic Nd:YAG laser (355 nm) was employed to produce stent on a stainless steel tube with the slit width of 0.05–0.1 mm. Reduced HAZs were observed with UV laser ablation compared to IR and MIR laser ablation. In producing the microchannels over stainless steel substrates with an IR Nd:YAG laser (1064 nm), the recast layer thickness was found to be as high as 30 μm (Mohammed, Al-Ahmari, & Umer, 2015). Recast layer was found to be

Figure 6.7 Nd:YAG laser (1064 nm) fabricated microchannels on steel, rectangular (top) and semicircular (bottom) (Mohammed et al., 2015).

inversely proportional to cutting speed. Fig. 6.7 shows the scanning electron microscopic (SEM) image of microchannel cross-sections.

Femtosecond laser micromachining of AISI 306 L steel was performed by Choi, Bong, Farson, Lu, and Lee (2009). The beams in this femtosecond laser processing were focused with a beam diameter as small as 1.6 μm. The channels were found to be clean with no traces of HAZ. Pulsewidth effect on drilling process on stainless steel was observed by Chichkov, Momma, Nolte, Alvensleben, and Tunnermann (1996) (Fig. 6.8). Holes drilled with 200 fs laser were found to be clean compared to 3.3 ns lasers for the same wavelength of laser. Weck, Crawford, Wilkinson, Haugen, and Preston (2007) has compared the micro-drills on a copper substrate with nanosecond, picosecond, and femtosecond lasers. Fig. 6.9 shows the SEM image of middle of the drilled hole for various pulse times. The ripples are visible at 150 fs of pulsewidth and subsequently melting takes place which increases with increase in pulsewidth time. Due to molten state, the surface smoothness was also found to be increasing in later stages, however, femtosecond ablation was found to give accurate dimensional control.

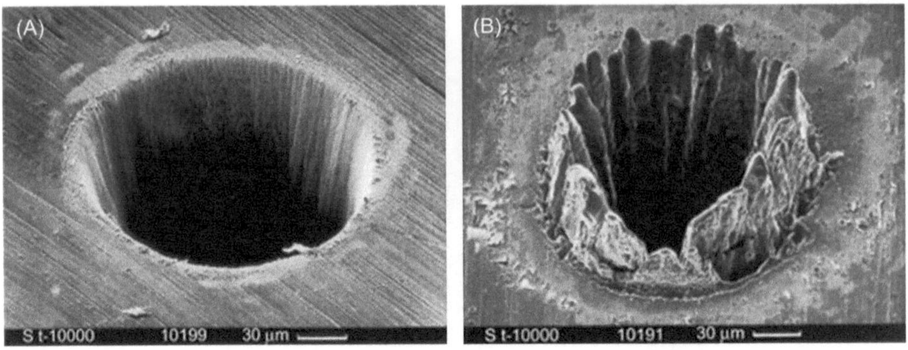

Figure 6.8 100 μm thick steel drilled by a 780 nm laser with (A) pulse width: 200 fs, (B) pulse width: 3.3 ns (Chichkov et al., 1996).

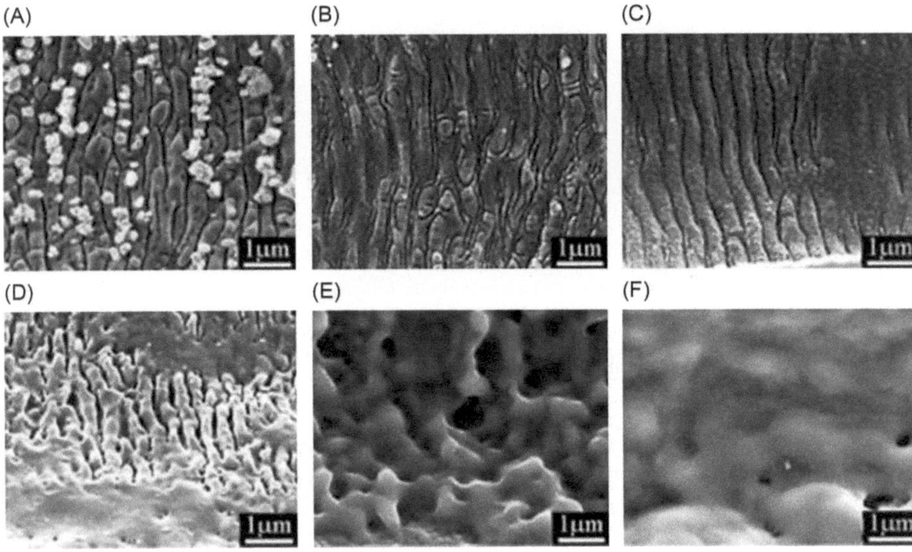

Figure 6.9 SEM image of middle of the drilled holes on copper (A) 150 fs, (B) 1 ps, (C) 10 ps, (D) 35 ps, (E) 220 ps, (F) 7 ns (Weck et al., 2007).

Polycrystalline aluminum oxide and aluminum nitride sleds were machined using a CO_2 laser and excimer lasers by Williams, Williams, Liu, Webb, and Firth (2009). Though the excimer laser was found to be producing repeatable good quality surfaces compared to CO_2 laser, both the laser machining processes were not found to be capable of producing parts suitable for use in opto-electronic assemblies due to extreme redeposition of melt material. CO_2 laser was found to be generating surface roughness as high as 4.5 μm which is not acceptable for such MEMS devices. Excimer laser was not found to be successful for large area micromachining. However, a femtosecond laser micromachining of aluminum was able to

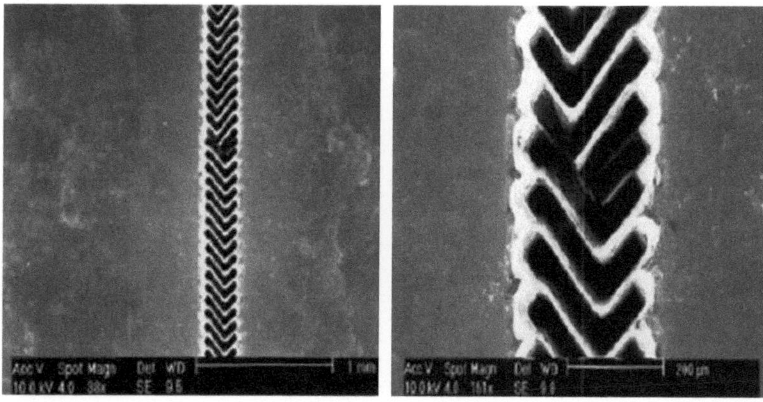

Figure 6.10 SEM image of staggered herringbone mixer viewed from the top. Right: close-up SEM image.
Source: Reproduced from Lim, D., Kamotani, Y., Cho, B., Mazumder, J., Takayama, S. (2003). Fabrication of microfluidic mixers and artificial vasculatures using a high-brightness diode-pumped Nd:YAG laser direct write method. *Lab on a Chip 3*, 318–323, (Lim et al., 2003) with permission from the Royal Society of Chemistry.

produce roughness of 1.1 μm (Perrie, Gill, Robinson, Fox, & ONeill, 2004). Kara and Kizil (2012) studied the femtosecond laser micromachining of titanium and found that the smoothest microchannels were obtained with 100 mm lens and 20 numbers of passes.

6.5.1.2 Semiconductors, composites, and specially developed materials

Diode pumped Nd:YAG lasers were applied to fabricate multilevel microchannels and herringbone mixers and artificial vascular on 660 μm thick silicon by Lim, Kamotani, Cho, Mazumder, and Takayama (2003). The channels were fabricated with depth of 125 μm and widths of 250, 200, 160, and 125 μm. It was observed that laser could fabricate the complex structures on silicon in one-step processing which otherwise could have taken more than several steps in other fabrication techniques. The staggered herringbone structures were found to achieve rapid mixing of liquids in fluorescence imaging test (Fig. 6.10).

Many times a postprocessing step is also required to improve surface roughness. Kam, Shah, and Mazumder (2008) fabricated a wide range of microchannels on silicon substrates using nanosecond and femtosecond lasers. Both types of lasers were found to be producing acceptable quality of microchannels. Surface roughness of the microchannels was found to be improved with defocused machining conditions (Fig. 6.11). Chemical wet etching was performed as a postprocessing for surface improvement in the chemical solution of 49% HF and 69% HNO_3 in the ratio of 10:1. A minimum 30-s etching time is required to achieve reasonable surface quality in nanosecond ablation. Femtosecond laser ablation was found to be

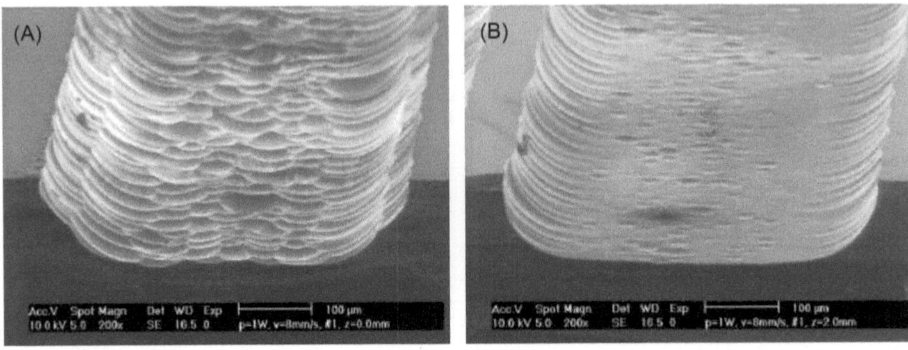

Figure 6.11 Nanosecond laser micromachining of silicon in (A) focused and (B) defocused conditions (Kam et al., 2008).

able to produce submicron features with less debris, collateral damage, and less etching time. Third harmonic Nd:YAG laser (355 nm) was utilized to fabricate microchannels inside photostructurable glass composite (Foturan) followed by thermal treatment and wet etching (Fernandez-Pradas et al., 2009). Larger pulse energies were found to be producing nonsymmetric channels when subjected to 10% ultrasonic HF bath.

6.5.1.3 Glass and polymers

Glasses are tough to be machined using conventional machining techniques but most suitable candidates for many microfluidic applications. Being a transparent material, it does not absorb normal IR radiations. This means that transparent glass can only be machined either by CO_2, UV lasers, or ultrashort lasers. However, a very small size microchannel can be generated on soda-lime surface with 1064 nm Nd:YVO_4 laser (Nieto, Delgado, & Flores-Arias, 2014). Such kind of machining normally requires a lot of thermal treatments and wet etching for practical applications. Femtosecond laser ablation of glasses was performed by Maselli et al. (2006) with astigmatically shaped beam for bio-photonic sensor applications. Microchannels were etched inside the microchannel in presence of HF solution.

Among all the materials, polymers were found to be etched by lasers utmost. The cost of the CO_2 laser system is at least 10 times lower than excimer laser, the femtosecond laser and wet etching equipment. It takes only a few minutes to fabricate microfluidic chip (Fu, Ju, Yang, & Wang, 2012). Hong et al. (2010) fabricated microchannels on PMMA with a CO_2 laser. It was found that direct laser etching when material resides at focal point produces rough channel surfaces and may not be suitable for many applications. Machining at a defocused condition may produce significantly improved surface roughness without the need of postannealing. PMMA ablation by a CO_2 laser is the most studied phenomenon and many authors have used this method to produce microchannels for various microfluidic devices. Since the beam diameter of a CO_2 laser may not be reduced below a

Figure 6.12 CO_2 laser micromachined microchannels on PMMA (A) before annealing, (B) after annealing (Cheng et al., 2004).

certain level (usually greater than 100 μm and most commonly lying around 250 μm (Snakenborg et al., 2004)), producing narrow width of the produced channel is also a constraint for such lasers. However, high power density focusing optics (HPDFO) can be used to focus the laser beam with focal spot size of 38 μm in diameter. The HPDFO involves two parts, the beam expander and high quality optics (Wang, Zheng, Lim, Wang, & Lam, 2011). CO_2 lasers can produce thousands of PMMA-based devices per day on a single system resulting in the possibility of mass production of portable microfluidic products (Cheng, Wei, Hsu, & Young, 2004). Cheng et al. (2004) also demonstrated that thermal annealing of the produced channels could enhance the surface smoothness of the walls significantly (Fig. 6.12).

Polycarbonate was also micromachined by CO_2 laser by Qi, Tao, Yao, and Tiechuan (2009). The ablation resulted in a deeper and wider molten pool, and the surface produced was found to be rough (Fig. 6.13). Ultraviolet lasers (UV lasers) were also found to be producing good quality surfaces on PMMA. Heng, Tao, and Tie-chuan (2006) and Waddell et al. (2002) fabricated the microchannels on PMMA substrates with 248 nm excimer laser. Cycloolefin copolymers (COC) were also successfully micromachined by Sabbert, Landsiedel, Bauer, and Ehrfeld (1999) using ArF excimer laser (193 nm). Femtosecond lasers were found to be extremely good in fabricating microchannels of small sizes on PMMA substrates. Circular

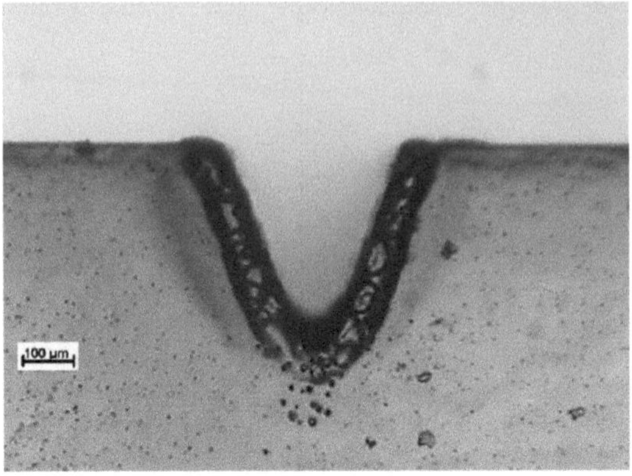

Figure 6.13 Cross-section of microchannel fabricated on polycarbonate substrate by a CO_2 laser (Qi et al., 2009).

microchannels of 2–20 μm inside PMMA were fabricated with the assistance of high pressure argon (374 KPa). Assist gas helps in fabricating clog-free microchannels for microfluidic applications.

6.5.2 Direct laser micromachining in different surrounding conditions

Since most of the laser machining process involves heat in some form, heat related defects are imperative. In order to reduce the HAZ, various surrounding conditions have been applied by various authors. Nd:YAG laser microchanneling of carbon doped PMMA was investigated in the presence of 1 mm of water above the surface (Prakash, Acherjee, Kuar, & Mitra, 2013). Microchannels were found to be clean and clog-free when fabricated in underwater condition. Nd: YAG laser micromachining of Inconel 718 was performed under static condition of water (Alahmari, Ahmed, & Darwish, 2015). It was found that two subchannels were generated on the surface instead of one microchannel due to redeposition of melt debris and disturbances in focus. Yan et al. (2011) machined alumina in underwater condition and found much better surface quality compared to open air condition.

Underwater machining minimizes crack and heat related defects. Water waves were captured on the surface in underwater machining condition. The water waves helps in removing the melt redeposition from the cutting zone. Bubbles were also found to be appearing on the surface (Fig. 6.14). Chung and Lin (2010) performed the underwater drilling of Pyrex (need registered trademark) glass utilizing CO_2 laser resulting in crack-free and scorch-less holes for inlet/outlet connections of microfluidic devices. Apart from water, other chemicals were also used as surrounding environment. Li and Achara (2004) found 300% increase in material removal rate when processing stainless steel in salt solution. Single crystal GaN

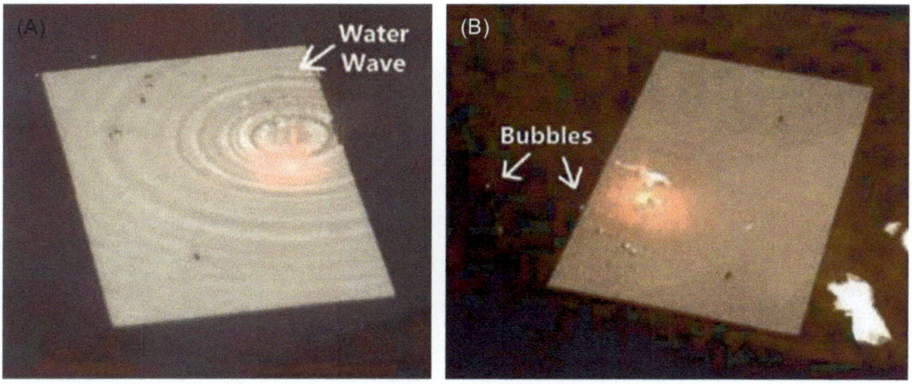

Figure 6.14 (A) Water waves and (B) bubbles on alumina surface in underwater CO_2 laser machining (Yan et al., 2011).

was machined in HCl solution by Nakashima, Sugioka, and Midorikawa (2009). A gas environment like argon was also used as a surrounding environment to improve the surface quality on silicon (Kam and Mazumder, 2008).

6.6 Challenges and future of laser processing

MEMS and microfluidics offer huge possibility for various technological advancements. The development of such devices lies in the ease of fabrication of such devices at the least possible cost. Lasers offer versatile solution for many types of substrates. IR nanosecond lasers present cost effective solutions for metallic and composite substrates. CO_2 lasers were found to be good for polymer and glass-based substrates. Apart from higher cost being the only constraint, ultrashort lasers present solution for almost all types of materials. Still there are few challenges for laser microfabrication of MEMS and microfluidic devices as listed below:

1. Majority of the laser machining processes suffer heat-related defects like HAZ, splashing, bulging, and cracks. These heat-related defects create problems in optical visibility, change in material properties, and bonding process.
2. The surface roughness of the generated profile is a big constraint for many materials. In most of the laser-based direct microfabrication process, the surface smoothness is not up to the mark due to uneven heat distribution. The surface roughness affects the fluid flow inside the microchannels and results in a complex flow pattern and is not desirable (Stroock et al., 2000). Although there are many postprocessing steps have been developed, the improvement in single-step processing is highly desirable to reduce time and cost consumption.
3. Lasers may not be able to produce high aspect ratio microchannels or microstructures on many substrates. Most of the lasers suffer from this drawback due to difficulty in material removal from deep cavities.
4. The initial cost of the laser system is very high. Although low power CO_2 laser comes cheaper than many other fabrication techniques, other types of lasers are still costly. A lot of effort and research is required to make it cost effective.

5. Since most of the laser ablation follows thermal cycle of heating, melting and vaporization, melt redeposition, and their effective removal from cutting zone is essentially required. Though some assist gases are used in removing the melt material, significant improvement in melt ejection is still required.
6. Laser beams generally follow Gaussian profile. This Gaussian nature of the beam hinders in producing exact rectangular or required shape and generally results in Gaussian shape feature only. This is one of the major drawbacks of the laser system. Although the beam shaping optics is being developed, it is still in its early stages and solutions for this problem are still pertinent.
7. A particular wavelength can only machine certain type of material and may not get absorbed into other types of materials. So lasers of different wavelengths are required to machine different types of substrate materials and only one laser source is not sufficient to machine all types of material.
8. Laser sources and associated optics are most important component of a laser machine and their maintenance is absolutely required. They age over the time and do not perform with original efficiency thereby reducing the performance of the laser.

Overall, lasers are wonderful tools for rapid fabrication of microfluidic and MEMS devices. A number of commercial firms are already using them for device fabrication. Although a number of improvements are required, the mass production of such devices is quite feasible by efficient and optimized use of laser machining systems.

References

Aimi, M. F., Rao, M. P., MacDonald, N. C., Zuruzi, A. S., & Bothman, D. P. (2004). High-aspect-ratio bulk micromachining of titanium. *Nature Materials*, *3*(2), 103–105.

Alahmari, A. M., Ahmed, N., & Darwish, S. (2015). Laser beam micro-machining under water immersion. *The International Journal of Advanced Manufacturing Technology*, 1–11.

Ali, S. Z., Udrea, F., Milne, W. I., & Gardner, J. W. (2008). Tungsten-based SOI microhot-plates for smart gas sensors. *Journal of Microelectromechanical Systems*, *17*(6), 1408–1417.

Banerjee, A., Ogale, A. A., Das, C., Mitra, K., & Subramanian, C. (2005). Temperature distribution in different materials due to short pulse laser irradiation. *Heat Transfer Engineering*, *26*(8), 41–49.

Barker, S. L. R., Tarlov, M. J., Branham, M., Xu, J., MacCrehan, W., Gaitan, M., & Locascio, L. E. (2000). Derivatization of plastic microfluidic devices with polyelectrolyte multilayers. *Micro Total Analysis Systems 2000* (pp. 67–70). Dordrecht: Springer Netherlands.

Barker, S. L. R., Tarlov, M. J., Canavan, H., Hickman, J. J., & Locascio, L. E. (2000). Plastic microfluidic devices modified with polyelectrolyte multilayers. *Analytical Chemistry*, *72*(20), 4899–4903.

Becker, H., & Locascio, L. E. (2002). Polymer microfluidic devices. *Talanta*, *56*(2), 267–287.

Bhatia, S. N., & Ingber, D. E. (2014). Microfluidic organs-on-chips. *Nature Biotechnology*, *32*(8), 760–772.

Brown, M. S., & Arnold, C. B. (2010). Fundamentals of laser-material interaction and application to multiscale surface modification. *Laser Precision Microfabrication* (pp. 91–120). Berlin, Heidelberg: Springer Berlin Heidelberg.

Buriak, J. M. (2002). Organometallic chemistry on silicon and germanium surfaces. *Chemical Reviews*, *102*(5), 1271–1308. PMID: 11996538.

Castano-Alvarez, M., Fernandez-Abedul, M. T., & Costa-Garcia, A. (2005). Poly(methylmethacrylate) and Topas capillary electrophoresis microchip performance with electrochemical detection. *Electrophoresis*, *26*(16), 3160–3168.

Castaño-Álvarez, M., Pozo Ayuso, D. F., García Granda, M., Fernández-Abedul, M. T., Rodríguez García, J., & Costa-García, A. (2008). Critical points in the fabrication of microfluidic devices on glass substrates. *Sensors and Actuators B: Chemical*, *130*(1), 436–448.

Cheng, J., sheng Liu, C., Shang, S., Liu, D., Perrie, W., Dearden, G., & Watkins, K. (2013). A review of ultrafast laser materials micromachining. *Optics & Laser Technology*, *46*, 88–102.

Cheng, J.-Y., Wei, C.-W., Hsu, K.-H., & Young, T.-H. (2004). Direct-write laser micromachining and universal surface modification of PMMA for device development. *Sensors and Actuators B: Chemical*, *99*(1), 186–196.

Chichkov, B. N., Momma, C., Nolte, S., Alvensleben, F., & Tunnermann, A. (1996). Femtosecond, picosecond and nanosecond laser ablation of solids. *Applied Physics A*, *63*(2), 109–115.

Choi, H. W., Bong, S., Farson, D. F., Lu, C., & Lee, L. J. (2009). Femtosecond laser micromachining and application of hot embossing molds for microfluidic device fabrication. *Journal of Laser Applications*, *21*(4), 196–204.

Chung, C., & Lin, S. (2010). CO_2 laser micromachined crackless through holes of Pyrex 7740 glass. *International Journal of Machine Tools and Manufacture*, *50*(11), 961–968.

Courbat, J., Briand, D., & de Rooij, N. (2008). Reliability improvement of suspended platinum-based micro-heating elements. *Sensors and Actuators A: Physical*, *142*(1), 284–291.

Dahotre, N. B., & Harimkar, S. P. (2008). *Laser Fabrication and Machining of Materials* (1st ed.). NY: Springer Science + Business Media, LLC.

De Marco, C., Eaton, S. M., Suriano, R., Turri, S., Levi, M., Ramponi, R., ... Osellame, R. (2010). Surface properties of femtosecond laser ablated PMMA. *ACS Applied Materials & Interfaces*, *2*(8), 2377–2384.

Duley, W. W. (1983). *Materials processing. Laser processing and analysis of materials* (pp. 69–176). Boston, MA: Springer US.

Dyer, P. (2003). Excimer laser polymer ablation: Twenty years on. *Applied Physics A*, *77*(2), 167–173.

Erickson, D., Li, D., & Krull, U. J. (2003). Modeling of DNA hybridization kinetics for spatially resolved biochips. *Analytical Biochemistry*, *317*(2), 186–200.

Fernandez-Pradas, J., Serrano, D., Serra, P., & Morenza, J. (2009). Laser fabricated microchannels inside photo-structurable glass-ceramic. *Applied Surface Science*, *255*(10), 5499–5502.

Fleury, G., Malhaire, C., Populaire, C., Verdier, M., Devos, A., Charvet, P.-L., & Polizzi, J.-P. (2007). Mechanical cross-characterization of sputtered inconel thin films for MEMS applications. *Sensors and Actuators B: Chemical*, *126*(1), 48–51.

Fu, L.-M., Ju, W.-J., Yang, R.-J., & Wang, Y.-N. (2012). Rapid prototyping of glass-based microfluidic chips utilizing two-pass defocused CO_2 laser beam method. *Microfluidics and Nanofluidics*, *14*(3), 479–487.

Fujii, T. (2002). PDMS-based microfluidic devices for biomedical applications. *Microelectronic Engineering*, *61-62*, 907–914.

Heng, Q., Tao, C., & Tie-chuan, Z. (2006). Surface roughness analysis and improvement of micro-fluidic channel with excimer laser. *Microfluidics and Nanofluidics*, *2*(4), 357–360.

Hong, T.-F., Ju, W.-J., Wu, M.-C., Tai, C.-H., Tsai, C.-H., & Fu, L.-M. (2010). Rapid prototyping of PMMA microfluidic chips utilizing a CO_2 laser. *Microfluidics and Nanofluidics*, *9*(6), 1125–1133.

Husain, A., & Kim, K.-Y. (2009). Thermal optimization of a microchannel heat sink with trapezoidal cross section. *Journal of Electronic Packaging*, *131*(2), 021005–021005.

Jiang, H., Lin, J., Jin, S., Li, J. (2002). Micro-size LED and detector arrays for minidisplay, hyper-bright light emitting diodes, lighting, and UV detector and imaging sensor applications, US Patent 6,410,940.

Joglekar, A. P., Liu, H.-h, Meyhöfer, E., Mourou, G., & Hunt, A. J. (2004). Optics at critical intensity: Applications to nanomorphing. *Proceedings of the National Academy of Sciences of the United States of America*, *101*(16), 5856–5861.

Kam, D. H., & Mazumder, J. (2008). Three-dimensional biomimetic microchannel network by laser direct writing. *Journal of Laser Applications*, *20*(3), 185–191.

Kam, D., Shah, L., Mazumder, J. (2008). Laser based micro- and nanopackaging and assembly Vol. 6880. *Proceedings of SPIE, Ch. Laser micromachining of branching networks* (pp. 68800S-1–168800S-8).

Kamitani, A., Morishita, S., Kotaki, H., & Arscott, S. (2011). Microfabricated microfluidic fuel cells. *Sensors and Actuators B: Chemical*, *154*(2), 174–180.

Kanlayasiri, K., & Paul, B. (2004). A nickel aluminide microchannel array heat exchanger for high-temperature applications. *Journal of Manufacturing Processes*, *6*(1), 72–80, multiscale Fabrica-tion/Integration.

Kara, V., & Kizil, H. (2012). Titanium micromachining by femtosecond laser. *Optics and Lasers in Engineering*, *50*(2), 140–147.

Kathuria, Y. (2005). Laser microprocessing of metallic stent for medical therapy. *Journal of Materials Processing Technology*, *170*(3), 545–550.

Klein, S., Seidel, H., Schmidt, U., Ziegler, V., Prechtel, U., & Würtz, A. (2008). *Tungsten-titanium as advanced material for RF-MEMS switches. Advanced Microsystems for Automotive Applications* (pp. 181–189). Berlin, Heidelberg: Springer Berlin Heidelberg.

Li, L., & Achara, C. (2004). Chemical assisted laser machining for the minimisation of recast and heat affected zone. *CIRP Annals-Manufacturing Technology*, *53*(1), 175–178.

Lim, D., Kamotani, Y., Cho, B., Mazumder, J., & Takayama, S. (2003). Fabrication of microfluidic mixers and artificial vasculatures using a high-brightness diode-pumped Nd:YAG laser direct write method. *Lab on a Chip*, *3*, 318–323.

Lin, C.-H., Lee, G.-B., Lin, Y.-H., & Chang, G.-L. (2001). A fast prototyping process for fabrication of microfluidic systems on soda-lime glass. *Journal of Micromechanics and Microengineering*, *11*(6), 726.

Malek, C. G. K. (2006). Laser processing for bio-microfluidics applications (part II). *Analytical and Bioanalytical Chemistry*, *385*(8), 1362–1369.

Maselli, V., Osellame, R., Cerullo, G., Ramponi, R., Laporta, P., Magagnin, L., & Cavallotti, P. L. (2006). Fabrication of long microchannels with circular cross section using astigmatically shaped femtosecond laser pulses and chemical etching. *Applied Physics Letters*, *88*, 19.

Mei, F., Jiang, J., & Meng, W. J. (2008). Evaluation of bond quality and heat transfer of Cu-based microchannel heat exchange devices. *Journal of Vacuum Science & Technology A*, *26*(4), 798–804.

Mela, P., van den Berg, A., Fintschenko, Y., Cummings, E. B., Simmons, B. A., & Kirby, B. J. (2005). The zeta potential of cyclo-olefin polymer microchannels and its effects on insulative (electrodeless) dielectrophoresis particle trapping devices. *Electrophoresis, 26*(9), 1792–1799.

Meng, H., Liao, J., Zhou, Y., & Zhang, Q. (2009). Laser micro-processing of cardiovascular stent with fiber laser cutting system. *Optics & Laser Technology, 41*, 300–302.

Mery, E., Ricoul, F., Sarrut, N., Constantin, O., Delapierre, G., Garin, J., & Vinet, F. (2008). A silicon microfluidic chip integrating an ordered micropillar array separation column and a nano-electrospray emitter for LC/MS analysis of peptides. *Sensors and Actuators B: Chemical, 134*(2), 438–446.

Mohammed, M. K., Al-Ahmari, A., & Umer, U. (2015). Multiobjective optimization of Nd: YAG direct laser writing of microchannels for microfluidic applications. *The International Journal of Advanced Manufacturing Technology, 81*(5), 1363–1377.

Molian, P., Pecholt, B., & Gupta, S. (2009). Picosecond pulsed laser ablation and micromachining of 4H-SiC wafers. *Applied Surface Science, 255*(8), 4515–4520.

Nakashima, S., Sugioka, K., & Midorikawa, K. (2009). Fabrication of microchannels in single-crystal GaN by wet-chemical-assisted femtosecond-laser ablation. *Applied Surface Science, 255*(24), 9770–9774.

Nieto, D., Delgado, T., & Flores-Arias, M. T. (2014). Fabrication of microchannels on soda-lime glass substrates with a Nd:YVO4 laser. *Optics and Lasers in Engineering, 63*, 11–18.

Ogonczyk, D., Wegrzyn, J., Jankowski, P., Dabrowski, B., & Garstecki, P. (2010). Bonding of microfluidic devices fabricated in polycarbonate. *Lab on a Chip, 10*, 1324–1327.

Pal, P., Sato, K. (2009) Silicon microfluidic channels and microstructures in single photolithography step. In *Design, test, integration packaging of MEMS/MOEMS, 2009. MEMS/MOEMS '09. Symposium on. IEEE* (pp. 419–423). Rome, Italy.

Parida, P. R., Mei, F., Jiang, J., Meng, W. J., & Ekkad, S. V. (2010). Experimental investigation of cooling performance of metal-based microchannels. *Heat Transfer Engineering, 31*(6), 485–494.

Paulraj, P., Paul, B. K., & Peterson, R. B. (2012). Development of an adhesive-bonded counterflow microchannel heat exchanger. *Applied Thermal Engineering, 48*, 194–201.

Perrie, W., Gill, M., Robinson, G., Fox, P., & ONeill, W. (2004). Femtosecond laser microstructuring of aluminium under helium. *Applied Surface Science, 230*(1–4), 50–59.

Persano, A., Quaranta, F., Cola, A., Taurino, A., De Angelis, G., Marcelli, R., & Siciliano, P. (2010). Ta_2O_5 thin films for capacitive RF MEMS switches. *Journal of Sensors*, 5.

Philips, R. (1990). Microchannel heat sinksIn A. Bar-Cohen, & A. D. Kraus (Eds.), *Advances in thermal modeling of electronic components and systems* (Vol. 2) New York: ASME Press, Chapter 3.

Powell, J. (1998). *CO_2 laser cutting* (2nd ed.). London: Springer-Verlag.

Prajapati, Y. K., Pathak, M., & Khan, M. K. (2015). A comparative study of flow boiling heat transfer in three different configurations of microchannels. *International Journal of Heat and Mass Transfer, 85*, 711–722.

Prakash, S., Acherjee, B., Kuar, A. S., & Mitra, S. (2013). An experimental investigation on Nd:YAG laser microchanneling on polymethyl methacrylate submerged in water. *Proceedings of the Institution of Mechanical Engineers, Part B: Journal of Engineering Manufacture, 227*(4), 508–519.

Prakash, S., & Kumar, S. (2015). Fabrication of microchannels: A review. *Proceedings of the Institution of Mechanical Engineers, Part B: Journal of Engineering Manufacture*, *229*(8), 1273–1288.

Pruessner, M. W., Siwak, N., Amarnath, K., Kanakaraju, S., Chuang, W.-H., & Ghodssi, R. (2006). End-coupled optical waveguide MEMS devices in the indium phosphide material system. *Journal of Micromechanics and Microengineering*, *16*(4), 832.

Qi, H., Tao, C., Yao, L., & Tiechuan, Z. (2009). Micromachining of microchannel on the polycarbonate substrate with CO_2 laser direct-writing ablation. *Optics and Lasers in Engineering*, *47*(5), 594–598.

Qin, S.-J., & Li, W. (2002). Micromachining of complex channel systems in 3D quartz substrates using Q-switched Nd:YAG laser. *Applied Physics A*, *74*(6), 773–777.

Rais-Zadeh, M., Kohl, P. A., & Ayazi, F. (2008). MEMS switched tunable inductors. *Journal of Microelectromechanical Systems*, *17*(1), 78–84.

Rubehn, B., Bosman, C., Oostenveld, R., Fries, P., & Stieglitz, T. (2009). A MEMS-based flexible multichannel ECoG-electrode array. *Journal of Neural Engineering*, *6*(3), 036003.

Sabbert, D., Landsiedel, J., Bauer, H.-D., & Ehrfeld, W. (1999). ArF-excimer laser ablation experiments on Cycloolefin Copolymer (COC). *Applied Surface Science*, *150*(1–4), 185–189.

Snakenborg, D., Klank, H., & Kutter, J. P. (2004). Microstructure fabrication with a CO_2 laser system. *Journal of Micromechanics and microengineering*, *14*(2), 182.

Sokolowski-Tinten, K., Bialkowski, J., Cavalleri, A., von der Linde, D., Oparin, A., Meyer-ter Vehn, J., & Anisimov, S. I. (1998). Transient states of matter during short pulse laser ablation. *Physical Review Letters*, *81*, 224–227.

Stroock, A. D., Weck, M., Chiu, D. T., Huck, W. T. S., Kenis, P. J. A., Ismagilov, R. F., & Whitesides, G. M. (2000). Patterning electro-osmotic flow with patterned surface charge. *Physical Review Letters*, *84*, 3314–3317.

Tanaka, Y., Sato, K., Shimizu, T., Yamato, M., Okano, T., & Kitamori, T. (2007). Biological cells on microchips: New technologies and applications. *Biosensors and Bioelectronics*, *23*(4), 449–458.

Terry, S., Jerman, J., & Angell, J. (1979). A gas chromatographic air analyzer fabricated on a silicon wafer.. *IEEE Transactions on electron devices*, *26*(12), 1880–1886.

Tuckerman, D., & Pease, R. (1981). High-performance heat sinking for VLSI. *IEEE Electron Device Letters*, *2*(5), 126–129.

Vasilyev, N. V., Gosline, A. H., Veeramani, A., Wu, M. T., Schmitz, G. P., Chen, R. T., ... Dupont, P. E. (2015). Tissue removal inside the beating heart using a robotically delivered metal MEMS tool. *The International Journal of Robotics Research*, *34*(2), 236–247.

Voskerician, G., Shive, M. S., Shawgo, R. S., von Recum, H., Anderson, J. M., Cima, M. J., & Langer, R. (2003). Biocompatibility and biofouling of {MEMS} drug delivery devices. *Biomaterials*, *24*(11), 1959–1967.

Waddell, E. A., Locascio, L. E., & Kramer, G. W. (2002). UV laser micromachining of polymers for microfluidic applications. *Journal of the Association for Laboratory Automation*, *7*(1), 78–82.

Wang, Y.-H., Lee, C.-Y., & Chiang, C.-M. (2007). A MEMS-based air flow sensor with a free-standing micro-cantilever structure. *Sensors*, *7*(10), 2389–2401.

Wang, Z. K., Zheng, H. Y., Lim, R. Y. H., Wang, Z. F., & Lam, Y. C. (2011). Improving surface smoothness of laser fabricated microchannels for microfluidic application. *Journal of Micromechanics and Microengineering*, *21*(9), 095008.

Weck, A., Crawford, T., Wilkinson, D., Haugen, H., & Preston, J. (2007). Laser drilling of high aspect ratio holes in copper with femtosecond, picosecond and nanosecond pulses. *Applied Physics A*, *90*(3), 537–543.

Williams, O., Williams, M., Liu, C., Webb, P., Firth, P. (2009). Laser micromachining of poly-crystalline alumina and aluminium nitride to enable compact optoelectronic interconnects. In: *Electronics packaging technology conference, 2009. EPTC '09, 11th* (pp. 920–925).

Wisitsoraat, A., Patthanasetakul, V., Lomas, T., & Tuantranont, A. (2007). Low cost thin film based piezoresistive MEMS tactile sensor. *Sensors and Actuators A: Physical*, *139* (1–2), 17–22.

Yan, Y., Li, L., Sezer, K., Wang, W., Whitehead, D., Ji, L., ... Jiang, Y. (2011). CO_2 laser underwater machining of deep cavities in alumina. *Journal of the European Ceramic Society*, *31*(15), 2793–2807.

Yao, L. Y., Chen, H., & Zhang, W. (2004). Time scale effects in laser material removal: A review. *The International Journal of Advanced Manufacturing Technology*, *26*(5), 598–608.

Yuan, D., & Das, S. (2007). Experimental and theoretical analysis of direct write laser micromachining of poly-methyl methacrylate by CO_2 laser ablation. *Journal of Applied Physics*, *101*(2).

Underwater pulsed laser beam cutting with a case study

7

N. Roy, A.S. Kuar and S. Mitra
Jadavpur University, Kolkata, West Bengal, India

7.1 Introduction

Diverse application of the laser in various fields of science, technology, art and entertainment makes it an invaluable tool for mankind. The invention of the laser has made a very significant contribution to the advancement of science and technology for new materials and process techniques to meet global competition. The laser is a device that emits electromagnetic radiation through the process of stimulated emission. Neodymium-doped yttrium aluminum garnet (Nd:YAG) laser, CO_2 laser, fiber laser, diode laser and excimer lasers are widely used for various material processing applications in industries. Nd:YAG laser is one of the most commonly used lasers for drilling, marking, cutting and welding applications. The lasing medium is Neodymium (Nd^{3+}) doped Yattrium Aluminum Garnet (YAG) crystal which is pumped by light source such as arc lamp or flash lamp. The wavelength of the Nd: YAG laser radiation is 1.06 μm which lies in the near-infrared region of the electromagnetic spectrum (Küper & Stuke, 1992).

7.2 Laser as a machine tool

Laser can be considered as one of the nonconventional machining processes because of their precision of operation, low cost, localized processing, flexibility and high speed of operation. Laser beam machining is the machining process involving a beam of laser as machine tool. It is a thermal process where a laser beam is used as a heat source to remove material without mechanical engagement with workpiece. The main properties of laser radiation are high spatial and temporal coherence along with high collimation, high monochromaticity, high brightness and minimal divergence which results in producing a very high intensity laser beam. A wide range of materials can be processed by laser due to its large irradiances (up to 10^{21} W/cm^2) at the surface, with high focusability which is enough to evaporate any material or even to start a nuclear fusion reaction (Küper & Stuke, 1992). In the laser beam machining process, material removal is not dependent on mechanical, physical or electrical properties of the workpiece but on thermo-optical properties of the material (Davim, 2013) (Fig. 7.1).

Microfabrication and Precision Engineering. DOI: http://dx.doi.org/10.1016/B978-0-85709-485-8.00007-3

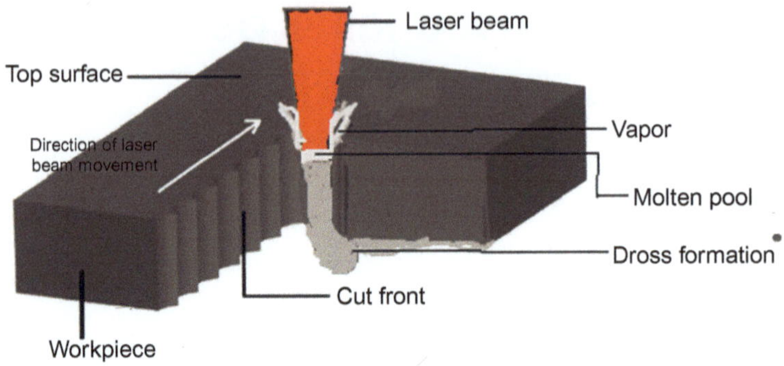

Figure 7.1 Laser beam cutting.

The beam wavelength has a great impact on absorptivity of laser light on a workpiece surface, which is relatively important for laser material processing. A discrete spatial profile, termed as transverse electromagnetic mode, is exhibited by a cross-section of laser beam. Generally fundamental TEM_{00} mode is used for laser beam machining operations, having Gaussian spatial distribution. The output of laser can either be constant amplitude or periodic, commonly known as continuous wave (CW) mode and pulsed beam mode respectively. Normal pulsing, Q-switching and mode locking are the various pulsing modes during pulsed laser operation. A high energy density is achieved in short pulse duration laser operation after threshold energy is reached. Pulse duration and pulse repetition rate are the most influencing process parameters for pulsed laser machining (Ready, 1997). During laser beam machining one of two geometrics are generally used i.e. either a moving target or a moving beam.

7.3 Laser material interaction

Reflection, refraction, absorption, scattering and transmission are the physical phenomenans that take place when the electromagnetic wave (laser beam) interacts with a workpiece surface. Linear and nonlinear absorption are the most important phenomena of the laser matter interaction. Fig. 7.2 shows a schematic of interaction of a laser beam with work material.

Absorption at irradiate surface not only depends on the wavelength of the laser radiation but also on other factors such as incident angle, surface roughness and temperature of the solid/machine zone. The absorption coefficient of material is proportional to electrical resistivity whereas the absorption by target substrate increases with decrease in wavelength. The probability of nonlinear absorption increases significantly in proportion to growing laser intensity. This aforesaid relationship is the reason behind multiphoton excitation process at constant laser fluence for a shorter laser–material interaction time. Absorption of laser energy can be described as interaction between electrons and nuclei lattice of target material.

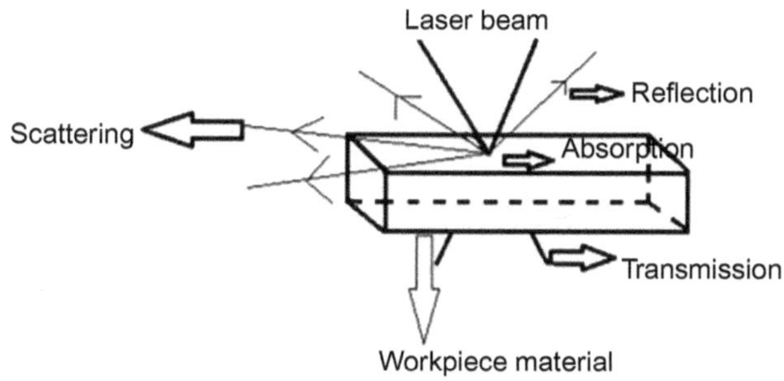

Figure 7.2 Interaction of laser beam with workpiece material.
Source: Reprinted from Roy, N., Kuar, A.S., Mitra, S., Acherjee, B. (2015). *Nd:YAG laser microdrilling of SiC-30BN nanocomposite: Experimental study and process optimization*. In: U. S. Dixit & S. N. Joshi (Eds.), *Lasers based manufacturing. Topics in mining, metallurgy, materials engineering* (pp. 317–341) Springer, India. Chapter 17, ISBN 978-81-322-2352-8, 1 with permission. Copyright Springer (Roy, Kuar, Mitra, and Acherjee, 2015).

A force is exerted by an optoelectric field of incident electromagnetic radiation by which the nuclei lattice and electrons get vibrational motion. Thus the absorbed radiation results in the excess energy of the electrons, such as kinetic energy of the free electrons, excitation energy of the bound electrons (Von, 1987). Excess energy of the electrons is damped by collision with vibrating lattice of nuclei and some incident laser energy is transferred to the lattice leads to generation of heat. By Beer-Lambert law (Steen, 1991), the absorption of laser radiation in the target substrate is generally expressed as,

$$I(Z) = I_0 e^{-\mu z}$$

where I_0 is the incident intensity, $I(z)$ is the intensity at depth z and μ is the absorption coefficient. The intensity of the laser radiation gets faded along with the depth inside the material. The length over which the significant reduction of intensity of the laser radiation takes place is known as attenuation length (L) and expressed as reciprocal of the absorption coefficient (μ) of target substrate (Welch & Gardner, 2002),

$$L = \frac{1}{\mu}$$

Generally for opaque material, the absorptivity can be defined as a fraction of incident radiation, absorbed at normal incidence which can be expressed (Duley, 1983) as,

$$A = 1 - R$$

where, A is absorptivity of irradiate substrate and R is the reflectivity of irradiate substrate.

Various physical effects in the material include heating, melting and vaporization of the substrate has occurred, depending on the temperature distributions in the material. Thermal energy is transported through the substrate by thermal conduction, convection and radiation from the surface of the substrate. Phase transformations such as surface melting and evaporation are taking place by sufficiently high laser intensity which is generally referred as melting (I_m) and vaporization (I_v) threshold intensity. During laser irradiation, the melting starts at a certain temperature when laser intensity reaches melting threshold energy and creates a solid–liquid interface at the surface. The solid–liquid interface moves away from the surface along the depth with increase in laser energy density and a melt pool is formed. The depth of melting is decreased during pulsed off time (Allmen, 1983). The depth of melting (Z max) reaches maximum when the laser irradiates surface temperature reaches at boiling point. Evaporation takes place with a further increase in laser power density and/or pulse on time, resulting in material removal from irradiate region. The liquid–vapor interface has to move inside the material with continuous laser irradiation after the vaporization is initiated. Depth of vaporization (Steen, 1991), mass of material removed per unit time and velocity of the liquid–vapor interface can be calculated by

$$\dot{m} = V_s \rho$$

where $\dot{m}$ the mass of material removed per unit time, V_s is the velocity of the liquid–vapor interface and ρ is the density.

$$d = V_s t_p$$

where, t_p is the pulse time and d is the depth of vaporization.

$$V_s = \frac{H}{\rho(cT_b + L_v)}$$

where, H is the absorbed laser power, T_b is the boiling temperature at surface, and L_v is the latent heat of vaporization. Evaporation-induced recoil pressure exceeds the highest possible value of surface tension pressure during laser beam processing, i.e., drilling, cutting, marking, welding, cladding, etc. During laser material processing, recoil pressure plays an important role in melt expulsion from machining zone (Steen, 1991). At high laser irradiance ($I \geq 109$ W/cm^2) the vapor or the ambient gas becomes ionized due to the interactions between the resulting vapor and the incident laser beam, termed as plasma. Plasma plume forms a shield over the machining area and reduces the energy available to the workpiece. When the incident laser energy on the target surface is sufficiently large to exceed the boiling temperature, rapid vaporization starts. In the photo-thermal ablation process, material is removed by thermal stresses and surface vaporization. Whereas in photo-chemical ablation, material removal takes place by molecular fragmentation without significant thermal damage by the energy of the incident photon causes the direct bond breaking of the molecular chains in the organic materials (Bäuerle, 2000). The physical phenomena of material removal mechanism are shown in Fig. 7.3.

Utilization of the laser in various manufacturing process is increasing rapidly. The applications of lasers have been demonstrated in many casting, forming, joining, and machining processes. The laser beam machining process is widely used in industries. Laser beam cutting, drilling and welding are the most common applications in automotive, medical and aircraft industries.

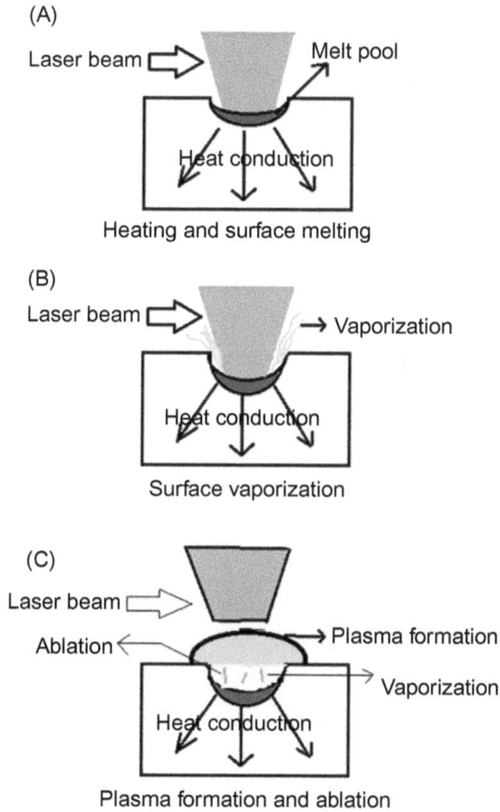

Figure 7.3 Various governing physical phenomena of material removal mechanism. (A) Heating and surface melting, (B) Surface vaporization, and (C) Plasma formation and ablation.
Source: Reprinted from Roy, N., Kuar, A.S., Mitra, S., Acherjee, B. (2015). *Nd:YAG laser microdrilling of SiC-30BN nanocomposite: Experimental study and process optimization.* In: U. S. Dixit & S. N. Joshi (Eds.), *Lasers based manufacturing. Topics in mining, metallurgy, materials engineering* (pp. 317–341) Springer, India. Chapter 17, ISBN 978-81-322-2352-8, 1 with permission. Copyright Springer.

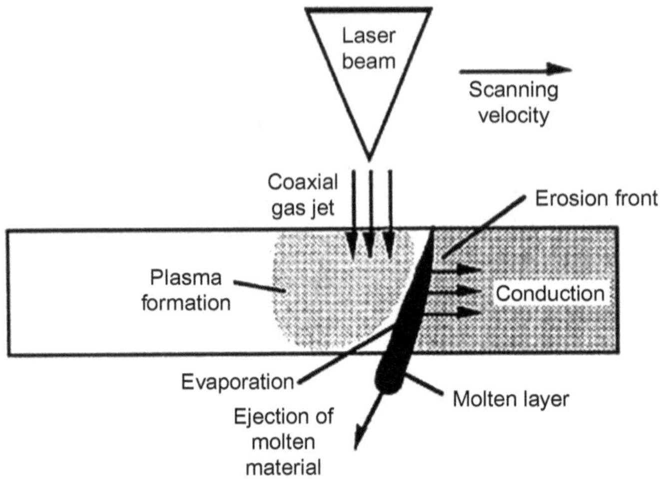

Figure 7.4 Schematic of laser through-cutting.
Source: Reprinted from Chryssolouris, G. (1991). *Laser machining: Theory and practice*. Springer, New York, NY with permission. Copyright Springer (Chryssolouris, 1991).

7.4 Laser beam cutting

Nowadays the laser beam cutting technique is the most established laser material processing technology to obtain desired geometry. It is the two-dimensional laser beam machining technique where the motion of either focused beam and/or the workpiece are relative to each other. Advantages of laser beam cutting can be divided into two categories (Powell, 1993), i.e., process characteristics and cut quality. The schematic diagram of laser beam cutting is illustrated in Fig. 7.4.

7.4.1 Process characteristics

It is a noncontact, thermal, high speed, and repeatable method to process a wide range of materials with different thickness.

1. There is no tool wear and mechanical damages on and adjacent to machining zone.
2. The workpiece does not need to be clamped or accurately positioned. Instead of that it must be properly focused or marginally de-focused as per machining requirement.
3. Most of the laser beams cutting machines are CNC-controlled to ensure the desired geometry and cutting speed.

7.4.2 Cut quality characteristics

1. A very narrow kerf width ($\sim$0.1 mm) with clean and smooth edges can be achieved during laser beam cutting.
2. Smaller heat affected zone (HAZ) with the least thermal stress can be obtained due to localized heating and melting during laser beam cutting.

7.4.3 Principles of laser beam cutting

During laser beam cutting, the laser beam interacts with workpiece material. The laser energy absorbed by the workpiece material transforms into thermal energy which heats up, subsequently melts and vaporizes the irradiate region of the workpiece locally/throughout the thickness, thus creating a cutting front. Melt pool or vapor is expelled from the machining zone by the hydrodynamic nature of melt pool, vapor and by assist gas axially/coaxially flown. Assist gas also cools the machining zone as well as takes part in chemical reactions such as oxidation or carburization of material to help in enhancement of material removal if reactive in nature (Chryssolouris, 1991).

7.4.3.1 Different types of laser beam cutting

Based on the role of assist gas and interaction between the laser beam and the workpiece, the substrate laser beam cutting operation can be classified into four groups, i.e., laser sublimation cutting, controlled fracture technique, fusion cutting, and reactive fusion cutting. Besides these four major laser cutting application/technique, there are few relatively new laser cutting applications evolved in recent times e.g. laser micro cutting, underwater laser cutting etc.

7.4.3.1.1 Laser sublimation cutting

During laser sublimation cutting, latent heat provided by laser energy transforms the target material to the vaporization point and ablates into a vapor state along the cutting seam. Materials with low thermal conductivity and lesser vaporization temperatures are suitable to process by this method. Extremely clean cut edges are achieved due to material removal takes place by direct phase change to vapor state. In this process all the thermal energy is used in evaporation process without any heat conduction in the target region to create a melt pool. Generally nonreactive assist gas flows coaxially to remove the vapor from machining zone (Biyikli & Modest, 1988).

7.4.3.1.2 Controlled fracture technique

In controlled fracture technique, incident laser energy produces mechanical stress on the impingement region of workpiece material which causes the separation of material from the surface of workpiece in a controlled manner along the path of laser beam movement. Less laser energy is required in controlled fracture technique, as material is removed by crack propagation and not by melting or evaporation. A high plastic compressive stress is generated around the laser spot diameter which relaxed with movement of laser beam, results in generation of residual tensile stress on the upper surface of irradiate region. This tensile stress propagates the crack from upper to lower surface of the substrate (Caristan, 2003).

7.4.3.1.3 Laser fusion cutting

Movement of laser beam relative to workpiece makes a straight or curved cut profile during laser fusion cutting. High intensity laser energy from impinging beam absorbed by workpiece substrate melts the metarial throughout the thickness to create a cutting front. Generally, high pressure, nonreactive assist gas, such as argon, or helium is used coaxially to transform momentum to the thin melt pool. Melt film is ejected from the bottom of the kerf as droplets due to vertical acceleration when momentum of the thin melt film exceeds the surface tension force. Otherwise dross formation occurs (Schulz, Simon, Urbassek, & Decker, 1987).

7.4.3.1.4 Reactive fusion cutting

It is a variation of fusion cutting where high pressure reactive assist gas is used instead of inert gas, resulting in exothermic reaction with reactive assist gas (mainly oxygen), when the top surface of workpiece substrate is heated. Exothermic reaction adds another source of heat along with laser irradiation to enhance the rise of molten layer temperature to evaporation, resulting in faster material removal from the machining zone by evaporation. In addition, the melt pool from the bottom of the machining zone removed by the frictional force works between reactive assist gas and the molten layer. The cutting front propagates through the thickness of the workpiece material by further melting at solid–liquid interface below the cutting front in the direction of laser beam movement. Oxidation plays an important role during oxygen-assisted reactive laser cutting. Enormous exothermic heat generation during oxygen-assisted laser cutting operation nullifies the conductive thermal loss of base material. Higher cutting speed is achieved by production of oxide film which has higher absorptivity and higher fluidity of oxide slag in erosion front. This laser cutting technique is mainly used to cut thick metal sheets (O'Neill & Gabzdyl, 2000).

7.4.3.1.5 Laser cutting at different assisted medium

Sometimes the laser cutting process is performed at various atmospheric conditions to obtain the desired cut quality. Laser cutting technique is sometimes performed in a submerged condition or in a preheated condition, or in a vacuum or sometime in a mixture of various conditions to get better cut characteristics.

7.4.3.1.6 Laser beam microcutting

In modern technologies, miniaturization is an important trend to fabricate microparts for biotechnological, microelectronics, telecommunication, MEMS, and medical applications. Laser beam micromachining is used to get microproducts with better flexibility in dimensional design. Ablation and laser-assisted chemical etching are the mechanisms behind material removal in micron or in submicron range during laser beam micromachining operation. Generally laser in NIR (1.06 μm) to deep UV(150 nm) region wavelengths with short pulse duration (<100 ns to few fs) is used for laser beam micromachining operation. Direct writing and mask projection are the most commonly used laser beam micromachining techniques used in industries. Resistor trimming, coronary stent fabrication are some of the applications of laser beam microcutting (Gower, 2000).

7.4.4 Application of laser beam machining

In the manufacturing industry, from automotive to medical, laser beam machining is used to process a wide range of materials with precision in a cost effective way. In the automobile industry, the LBM process is used to manufacture products from airbags to body frames. The laser beam hardening process is used for hardening gear teeth. In aerospace industries various laser beam machining processes are used to produce jet engine components, polymer cabling insulation, titanium ducting, fuselage skin stiffeners, fabrication of body parts, etc. In medical or biomedical industries laser material process is utilized in the fabrication of microfluidic devices, coronary stents, and for manufacturing of various surgical equipment.

7.5 Underwater laser beam machining

Undesired quality aspects, i.e., kerf taper, dross formation, HAZ, recast layer, burr formation, difficulties in debris removal and desired surface morphology during laser cutting are sometimes totally removed or partially corrected by irradiating the workpiece material at submerged condition.

7.5.1 Advantages of laser beam cutting at submerged condition

1. Narrow HAZ is generated due to low thermal load on workpiece.
2. Strong hydrodynamic force generated by bubble dynamics, carries away the debris from machining zone.
3. Shorter wavelength of irradiate laser beam at liquid helps to increase the absorptivity of workpiece at submerged condition.
4. Risk of atmosphere contamination is reduced at submerged condition.

Sometimes water solution of salts or bases or acids are used during submerged laser beam machining to improve the etch rate and finish quality. To improve the wetting quality of water film on workpiece surface, organic additives are added.

7.5.2 Material removal mechanism of nanosecond pulsed laser beam cutting at submerged condition

Along with the absorption, heating, melting/vaporization, and ablation, some other phenomenon also occur during nanosecond pulsed laser beam cutting at submerged condition.

High drag force and low settling velocity in liquid help quick removal of debris from machining zone. Formation of bubbles and thermal gradient generated by laser induced heat cause convection in liquid. In some circumstances fluid in dynamic condition is recommended to avoid scattering and absorption of laser light by debris. Liquid molecules may be excited, ionized, and dissociated by laser-induced plasma temperatures and plasma UV/IR radiation, thus become chemically active (Kruusing, 2004). Restriction of plasma expansion in confined condition enhances the action of laser irradiation. Dissolution of workpiece and debris takes place in supercritical water generated by high intensity laser beam. Collapse of the vapor bubbles generates mechanical impact of microjets resulting in easy removal of debris from machining zone (Li & Achara, 2004).

7.5.3 Development of different types of liquid-assisted laser beam machining

Current research on different types of liquid-assisted laser beam machining, i.e., laser ablation in submerged condition, underwater waterjet guided laser beam machining, underwater gas-assisted laser beam machining, chemical-assisted laser beam machining is carried out by researchers. Some researchers have found that laser beam cutting in chemical solution produces better surface quality than processing in pure water (Datta, Romankiw, & Vigliotti, 1987). Electrochemical dissolution/localized breakdown is enhanced by laser beam in neutral salt solution. Optical absorptivity of workpiece substrate increases in water due to higher refractive index of water (Kim & Lee, 2001).

7.5.3.1 Laser beam cutting in submerged condition

In many underwater laser beam cutting techniques, the laser ablation at submerged condition can be said to be the simplest process to cool the workpiece by water during the cutting induced by laser. A schematic view is illustrated in Fig. 7.5. In this technique layer or thin film of stagnant water creates better natural convection than air, which results in reduction in temperature gradient and thermal stress on the top surface of workpiece substrate, thereby reducing the possibility of subsurface crack formation, results in micro crack free machined zone with smaller heat affected zone.

7.5.3.2 Underwater assist gas jet/waterjet assisted laser beam cutting

In this process a high pressure assist gas or waterjet coaxially flows along with high intensity laser beam from a specially designed nozzle for impingement on the top surface of submerged workpiece. Sometime an off axial waterjet is impinged on the irradiation spot by a specially designed delivery nozzle. A schematic view is illustrated in Fig. 7.6. Removal of molten pool from the bottom of the kerf is easier in this process (Mullick et al., 2013). Sometimes a waterjet is directed into the cutting region to increase the thermal efficiency of the laser beam in underwater conditions. Here high intensity laser beam is generally used. The direction of the waterjet in

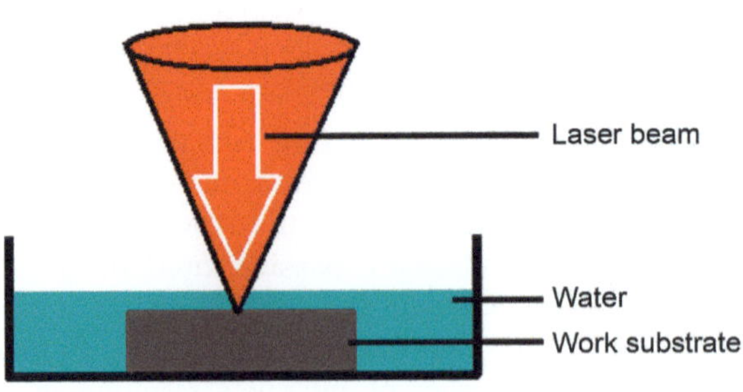

Figure 7.5 Underwater laser ablation process.

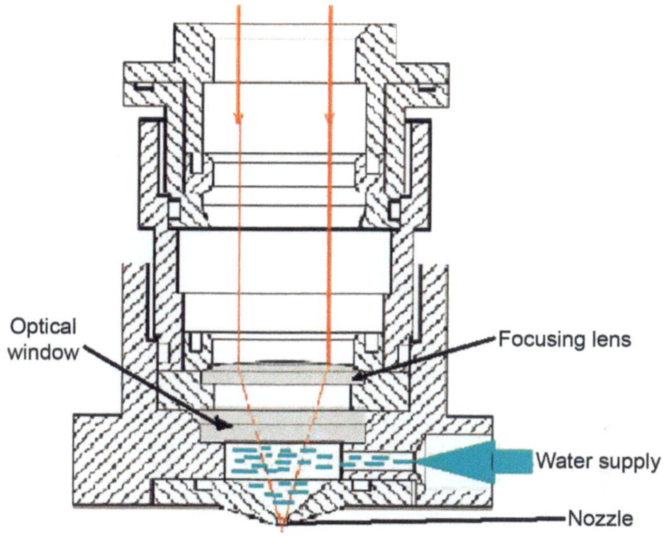

Figure 7.6 Schematic of the underwater cutting head and transmission of laser beam through water.
Source: Reprinted from Mullick, S., Madhukar, Y.K., Roy, S., Kumar, S., Shukla, D.K., & Nath, A.K. (2013). Development and parametric study of a water-jet assisted underwater laser cutting process. *International Journal of Machine Tools & Manufacture*, *68*, 48–55 with permission. Copyright Elsevier Ltd.

the fusion region results in desired cut quality with higher cutting speed (Owaki, Uehara, & Tsuchiya, 1999).

7.5.3.3 Molten salt-jet-guided/chemical laser beam

In this process, instead of a waterjet a molten salt jet or chemical is used as assisted medium which is directed in fusion zone. Molten salt jet/chemical is used for laser reactive cutting due to higher heat capacity and higher oxygen content by molten salt, which facilitates the removal of the material rapidly. In that instance, high temperature chemical etching of the workpiece surface also takes place, which increases the rate of material removal.

7.5.3.4 Water jet following the laser beam

In this process, the purpose of the waterjet is to confine the assist gas in order to increase the assist gas pressure in kerf which results in improvement in melt removal. This process is suitable for both the high and low intensity laser beam due to absence of the waterjet in the fusion zone (Stephen, 2011). Schematic diagram of this process given below in Fig. 7.7.

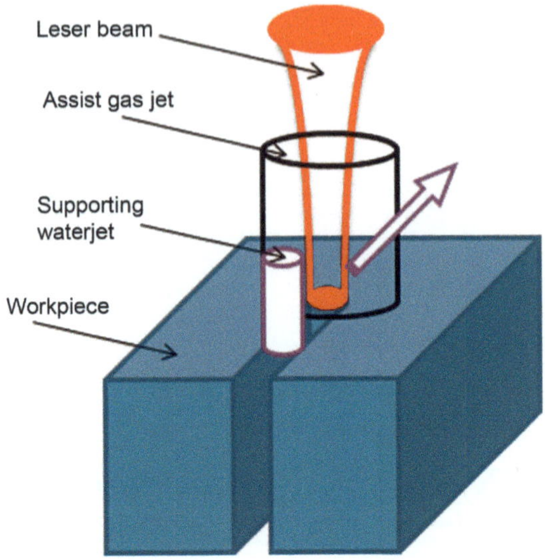

Figure 7.7 Method of laser cutting supported with a water jet.

7.5.3.5 Laser beam cutting of opaque material at partially submerged condition

In this process during cut melt redeposition on the cutting front at backside is reduced or absent. Liquid prevents the adhesion of debris and dross at the backside of the workpiece.

7.5.3.6 Laser beam cutting of transparent material at partially submerged condition

Because of low absorption of the light, inorganic optical materials are difficult to machine with conventional laser cutting techniques. Desired microstructuring can be achieved by this technique. The damage and the optical breakdown threshold at a glass plate in contact with water were 2.5 times higher compared to those in air. Debris is removed efficiently by liquid motion, gravity force and debris dissolution. In this process, the etching threshold value is comparatively less than in air, and etching occurs well below the optical damage threshold of the materials. Absence of plasma shielding results in increased laser cutting efficiency (Kawaguchi, Sato, and Narazaki, 2005). Schematic diagram of this process is shown in Fig. 7.8.

7.5.3.7 Hybrid waterjet laser cutting

In this process a defocused laser beam travels along the workpiece followed by a high pressure waterjet. Here the waterjet flows off axially along the laser beam but enters in the fusion zone. In this process, the laser in defocused position is used to heated up the soft solid state and the material is removed by the impact of the following high

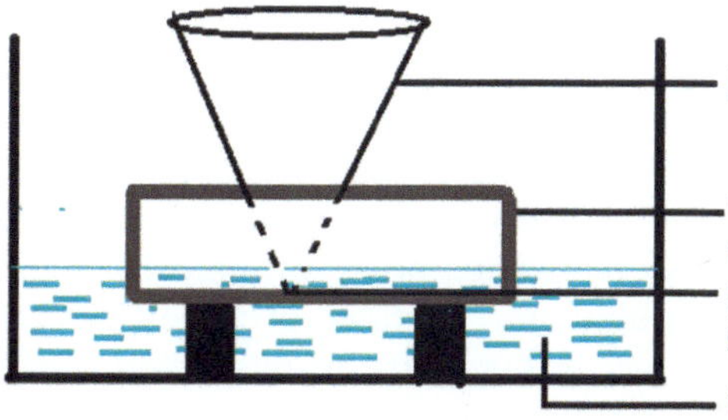

Figure 7.8 Schematic of laser back side etching of transparent material.

pressure waterjet. For that reason, the amount of thermal damage on the workpiece surface and sub-surface regions is minimized. A specially designed mechanism is attached to the workpiece nozzle to change its inclination angle with the laser beam and standoff distance from the workpiece. During this process, compressive stress is generated on the laser-induced heated zone. Forced convection due to waterjet flow generates the quenching effect to the heated region, resulting in formation of tensile stress on the top surface of the workpiece. That tensile stress is the reason behind propagation of cracks leading to removal of material from workpiece specimen with flow of water as microchips (Kalyanasundaram, Shrotriya, & Pal, 2010).

7.6 Pulsed IR laser ablation of Inconel 625 superalloy at submerged condition: A case study

A parametric study on laser beam cutting of Inconel 625 superalloy at submerged condition has been carried out. The effect of the height of water column along with the other laser beam cutting parameters such as lamp current, pulse frequency, pulse width and cutting speed, on kerf width, HAZ width and depth of cut is investigated. Height of the water column is varied from 1 mm to 5 mm to see the effect on machining responses. In this study lamp current is consider as an function of average power or working power. Design of experiment based on response surface methodology (RSM) has been implemented here to carry out the experiments.

7.6.1 Experimental setup

Inconel 625 with 80 mm × 80 mm × 0.9 mm dimension has been considered as workpiece material for experimental purposes. In this study water is used as assisted medium at stagnant condition. A pulsed Nd:YAG laser-based CNC machining system (SLT-SP- 2000), manufactured by M/s Sahajanand Laser Technology, India, is used for the experimental study. The CNC controller consists of x–y–z axis and a controlling unit. One stepper motor is attached to each of the axes and connected to the controlling unit. A CCD camera is attached over the CNC z-axis controller unit which, along with the CCTV monitor, is used for viewing the location of the workpiece and also for checking the proper focusing condition of surface of the workpiece. The schematic diagram of the underwater pulsed Nd:YAG laser-based CNC machining system is shown in Fig. 7.9.

The workpiece holding unit for underwater laser beam cutting at static condition has been designed and fabricated at the workshop. Transparent PMMA sheets with different thicknesses are used for fabrication of the workpiece holding device. This specially designed workpiece holding unit is placed over the CNC controlled work table. Image of this unit is shown in Fig. 7.10. In this experimental procedure a water film from 1 to 5 mm is used to carry out the experiments. For the study, a laser beam comes from outside, first interacts with water film then travels a distance through the water to strike the workpiece substrate. The workpiece is held in submerged condition by pouring water externally. The height of the water column over the workpiece is maintained precisely by placing the slip gauge over the workpiece material. From the dimension of the slip gauge, the height of the water column over the surface of the workpiece material is controlled as required. A steady state of water column is maintained during the experiment. De-ionized water at room temperature (20–22°C) has been used as water medium because of its nonreactive nature to workpiece material even at elevated temperatures and easy availability. The bubbles accumulated near the fusion region are removed from the machining zone by a thin stick after each experimental run, followed by a few minute wait for stabilization of stagnant water. Jammer is used for precisely fixed the

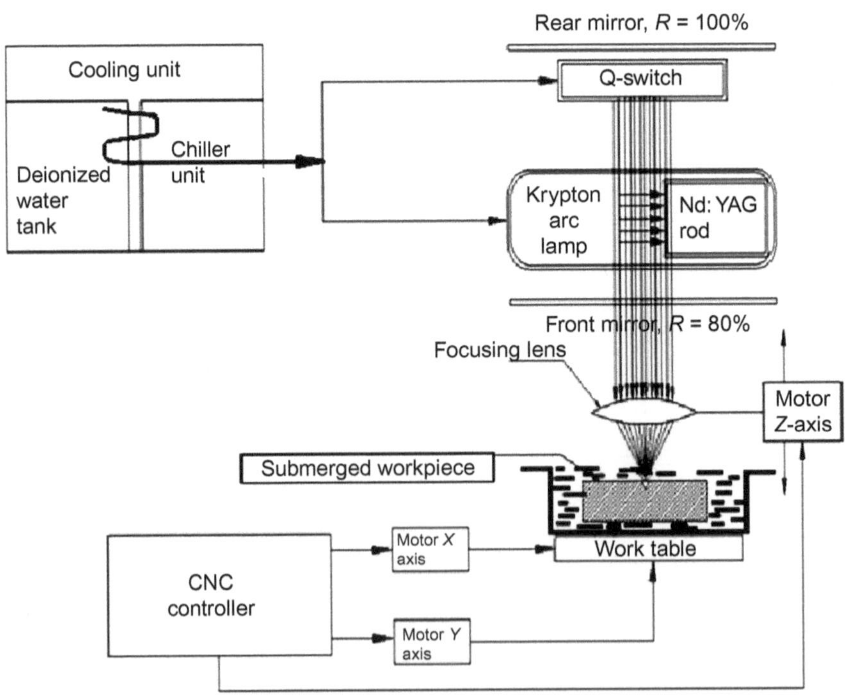

Figure 7.9 Schematic diagram of Pulsed Nd:YAG laser beam machining setup at submerged condition.

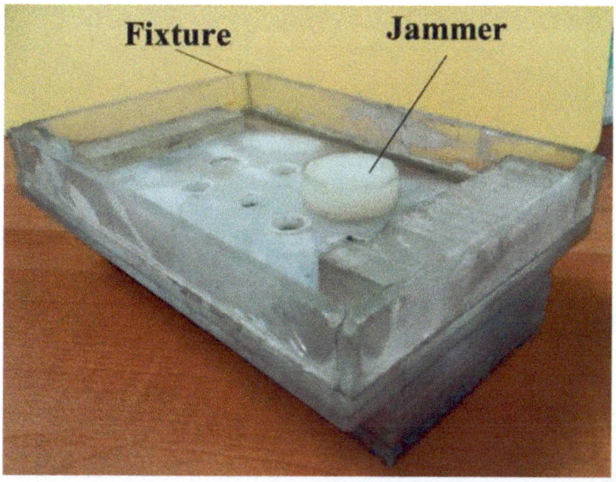

Figure 7.10 Specially designed tool holding unit for underwater machining.

workpiece with proper alignment. The schematic diagram of the underwater pulsed Nd:YAG laser-based CNC machining system is shown in Fig. 7.10.

Experiments have been carried out according to the central composite rotatable second-order design based on RSM. Response surface modeling is used to establish the mathematical relationship between the response, y_u and the various machining parameters, with the eventual objective of determining the optimum operating conditions for the system. The ranges of these controllable process variables are selected on the basis of trial experiments conducted by using a one factor at a time approach.

Usually, a second order polynomial equation is used in RSM,

$$y_u = \beta_0 + \sum_{j=1}^{k} \beta_j x_j + \sum_{j=1}^{k} \beta_{jj} x_j^2 + \sum_{i<j} \sum_{j=2}^{k} \beta_{ij} x_i x_j \qquad (7.1)$$

Here y_u is the corresponding response, x_i is the coded value of the ith machining parameter, k is the number of machining parameters and β_i, β_{ii}, β_{ij} are the second-order regression coefficients. Ranges of all controllable input process variables for underwater laser machining, i.e., lamp current, pulse frequency, pulse width, cutting speed, and height of water level as selected are listed in Table 7.1.

Images of cut region have been captured by Olympus (Model STM 6-F10-3; No. OH13697) optical measuring microscope at 20 × magnification. Here heat affected zone (HAZ) is measured by measure the area of change in colour adjacent to machined zone. Then the machining responses are evaluated by image analysis software. Calculation of HAZ width is governed by the formula given below,

$$HAZ \text{ width } (mm) = \frac{HAZ_{\text{Top}} - Kerf \text{ width}_{\text{Top}}}{2} \qquad (7.2)$$

Machining responses are measured at five different positions along the kerf for all experiments and an average is taken for further analysis to reduce the measurement error. Experimental results are listed in Table 7.2.

Table 7.1 Process parameters levels

Process parameter and symbol	Unit	Value of α	Levels	
			Lowest (−1)	Highest (+1)
Lamp current (X_1)	Amp	2.00	22	26
Pulse frequency (X_2)	kHz		4	8
Pulse width (X_3)	%		4	8
Cutting speed (X_4)	mm/sec		1.5	2.5
Height of water level (X_5)	mm		2	4

7.6.2 Development of mathematical model

Response surface modeling (central composite design) is used to establish the mathematical relationship between the response and variable process parameters. The second order polynomial equations are given below:

$$\text{Kerf width} = -0.925 + 0.0605\,X_1 + 0.0862\,X_2 + 0.0082\,X_3 + 0.0031\,X_4 - 0.0052\,X_5 - 0.000645\,X_1 * X_1 - 0.000417\,X_2 * X_2 - 0.000413\,X_3 * X_3 + 0.00291\,X_4 * X_4 - 0.00112\,X_5 * X_5 - 0.003672\,X_1 * X_2 - 0.000105\,X_1 * X_3 - 0.00152\,X_1 * X_4 + 0.001680\,X_1 * X_5 - 0.000213\,X_2 * X_3 + 0.00106\,X_2 * X_4 - 0.003453\,X_2 * X_5 + 0.00037\,X_3 * X_4 - 0.000627\,X_3 * X_5 + 0.00141\,X_4 * X_5 \tag{7.3}$$

$$\text{Depth of cut} = -0.1572 + 0.01547\,X_1 + 0.00207\,X_2 + 0.00202\,X_3 + 0.01059\,X_4 - 0.00984\,X_5 - 0.000391\,X_1 * X_1 - 0.000107X_2 * X_2 - 0.000585\,X_3 * X_3 - 0.000897\,X_4 * X_4 - 0.000114\,X_5 * X_5 + 0.000076\,X_1 * X_2 + 0.000251\,X_1 * X_3 + 0.000138\,X_1 * X_4 + 0.000293\,X_1 * X_5 - 0.000054\,X_2 * X_3 - 0.001755\,X_2 * X_4 + 0.000109\,X_2 * X_5 - 0.000417\,X_3 * X_4 + 0.000109\,X_3 * X_5 + 0.000404\,X_4 * X_5 \tag{7.4}$$

$$\text{HAZ width} = 0.63058 + 0.007893\,X_1 - 0.001267\,X_2 + 0.001022\,X_3 - 0.002791\,X_4 - 0.005027\,X_5 - 0.001270\,X_1^2 - 0.000626\,X_2^2 + 0.001061\,X_3^2 - 0.000879\,X_4^2 + 0.000830\,X_5^2 + 0.003362\,X_1 * X_2 + 0.000316\,X_1 * X_3 + 0.000607\,X_1 * X_4 + 0.000820\,X_1 * X_5 + 0.000844\,X_2 * X_3 - 0.002134\,X_2 * X_4 + 0.001911\,X_2 * X_5 + 0.000410\,X_3 * X_4 - 0.01027\,X_3 * X_5 - 0.000711\,X_4 * X_5 \tag{7.5}$$

7.6.3 ANOVA analysis

Analysis of variance (ANOVA) and subsequent *f*- and *p*-value tests have been carried out on machining responses, i.e., kerf width, depth of cut, and HAZ width to test the adequacy of the corresponding developed mathematical models. From the Table 7.3 it is observed that *p* value of the model for all the machining responses is less than 0.05 (i.e., $\alpha = 0.05$, or at 95% confidence level) indicating a statistical significance. From the *p* value it is also observed that all the machining parameters are significant individually along with their square terms and 2-way

Table 7.2 **Experimental results**

Lamp current (Amp)	Pulse frequency (kHZ)	Pulse width (%)	Cutting speed (mm/sec)	Height of water level (mm)	Kerf width (mm)	Depth of cut (mm)	Heat affected zone (HAZ; mm)
22	4	4	1.5	2	0.1346	0.0271	0.06582
26	4	4	1.5	2	0.1836	0.0243	0.07205
22	8	4	1.5	2	0.0933	0.0274	0.05652
26	8	4	1.5	2	0.1039	0.0230	0.07403
22	4	8	1.5	2	0.1141	0.0272	0.06717
26	4	8	1.5	2	0.1886	0.0275	0.07583
22	8	8	1.5	2	0.0860	0.0274	0.06133
26	8	8	1.5	2	0.0963	0.0257	0.07983
22	4	4	2.5	2	0.1193	0.0315	0.06440
26	4	4	2.5	2	0.1571	0.0255	0.07280
22	8	4	2.5	2	0.0912	0.0232	0.04652
26	8	4	2.5	2	0.0910	0.0189	0.06787
22	4	8	2.5	2	0.1052	0.0274	0.06833
26	4	8	2.5	2	0.1640	0.0280	0.07682
22	8	8	2.5	2	0.0843	0.0198	0.05060
26	8	8	2.5	2	0.0854	0.0216	0.07402
22	4	4	1.5	4	0.1355	0.0220	0.05582
26	4	4	1.5	4	0.2294	0.0204	0.06347
22	8	4	1.5	4	0.0941	0.0233	0.05132
26	8	4	1.5	4	0.1156	0.0225	0.07473
22	4	8	1.5	4	0.1477	0.0243	0.05177
26	4	8	1.5	4	0.1997	0.0241	0.06120
22	8	8	1.5	4	0.0866	0.0229	0.05150
26	8	8	1.5	4	0.0949	0.0248	0.07607
22	4	4	2.5	4	0.1301	0.0265	0.05097
26	4	4	2.5	4	0.2176	0.0231	0.06157
22	8	4	2.5	4	0.0919	0.0187	0.03750

(*Continued*)

Table 7.2 **(Continued)**

Lamp current (Amp)	Pulse frequency (kHZ)	Pulse width (%)	Cutting speed (mm/sec)	Height of water level (mm)	Kerf width (mm)	Depth of cut (mm)	Heat affected zone (HAZ; mm)
26	8	4	2.5	4	0.0942	0.0188	0.06297
22	4	8	2.5	4	0.1270	0.0257	0.04867
26	4	8	2.5	4	0.2043	0.0268	0.06110
22	8	8	2.5	4	0.0756	0.0161	0.04193
26	8	8	2.5	4	0.0824	0.0213	0.06718
20	6	6	2	3	0.0945	0.0225	0.04162
28	6	6	2	3	0.1526	0.0222	0.07380
24	2	6	2	3	0.1837	0.0307	0.06198
24	10	6	2	3	0.0707	0.0231	0.05858
24	6	2	2	3	0.1255	0.0189	0.06557
24	6	10	2	3	0.1291	0.0196	0.06850
24	6	6	1	3	0.1358	0.0299	0.06588
24	6	6	3	3	0.1378	0.0255	0.05267
24	6	6	2	1	0.1188	0.0309	0.07733
24	6	6	2	5	0.1400	0.0254	0.05488
24	6	6	2	3	0.1428	0.0286	0.06233
24	6	6	2	3	0.1349	0.0287	0.06232
24	6	6	2	3	0.1311	0.0289	0.06472
24	6	6	2	3	0.1359	0.0278	0.06167
24	6	6	2	3	0.1289	0.0301	0.06405
24	6	6	2	3	0.1371	0.0270	0.06360
24	6	6	2	3	0.1238	0.0301	0.06412
24	6	6	2	3	0.1277	0.0279	0.06410
24	6	6	2	3	0.1264	0.0291	0.06292
24	6	6	2	3	0.1328	0.0281	0.06130

Table 7.3 ANOVA analysis of machining responses

Machining responses		Kerf width		Depth of cut		HAZ width	
Source	DF	*f* value	*p* value	*f* value	*p* value	*f* value	*p* value
Model	20	45.47	0.000	37.49	0.000	196.8	0.00
Linear	5	156.67	0.000	52.05	0.000	641.99	0.00
Lamp current	1	166.32	0.000	5.98	0.020	2040.49	0.00
Pulse frequency	1	576.06	0.000	139.23	0.000	52.53	0.00
Pulse width	1	5.89	0.021	6.91	0.013	34.19	0.00
Cutting speed	1	10.69	0.003	24.47	0.000	255.17	0.00
Height of water level	1	24.40	0.000	83.65	0.000	827.58	0.00
Square	5	1.32	0.282	63.39	0.000	24.22	0.00
2-Way interaction	10	11.95	0.000	17.25	0.000	60.49	0.00
Error	31						
Lack of fit	22	2.85	0.053	2.85	0.603	0.85	0.645
Pure error	9						
Total	51						

Machining responses	Value of *S*	*R*-sq	*R*-sq(adjusted)	*R*-sq(predicted)
Kerf width	0.0086769	96.70%	94.58%	88.71%
Depth of cut	0.0009534	96.03%	93.47%	88.72%
HAZ width	0.0011052	99.22%	98.71%	97.72%

Figure 7.11 Microscopic view of HAZ width and kerf width.

interaction terms, except the square terms for kerf width. From the *f* value it is observed that lamp current is the most dominating factor for HAZ width but moderate in terms of kerf width and lesser dominant for depth of cut. Pulse frequency has the most dominant factor both the kerf width and depth of cut but a lesser dominant for HAZ width. Pulse width is the least dominant for all the responses. Cutting speed is more dominant factor depth of cut and HAZ width but lesser dominant for kerf width. Height of water level or column shows moderate dominancy for all the responses in which it has most dominant for HAZ width followed by depth of cut and kerf width. Values of other adequacy measures *R*-sq, adjusted *R*-sq, and predicted *R*-sq are in reasonable agreement and are close to 100%, which indicate adequacy of the model. Microscopic view of HAZ width is given in Fig. 7.11. A varying HAZ width along the kerf shown in microscopic view.

7.6.4 Effects of different process parameters on machining responses

7.6.4.1 Effect of different process parameters on kerf width

From the Fig 7.12 it is observed that kerf width marginally increases with an increase in height of water level, then decreases with further increase in height of water level. With increase in water column bubbles form and their collapsing is moved away from fusion zone results in increase in material removal. But further increase in water column results in increase of travel distance for laser beam to interact with work substrate. For that absorption of laser energy by water increase, results in decrease in material removal. With increase in lamp current more laser power generates, which travels through the thin water column to interact with the workpiece which may result in more material being removed from fusion zone due to hydrodynamic mechanism.

7.6.4.2 Effect of different process parameters on depth of cut

Here from Fig. 7.13 it is observed that depth of cut decreases with the increase in water level height, and decreases with increase in cutting speed. Increase in cutting speed means less interaction time of the laser beam with the workpiece per pulse.

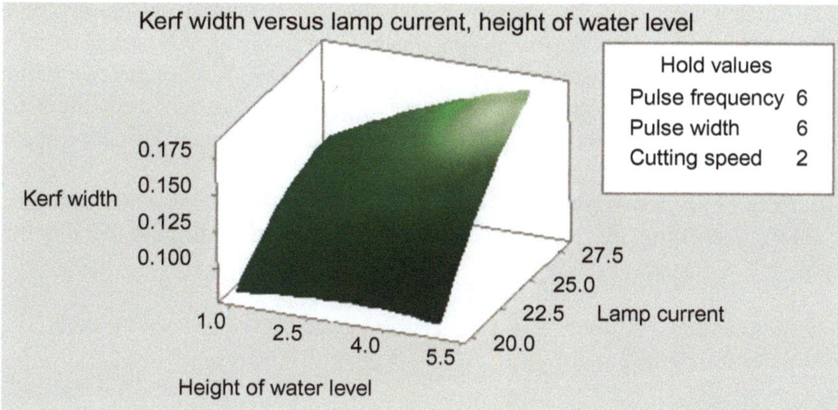

Figure 7.12 Surface plot of kerf width versus height of water column versus lamp current.

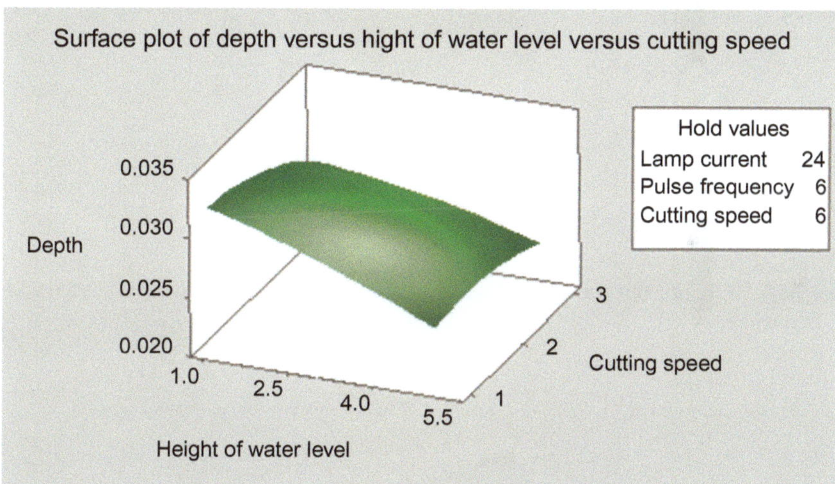

Figure 7.13 Surface plot of depth versus height of water column versus cutting speed.

For that reason less penetration may be achieved, resulting in decrease in depth of cut with increase in cutting speed. In underwater machining, laser energy is absorbed by water resulting in a lesser amount of laser fluence to interact with top surface of workpiece which also produces vapor bubbles and superheated water. These vapor bubbles sometimes may scatter the laser beam from its irradiation point, which may be the reason behind the decrease in depth of cut with increase in height of water level.

7.6.4.3 Effect of different process parameters on HAZ width

At lower laser pulse frequency, more laser power is generated than at higher pulse frequency, which could account for the amount of material removal. During underwater laser beam machining, settling time of the debris material removed from cut zone is much longer and the scattering and absorption of thermal energy of laser beam is greater. For that reason during laser machining at submerged condition, HAZ width is decreased with increase in pulse frequency. With change in height of water column, the refractive index changes linearly result in less heat input to the top surface of the workpiece. For that reason HAZ width may be decreased with increase in height of water column (Fig. 7.14).

From Fig. 7.15 it is observed that, with increase in height of water column, the laser energy reaching the workpiece surface is reduced; this may be due to the

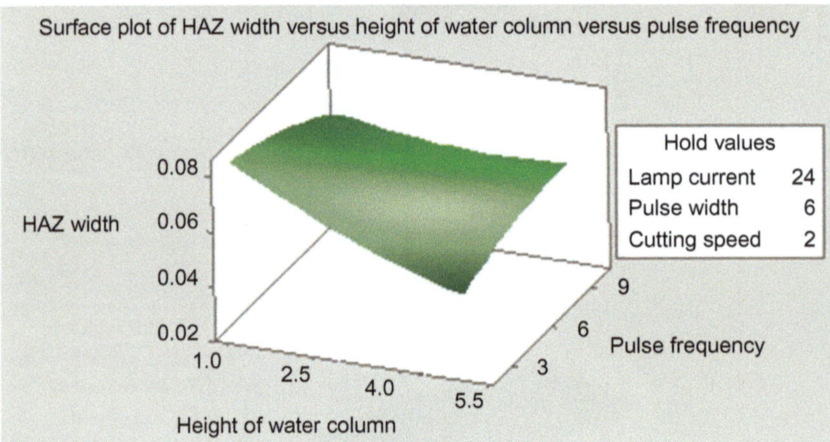

Figure 7.14 Response surface of HAZ width with respect to height of water column and pulse frequency.

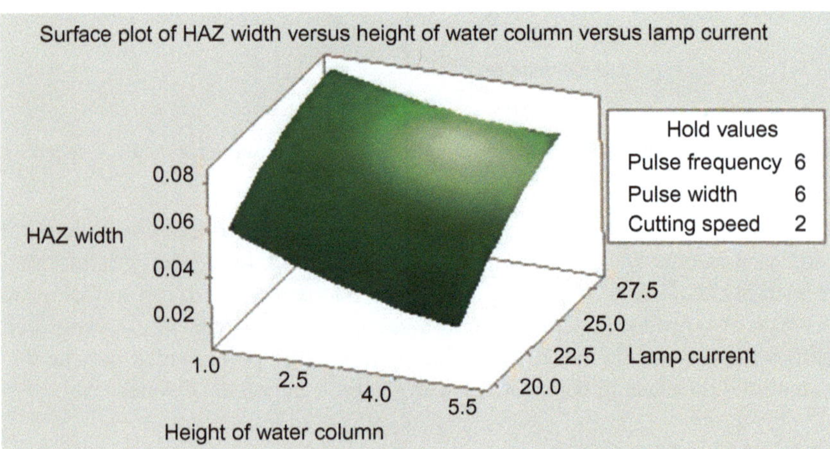

Figure 7.15 Response surface of HAZ width with respect to Height of water column and Lamp current.

absorption of laser energy by ionization of water and laser beam scattering by water vapor formed in the laser–material–water interaction zone which results in a lesser amount of HAZ width along the kerf. Higher energy density of laser beam causes by increase in lamp current, resulting in more heat input to the fusion zone to increase the HAZ width.

Conclusion

In this chapter a overview on Nano second pulsed laser beam machining has been discussed. Illustration of laser matter interaction and material removal of short pulsed laser beam machining is given. Difficulties of pulsed laser ablation is discussed. Remedy of this difficulties in form of laser machining at different environments, specially in underwater is illustrated. Developments of laser material interaction in submerged condition is exemplified.

Laser beam micromachining of Inconel 625 superalloy at submerged condition has been experimentally studied here. Center composite design-based DOE technique is used here to carry out the experiments. The effect of height of a water column along with other controllable process parameters, i.e., lamp current, pulse frequency, pulse width and cutting speed on kerf width, depth of cut, and HAZ width are investigated successfully. According to result of ANOVA during experimental study, carried out within the selected range, it has to be seemed that height of water column has a great effect on all the machining responses. Lamp current which is used here as an function of average power is the most influencing process variable whereas pulse width is the least affecting factor for all the machining responses. HAZ width, kerf width and depth of cut is decreased with increase in height of water.

Acknowledgment

The authors acknowledge University Grants commission (UGC), New Delhi, for financial support under the scheme of Rajiv Gandhi National Fellowship Programme and CAS Ph-IV (UGC) programme of Production Engineering Department, Jadavpur university, Kolkata, INDIA for technical equipment support.

References

Allmen, M. F. (1983). In M. Bertolotti (Ed.), *Physical processes in laser-materials interactions* (pp. 49–75). New York: Plenum, Chapter 2.

Bäuerle, D. (2000). *Laser processing and chemistry*. Heidelberg: Springer.

Biyikli, S., & Modest, M. F. (1988). Beam expansion and focusing effects on evaporative laser cutting. *ASME Journal of Heat Transfer, 110*, 529–532.

Caristan, C. (2003). *Laser cutting guide for manufacturing*. Dearborn: Society of Manufacturing Engineers.

Chryssolouris, G. (1991). *Laser machining: Theory and practice*. New York, NY: Springer.

Datta, M., Romankiw, L. T., & Vigliotti, D. R. (1987). Laser etching of metals in neutral salt solutions. *Applied Physics Letters, 51*(24), 2040–2042.

Davim, P. J. (2013). *Nontraditional machining processes, research advances*. London: Springer-Verlag. ISBN 978-1-4471-5178-4, http://dx.doi.org/10.1007/978-1-4471-5179-1.

Duley, W. W. (1983). *Laser processing and analysis of materials*. New York: Plenum Press.

Gower, M. C. (2000). Industrial applications of laser micromachining. *Optics Express, 7*, 56–67.

Kalyanasundaram, D., Shrotriya, P., & Pal, M. (2010). Fracture mechanics-based analysis for hybrid laser/waterjet (LWJ) machining of yttria-partially stabilized zirconia (Y-PSZ). *International Journal of Machine Tools & Manufacture, 50*, 97–105.

Kawaguchi, Y., Sato, T., & Narazaki, A. (2005). Etching a micro-trench with a maximum aspect ratio of 60 on silica glass by laser-induced backside wet etching (LIBWE). *Japanese Journal of Applied Physics, 44*, L176–L178.

Kim, D., & Lee, H. (2001). Enhanced ablation and photoacoustic excitation in near-threshold laser ablation of liquid-coated surfaces. *Journal of Applied Physics, 89*(10), 5703–5706.

Kruusing, A. (2004). Underwater and water-assisted laser processing: Part 2-etching, cutting and rarely used methods. *Optics and Lasers in Engineering, 41*, 329–352.

Küper, S., & Stuke, M. (1992). *Applied Physics Letters, 60*, 1633.

Li, L., & Achara, C. (2004). Chemical assisted laser machining for the minimisation of recast and heat affected zone. *Annals CIRP, 53*(1), 175–178.

Mullick, S., Madhukar, Y. K., Roy, S., Kumar, S., Shukla, D. K., & Nath, A. K. (2013). Development and parametric study of a water-jet assisted underwater laser cutting process. *International Journal of Machine Tools & Manufacture, 68*, 48–55.

O'Neill, W., & Gabzdyl, J. T. (2000). New developments in laser-assisted oxygen cutting. *Optics and Lasers in Engineering, 34*, 355–367.

Owaki, K., Uehara, M., Tsuchiya, K. (1999). *Laser water jet composite cutting device*. Patent JP11000780.

Powell, J. (1993). *CO_2 laser cutting*. London: Springer.

Ready, J. F. (1997). *Industrial applications of lasers*. San Diego: Academic Press.

Roy, N., Kuar, A.S., Mitra, S., & Acherjee, B. (2015). Nd:YAG laser microdrilling of SiC-30BN nanocomposite: Experimental study and process optimization. In: U. S. Dixit, & S. N. Joshi, (Eds.), *Lasers based manufacturing. Topics in mining, metallurgy, materials engineering* (pp. 317–341). Springer, India, pp. 317–341 Chapter 17, ISBN 978-81-322-2352-8, 1.

Schulz, W., Simon, G., Urbassek, H. M., & Decker, I. (1987). Laser fusion cutting of metals. *Journal of Physics D: Applied Physics, 20*, 481–488.

Steen, W. M. (1991). *Laser materials processing*. London: Springer.

Stephen, A. (2011). Mechanisms and applications of laser chemical machining. *Physics Procedia, 12*, 261–267.

Von, A. M. (1987). *Laser-beam interactions with materials*. Berlin: Springer.

Welch, A. J., & Gardner, C. (2002). Optical and thermal response of tissue to laser radiation. In R. W. Waynant (Ed.), *Lasers in medicine* (pp. 27–45). Boca Raton, FL: CRC Press.

Glass molding process for microstructures

8

T. Zhou[1] *and J. Yan*[2]
[1]Beijing Institute of Technology, Beijing, P.R. China, [2]Keio University, Yokohama, Japan

8.1 Application of microstructures

8.1.1 Optical imaging in an optical system

8.1.1.1 Refraction

Refraction is the change in direction of propagation of a wave due to a change in its transmission medium, and optical components with microstructures have many applications based on the refraction function in the optical system.

Microstructure arrays are configurations of a number of lenses or prisms in a micro-nanoscale and can provide a variety of optical functions due to their special geometrical features. The basic types of microstructure arrays are displayed in Fig. 8.1 according to the element shape. Microstructure arrays with large element size, for instance, 0.5 ~ 50 μm, mainly achieve optical performance by their refraction and reflection properties. They can raise light energy utilization ratio and realize miniaturization of optical systems through multiple imaging. Moreover, a combination of several types of microstructure arrays is able to contribute to beam guidance control, smart scan and other complex functions (shown in Fig. 8.2). Due to these functions, microstructure arrays are widely used in LCD, mobile phones, palm pilots, TVs and other electrical products.

An example of microstructures based on the refraction principle is a wave-front sensor, which is composed of microlens array and CCD (charge-coupled device) array. A Shack-Hartmann wave-front sensor can be used to characterize the performance of an optical system. In addition, they are increasingly being used to control the adaptive optical elements by real-time monitoring of wave-fronts in order to achieve the goal that eliminates the wave-front distortion before the imaging.

8.1.1.2 Diffraction

Diffraction, which is defined as the bending of light around the corners of an obstacle or aperture into the region of geometrical shadow of the obstacle, is another significant property of microstructures.

Microstructure arrays with small element size, for instance, 0.5 ~ 5 μm, mainly achieve optical performance by their interference and diffraction reflection properties. They can produce entirely different phenomena compared with macro lenses

Microfabrication and Precision Engineering. DOI: http://dx.doi.org/10.1016/B978-0-85709-485-8.00008-5

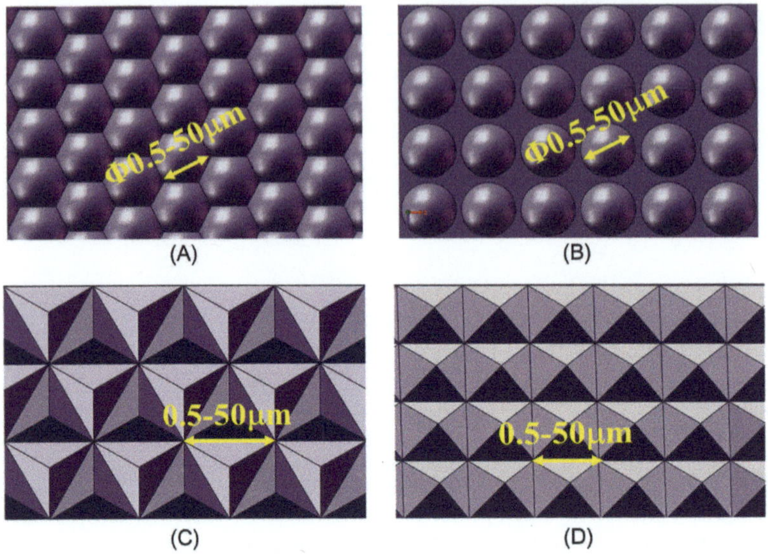

Figure 8.1 Element shape of microstructure arrays: (A) adjacent microlens arrays, (B) distributed microlens arrays, (C) triangular pyramid arrays, (D) rectangular pyramid arrays.

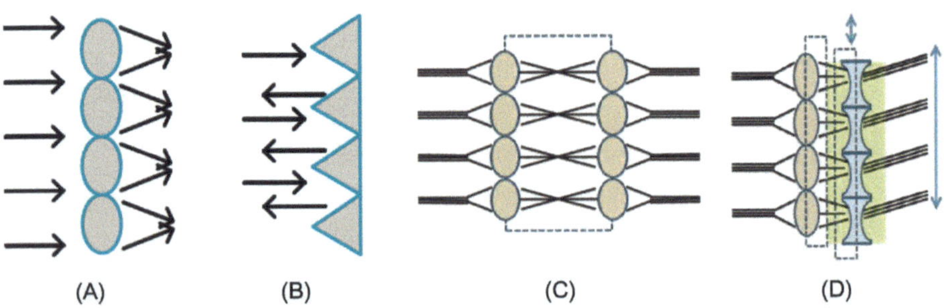

Figure 8.2 Function and application of microstructure arrays with large element size: (A) focus, (B) reflection, (C) beam guidance, (D) smart scan.

when light goes through microstructure arrays. As shown in Fig. 8.3, microstructure arrays are equipped with properties of one-dimensional diffraction and two-dimensional diffraction. They are probably able to accomplish antireflection and other goals such as polarization beam splitting, optical waveguide coupling, light beam transformation and integration if we focus on the complicated design of cycle structures and the shape of microstructure arrays.

The most common use of optical components with microstructures based on diffraction principle is diffraction grating, which is a component of optical devices consisting of a surface ruled with close, equidistant and parallel lines for the purpose of resolving light into spectra. As a typical application of diffraction grating,

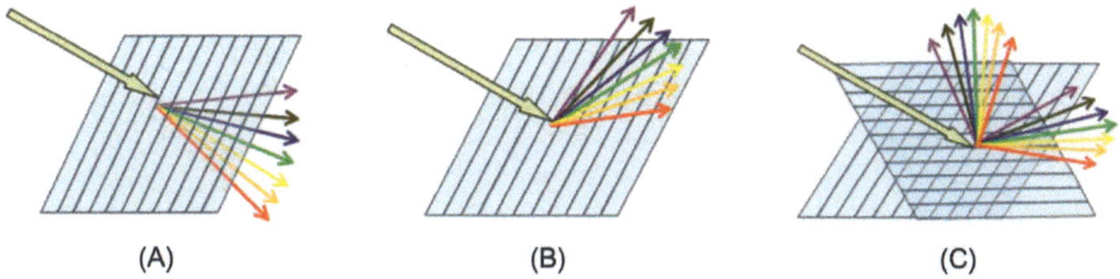

Figure 8.3 Function and application of microstructure arrays with small element size: (A) one-dimensional transmission diffraction, (B) one-dimensional reflection diffraction, (C) two-dimensional diffraction.

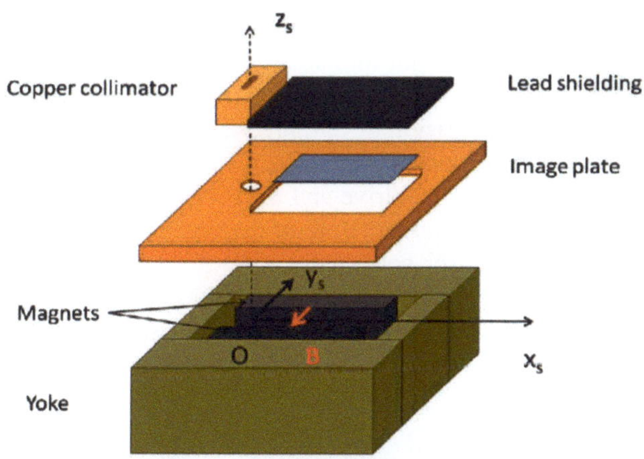

Figure 8.4 Drawing of the spectrometer setup (Gobet et al., 2016).

optical spectrometers are widely applied in the field of process control, plasma diagnostics, spectroscopic analysis of gases and liquids (Gruger, Wolter, Schuster, Schenk, & Lakner, 2003). The working principle of a spectrometer is shown in Fig. 8.4. The electrons enter into the magnet gap from the O point placed at the accelerator isocenter. An image plate shielded by lead is used to detect the electrons that are deflected between the two magnets.

The diffraction grating can also be used in the imaging system, fiber grating, and encoder. However, one of the most important usages of diffraction grating is the sensor that is used in machine tools and measurement equipment.

8.1.2 Positioning sensor in machine tools and measurement equipment

8.1.2.1 Linear grating

There are hundreds of types of positioning sensors in machine tools, and the simplest one is the grating ruler, which is used as the checkout gear in the numerical control machine tool.

In the installation process for a grating ruler, the line on the indicator grating will generate a small angle with the line on the ruler grating, and these lines on two gratings cross each other. The lines overlap and form black fringes near the intersection and bright fringes in other places when light goes through the grating. These fringes are called moiré fringes (shown in Fig. 8.5).

The moiré fringes change between the bright and the dark with the movement of grating. After signal processing, circuit amplification, shaping and differential, the system will output the pulse. Each output of a pulse is represented by the distance of a grid. By counting the pulses, the moving distance of the working table can be obtained.

8.1.2.2 Face grating

In optics, face grating is an optical component with a periodic structure on the surface, which splits and diffracts light into several beams traveling in different directions. With the development of ultraprecision numerical control manufacturing technology, the machining precision of ultraprecision machine tools has reached the submicron scale shape precision and nanoscale surface roughness, which directly dependends on strict control of the drive and ultraprecision testing technology.

Fig. 8.6 shows the fundamental structure of the three-axis surface encoder, which is composed of an optical sensor head and a scale *XY* planar grating. The scale *XY* planar grating has periodic grating structures with a period of *g* in the *X*- and *Y*-axes. The components of the optical sensor head are a laser source, a nonpolarizing beam splitter (BS), a detector unit and a reference *XY* planar grating. A laser

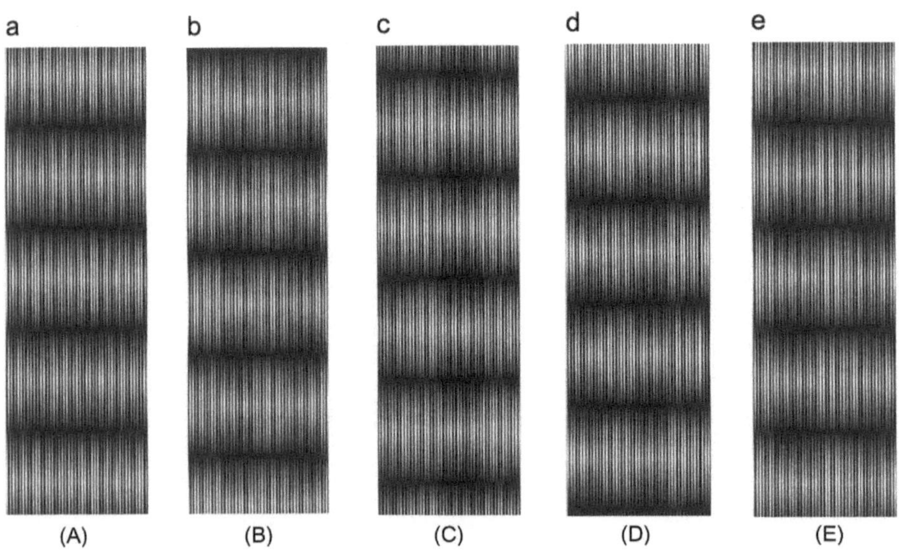

Figure 8.5 A typical moiré's pattern formed by two superimposed Ronchi gratings rotated by an angle with displacement equal to: (A) zero, (B) a quarter of the grating pitch, (C) half of the grating pitch, (D) 3 quarters of the grating pitch, (E) the grating pitch (Madanipour & Tavassoly, 2010).

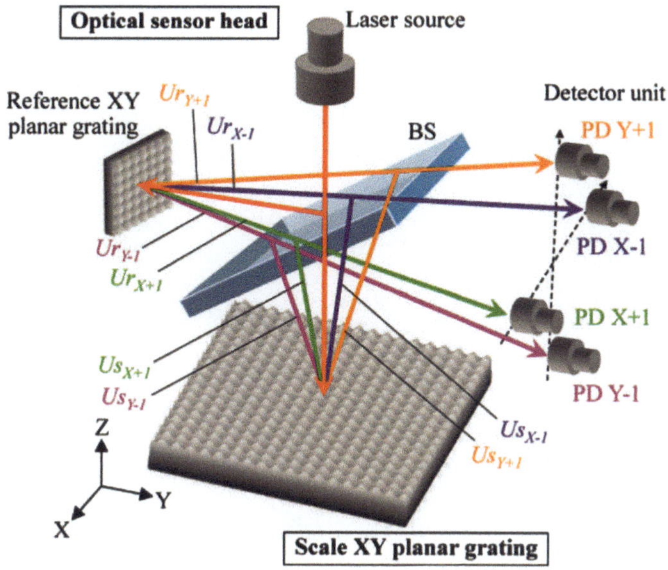

Figure 8.6 Basic principle of the three-axis surface encoder (Kimura et al., 2012).

beam from the optical sensor head is projected onto the moving scale grating. The *X*-directional positive and negative first-order diffracted beams from the scale grating interfere with each other to generate interference signals, from which the *X*-directional displacement can be obtained. Similarly, the *Y*-directional displacement can be obtained from the interference between the *Y*-directional positive and negative first-order diffracted beams from the scale grating. The *XY* planar encoder has been successfully used for two-axis *XY* position measurements of CNC machine tools and photolithography scanners.

8.1.3 Micro fluid control in a biomedical field

The microstructures are widely employed not only in optical imaging and mechanical engineering, but also in the biomedical field. For instance, they can be used in the measurement system of the deformability of red blood cells (RBC). The deformability of RBC is the ability of getting through the narrow vascular channels by their deformation when the RBC are in the flow process, which is one of the most significant blood rheology indexes. Therefore, the study of RBC deformability has played an important role in preventing and curing some diseases because of RBC morphological features.

A cell deformability analysis system, based on MEMS and microfluidic technology, is useful for simulating capillaries (shown in Fig. 8.7), improving flow path and driving mode. It also combines with the enhancement of a high-speed image acquisition system (shown in Fig. 8.8) and static and dynamic image processing when using a microchannel array.

(A) (B)

Figure 8.7 The structure map of microchannel with different sizes and shapes: (A) 12 × 4 × 4 (μm), (B) 3.3 × 4.9 × 0.5 (mm) (Sutton et al., 1997).

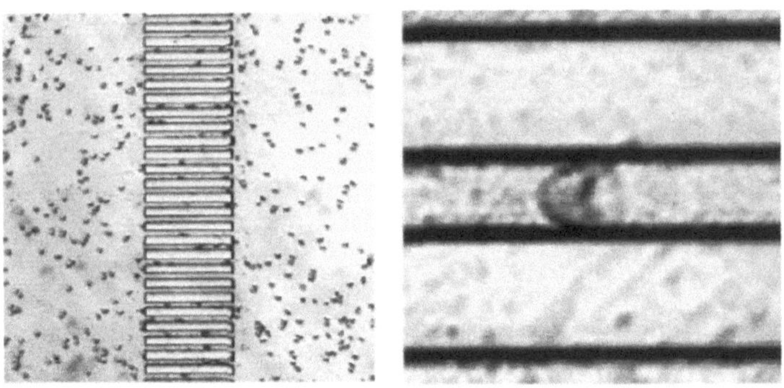

(A) (B)

Figure 8.8 The image photograph by high-speed image acquisition system: (A) images of flowing RBCs, (B) a RBC with the velocity of 1.07 mm/s (Tsukada, Sekizuka, Oshio, & Minamitani, 2001).

Compared with normal rheological detection technology, this method is advanced in many aspects, such as multiinformation, flexible, controllable structure, easy to handle and low cost. In addition, this approach is objective and can provide a valuable reference for the auxiliary diagnosis of relative diseases, as well as other applications in the fields of biomedicine and biomedical engineering.

8.2 Fundamental of glass molding technique

8.2.1 Introduction

Three-dimensional microsurface structures such as microgrooves, micropyramids, microprisms and microlenses are demanded more and more in recent optical, optoelectronic, mechanical and biomedical industries. Components with microsurface structures yield new functions for light operation, thereby improving significantly the imaging quality of optical systems. Microgrooves can also be used as fluid channels in biomedical and biochemical applications. Therefore, high-precision and high efficiency fabrication of microgrooves on flat or curved surfaces are receiving focused interests (Zhou, Yan, Masuda, Oowada, & Kuriyagawa, 2011).

Commercially, glass and plastic are two major substrate materials for microstructured components. Glass has predominant advantages over plastic in aspects of hardness, refractive index, light permeability, stability to environmental changes in terms of temperature and humidity, etc. A few microstructures on glass can be fabricated by material removal processes, such as sand blasting, photolithography, wet/dry etching and focused ion beam (FIB), and so on. These processes are effective for manufacturing microstructures with rectangular cross-sections (U-grooves), but it is difficult to fabricate microgrooves with sharp-angled cross-sections (V-grooves). Microcutting of glass using microendmills has also been reported, but the production efficiency is limited and the production cost is considerably high for mass production. As an alternative approach, glass molding process (GMP) is able to produce glass optical elements by replicating the shape of the mold to heated glass preforms without further machining process. Fig. 8.9 shows the glass molding process technology of the microstructure array.

From the viewpoint of fabrication cost and process time, the GMP is undoubtedly a better approach to produce precision optical elements, such as aspherical lenses, Fresnel lenses, diffractive optical elements (DOEs), microprism arrays and microlens arrays. In recent years, glass molding for microstructure, alternatively termed hot embossing or thermal imprinting, has also been reported.

The production of precision glass lenses with microgrooves by hot compression molding is a promising manufacturing method in terms of saving production costs and environment compatibility. It has shown that the molding process can produce lenses with form accuracy, surface finish and an optical performance that is comparable to lenses manufactured by conventional material removal techniques. Fig. 8.10 shows the process flow of microgroove forming. According to the thermal cycle, the process can be divided into four stages: heating, pressing, annealing, and

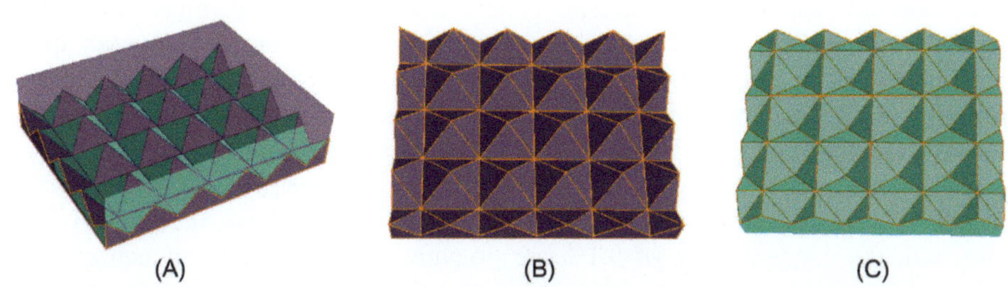

Figure 8.9 Glass molding technology of microstructure arrays: (A) forming of the microstructure array, (B) mold with microstructures, (C) glass with microstructures.

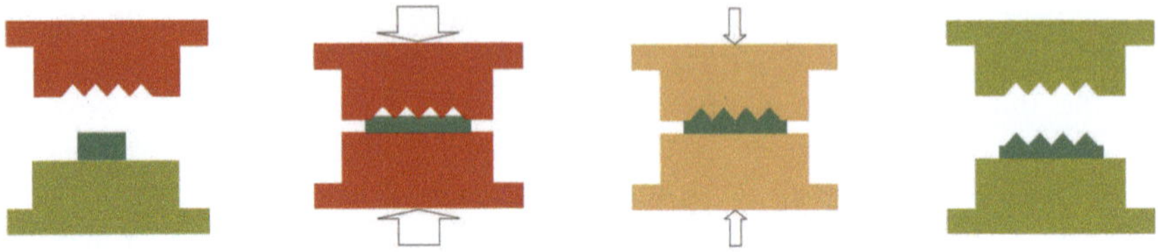

Figure 8.10 Process flow of microgroove forming: (A) heating, (B) pressing, (C) annealing, and (D) cooling.

cooling. Firstly, a glass preform is placed on the lower mold and inert gas, such as nitrogen (N_2), is flowed to purge the air in the machine chamber; then the molds and glass preform are heated to the molding temperature by a heat source, such as infrared lamps (Fig. 8.10A). Secondly, the glass preform is pressed by closing the two mold halves (Fig. 8.10B). Thirdly, while a small pressing load is maintained, the formed lens is slowly cooled down to release the internal stress, namely, annealing (Fig. 8.10C). Finally, the glass lens is cooled rapidly to ambient temperature and released from the molds (Fig. 8.10D). Through these four stages, the shapes of the mold cores are precisely replicated to the glass lens.

However, in the process of microgroove forming, there are several technical challenges associated with the molding process which have prevented it from being applicable in industry for high-volume lens production. These challenges include thermal shrinkage of the lens on cooling, optical surface finish and precise mold shape, mold life and the selection of process parameters.

8.2.2 Materials suited for optical microstructures molding

8.2.2.1 Polymethyl methacrylate

Polymethyl methacrylate (PMMA) is a transparent, tough and rigid plastic. Some detail characteristics are illustrated in Table 8.1. Its properties remain stable when exposed to ultraviolet radiation and terrible conditions. Therefore, it is an ideal substitute for glass. PMMA is widely employed in domed skylights, swimming pool enclosures, aircraft canopies, instrument panels and luminous ceilings. For these applications, the plastic is drawn into sheets that are machined or thermoformed, and it is also injection-molded into automobile lenses and lighting-fixture covers. PMMA is frequently made into optical fibers for telecommunication or endoscopy because of its unusual property of keeping a beam of light reflected within its surfaces.

Due to the outstanding performance in the optical system, PMMA can also be used for microstructure fabrication. In injection molding process, the PMMA solutions with the concentration of 9.1% are prepared in the solvents of benzene and methyl methacrylate (MMA) separately. The solution is spin-coated on the glass substrate at the speed of 1000 r/min for 30 s, and then gradually heated up to 100°C. The patterned surface of a polydimethylsiloxane (PDMA) stamp is

Table 8.1 Basic properties of PMMA (Smith & Hashemi, 2011)

Chemical formula	$(C_5O_2H_8)_n$
Molar mass	Varies
Density	1.18 g/cm^3
Melting point	160°C (320°F; 433K)
Refractive index (nD)	1.4905 (at 589.3 nm)

Figure 8.11 The molded low-melting optical aspheric lenses.

contacted closely with PMMA film, and kept at 120°C for about 4 h under a certain pressure. Having removed off the PDMS stamp gently, the cylindrical micropattern on the PMMA film, the same as that on the template, is obtained. The cylindrical microlens array of PMMA film on the glass substrate is kept at the temperature beyond its softened temperature for a suitable time period until every microcylinder is completely shrunk into a hemispherical lens.

8.2.2.2 Low-melting optical glass

The low-melting optical glass has a low melting point, which is suitable for molding. This kind of optical glass is made by mixing high-purity silicon oxide, boron, sodium, potassium, zinc, lead, magnesium, calcium, barium and other oxide in a specific recipe. The glass usually melts in a platinum crucible with high temperature at first. Then it is stirred well and gets away from the bubble in ultrasound. Finally internal stress is removed after a long, slow cooling and annealing process. At this point the manufacture of low-melting optical glass is completed, and the molding process can be carried out. The molded glass aspheric lens is shown in Fig. 8.11.

The glass with high softening point temperature needs a high forming point temperature. In high-temperature conditions the glass may involve with physical or chemical reaction with the mold which shortens the service life of the mold. From the viewpoint of prolonging the service life of the mold, we should develop the glass material which is suitable for low temperature (under 600°C).

The low-melting optical glass is one of the most important materials of the GMP. It can be used not only in the manufacture of the lens with normal size, but also to mold the microlens arrays. As a result, the products can get an excellent service performance due to its excellent optical and physical properties.

8.2.2.3 Infrared Materials

The Infrared Materials is a general term of several kinds of materials and the most common use is the chalcogenide glass, which contains one or more chalcogenide elements (not counting oxygen). Glass-forming abilities decrease with increasing molar weight of constituent elements, i.e., $S > Se > Te$.

Chalcogenide glass, as shown in Fig. 8.12, is widely used in various aspects of industrial systems. The semiconducting properties of chalcogenide glasses were revealed in 1955 by B.T. Kolomiets and N.A. Gorunova from Loffe Institute, USSR (Kolomiets, 1964a, 1964b). This discovery initiates numerous researches and applications of this new semiconducting material. Modern chalcogenide compounds such as AgInSbTe and GeSbTe, are in widespread application in rewritable optical disks and phase-change memory devices. However, the most important application is in the infrared field including infrared detectors, moldable infrared lenses and infrared optical fibers, with the main advantage that these materials transmit across a wide range of the infrared electromagnetic spectrum. The physical properties of chalcogenide glasses (high refractive index, low phonon energy and high nonlinearity) also make them ideal for incorporating into lasers, planar optics, photonic integrated circuits and other active devices especially if rare earth ions are added.

Chalcogenide glasses is also moldable, which means complex shapes can be developed without using traditional expensive processes such as single point diamond turning (SPDT) or polishing. A molding method significantly lowers cost and improves production, so nowadays many chalcogenide glasses lenses are made by the GMP.

Figure 8.12 Photographs of chalcogenide glass.

8.2.3 Mold material

8.2.3.1 Commonly used mold material

In the molding/embossing/imprinting process, mold fabrication is an important issue. Silicon or nickel molds, usually employed for polymer forming, cannot be used for glass forming because of their poor heat resistance and difficulty of removal arising from cohesion between the mold and glass. Super-hard materials, such as silicon carbide (SiC) and tungsten carbide (WC), are preferable mold materials for pressing continuous surfaces like aspherical lenses.

Silicon carbide (SiC), exceedingly hard, is synthetically produced by crystalline compound of silicon and carbon. Silicon carbide was the hardest synthetic material known until the invention of boron carbide in 1929. It has a Mohs hardness rating of 9, approaching that of diamond. In addition to hardness, silicon carbide crystals have fracture characteristics that make them extremely useful for grinding wheels, abrasive paper and cloth products. Its high thermal conductivity, together with its high-temperature strength, low thermal expansion, and resistance to chemical reaction, make it valuable in the manufacture of high-temperature glass molds and other refractories.

Tungsten carbide (WC) is a chemical compound containing equal parts of tungsten and carbon atoms. It can be pressed and formed into shapes for use in industrial machinery, cutting tools, abrasives, armor-piercing rounds, other tools and instruments, and jewellery. Tungsten carbide is approximately twice stiffer than steel, with a Young's modulus of approximately 530~700 GPa (Groover, 2007; Sloely, 2001), and is double the density of steel, which is nearly midway between that of gold. It is comparable with corundum (α-Al2O3) in hardness and can only be polished and finished with abrasives of superior hardness such as cubic boron nitride and diamond powder, wheels and compounds. The other properties of the tungsten carbide are shown in the Table 8.2.

But the chemical reaction, stress and repeated heat treatment on the surface of glass will affect the life of the mold core. The glass adhering to the mold surface, the oxidation, and the wear of the mold greatly reduce the service life of mold. The mold and glass materials at high temperature will produce quite an intense chemical reaction, ion exchange reaction and heat reaction, thereby leading mutual diffusion or generating new compounds in the interface. Therefore, it is highly necessary to deal with the surface of the mold. Ni-P coating (shown in Fig. 8.13) can effectively

Table 8.2 Properties of Tungsten carbide (Cardarelli, 2008)

Chemical formula	WC
Molar mass	195.85 g/mol
Appearance	Gray-black lustrous solid
Density	15.6 g/cm^3
Melting point	2785~2830°C (5045~5126°F; 3058~3103K)
Boiling point	6000°C (10,830°F; 6270K) at 760 mmHg
Solubility in water	Insoluble

Figure 8.13 (A) Ni-P coating structure and (B) Ni-P coated molding core.

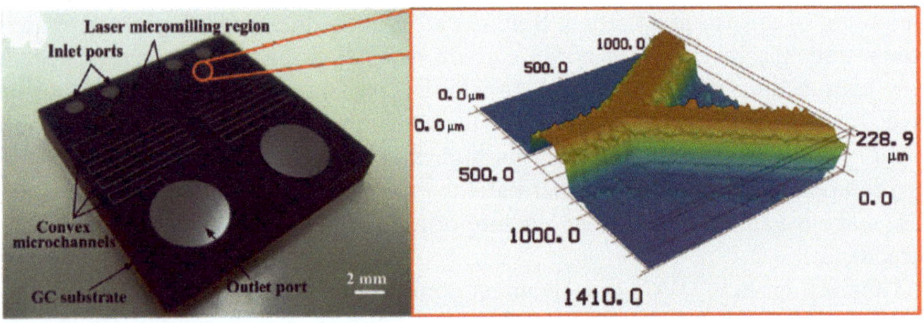

Figure 8.14 Photographs of (A) the glass molding die after laser micromilling onto the glassy carbon substrate with convex microchannels and (B) enlarged partial convex microchannels as measured by the 3D laser confocal microscope (Tseng et al., 2014).

solve the problem by retarding the surface reaction so as to reduce mechanical damage during hot pressing.

These materials have not been commonly used for molding microstructures, since they are very difficult to generate microstructures. In embossing/imprinting research, glassy carbon (GC) molds fabricated by FIB or dicing are used at present.

Glass-like carbon, often called GC or vitreous carbon, is a nongraphitizing carbon that combines glassy and ceramic properties. The most important properties are high-temperature resistance, high hardness (7 Mohs), low density, low electrical resistance, low friction, low thermal resistance, extreme resistance to chemical attack, and impermeability to gases and liquids. GC is widely used as an electrode material in electrochemistry, as well as for high-temperature mold (shown in Fig. 8.14). As a component of some prosthetic devices, it can be fabricated as different shapes, sizes and sections.

The reason why GC is used to make the mold of the GMP is that GC is suitable for a high temperature of 1400°C, comparable to the transition temperature of quartz glass. Another reason is that cohesion between GC and glasses is generally poor, and we can release the structure easily after embossing (Takahashi, Murakoshi, Maeda, & Hasegawa, 2007).

Normally the methods of the MEMS, like the FIB technique, are used to generate nanometer-level microstructures on a small-area GC mold, and the dicing technique is used for fabricating bigger microstructures on a large-area mold.

8.2.3.2 Mold machining method

There are various methods of MEMS to fabricate the mold of the microstructure of the glass, such as FIBs, femtosecond lasers and KrF eximer lasers (Youn, Takahashi, Goto, & Maeda, 2006). At present, the micro/nanomachining of GC is usually employed by a FIB to fabricate a mold for glass embossing (see Fig. 8.15). FIB machining characteristics are investigated with respect to accuracy, resolution, roughness and aspect ratio. Glass-embossed structures are successfully fabricated by hot molding using a GC mold (Takahashi, Sugimoto, & Maeda, 2005).

In above-mentioned text, micro/nano imprinting is developed for Pyrex glasses using a GC mold prepared by FIB machining. The disadvantage of FIB machining is the limited area of etching, the typical area of which is less than several hundred square micrometers. This is the reason that researchers tried the large area of embossing using GC mold fabricated by using a dicing machine as shown in Fig. 8.16.

Machining methods of mold consist of dicing and grinding, when the mold is used to fabricate the glass lens which is in macro size. It means this method is usually not amenable to the manufacture of the microstructure mold.

In the Cartesian coordinate system, the grinding area is the arc area of the grinding wheel side face and end face due to the high speed. In the process of

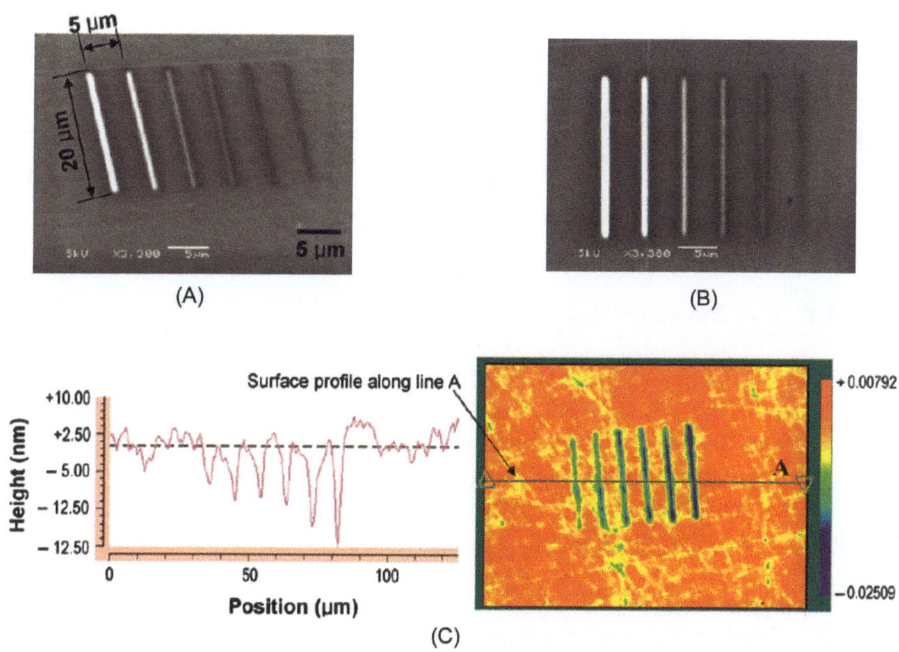

Figure 8.15 The image of the microstructure on the mold fabricated by the FIB: (A) SEM image of the line structures made by FIB milling, (B) corresponding top view image, (C) surface morphology measured by interferometric microscopy (Youn et al., 2006).

(A) (B) (C)

Figure 8.16 SEM images of patterns on GC surfaces machined by dicing: (A) array of 50 μm-high and 100 μm-wide squares with 50 μm spacing, (B) array of 50 μm-high and 300 μm-wide squares with 50 μm spacing, (C) array of pyramids with bottom width of 400 μm and side wall angle of 40° (Youn et al., 2006).

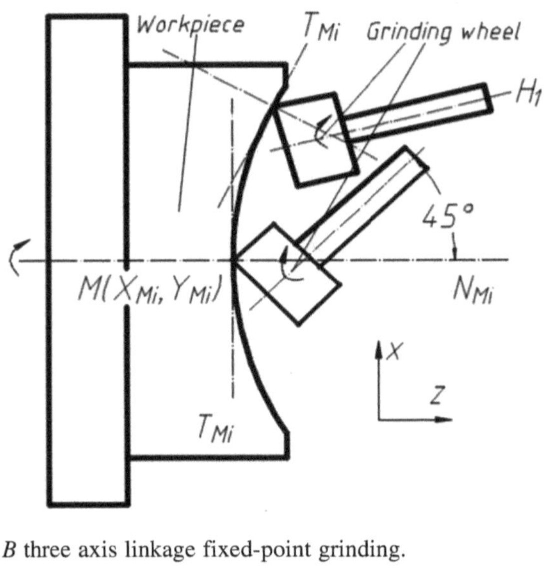

Figure 8.17 *X*, *Z*, *B* three axis linkage fixed-point grinding.

ultraprecision grinding, the lens die and the grinding wheel are in point contact as shown in Fig. 8.17, which is beneficial to the control of the machining track and the subsequent compensation. Furthermore, it is easy to improve the quality of the grinding process by repairing the grinding wheel. Especially in the processing of microaspheric lens die, we can effectively avoid the interference problem by controlling the rotation of the B axis.

8.2.3.3 New mold plating material

Nickel–phosphorous (Ni-P) electroless plating has been known as an important mold surface preparation technology for manufacturing plastic optical parts. Ni-P plating provides hard, wear- and corrosion-resistant surfaces at relatively high temperatures, and at the same time, maintains excellent precision micromachinability. In previous work, researchers have demonstrated that Ni-P plating can also be used for molding glass components, as shown in Fig. 8.18, such as aspherical and diffractive lenses, with a considerably long mold service life (Zhou, Liang, Wang, & Kuriyagawa, 2013).

Some microcutting experiments are conducted on electroless-plated Ni-P surfaces to fabricate microstructures such as microgrooves and micropyramid arrays (shown in Fig. 8.19) (Yan, Oowada, Zhou, & Kuriyagawa, 2009).

Burr formation behavior and cutting force characteristics are investigated experimentally and simulated by the finite element method (FEM) under various conditions. A simple two-step cutting process is proposed to improve the surface quality. The machined microstructure arrays are used as molds for hot-press glass molding experiments and good geometrical transferability is confirmed. The results verify that diamond-machined Ni-P microstructure molds are applicable to glass molding processes for mass production of precision microoptical components.

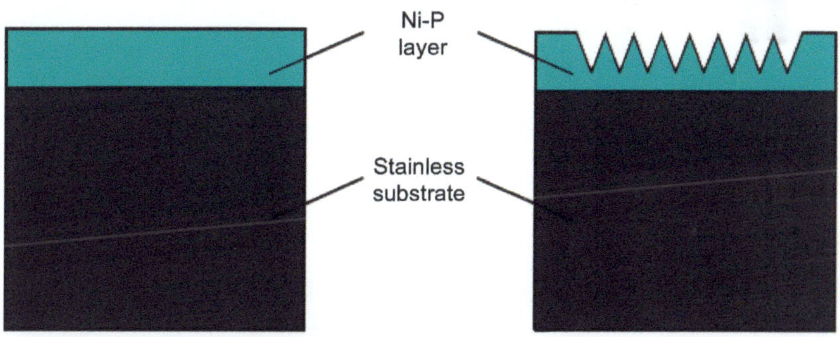

Figure 8.18 Ni-P mold for microgroove forming in glass molding process.

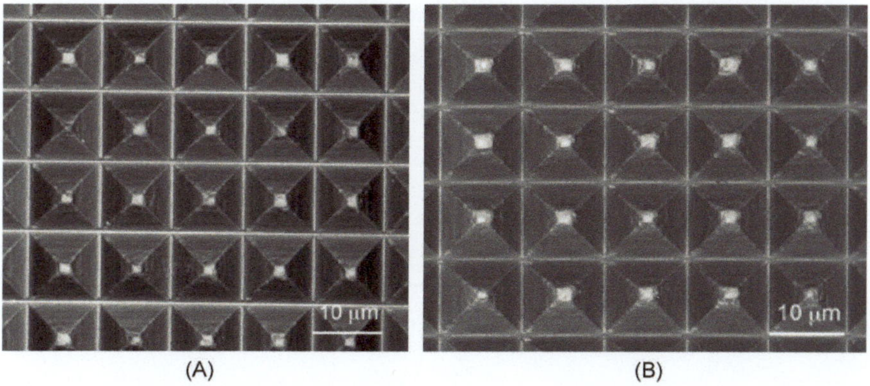

Figure 8.19 Microscope images of micropyramid arrays: (A) Ni-P mold, (B) molded glass plate.

8.3 Modeling and simulation of microstructure molding

8.3.1 Modeling of viscoelastic constitutive

To date, there have been many attempts to simulate the behavior of viscoelastic materials. This has been aimed at facilitating analysis of the behavior of glass products, assisting with extrapolation and interpolation of experimental data and reducing the need for extensive, time-consuming creep tests. Viscoelastic behavior is the time-dependent response of a material to a strain or stress. It can be illustrated best with the help of mechanical model-combinations of springs and dashpots. *A* spring represents elastic or hookean behavior, and a dashpot represents viscous or Newtonian behavior. These two elements cover both parts of the notion "viscoelasticity." In addition, a differential equation can be obtained describing the behavior of the model. The objective is to devise the model and obtain the equation that will describe the behavior of a real material (Rekhson, 1986).

Although there are no discrete molecular structures that behave like the individual elements of the models, they nevertheless do aid in the understanding and analysis of the behavior of viscoelastic materials. Several common models are shown in Fig. 8.20 (Crawford & Crawford, 1998).

8.3.1.1 The Maxwell model

The spring is the elastic component of the response and obeys the relation

$$\sigma_1 = E\varepsilon_1 \tag{8.1}$$

where σ_1 and ε_1 are the stress and strain respectively and E is a constant.

The dashpot is the viscous component of the response and in this case the stress σ_2 is proportional to the rate of strain $\dot{\varepsilon}_2$,

$$\sigma_2 = \eta\dot{\varepsilon}_2 \tag{8.2}$$

where η is a material constant.

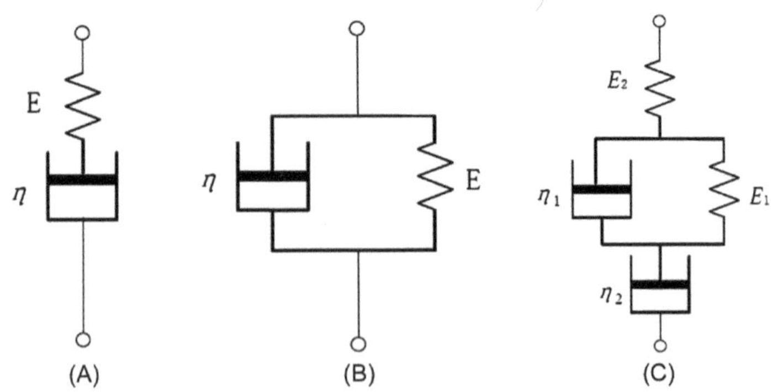

Figure 8.20 Common viscoelastic models: (A) the Maxwell model, (B) the Kelvin model, (C) the Burger model.

For equilibrium of forces, assuming a constant area. As shown in Fig. 8.20A, applied stress,

$$\sigma = \sigma_1 = \sigma_2 \tag{8.3}$$

The total strain, ε is equal to the sum of the strains in the two elements. So

$$\varepsilon = \varepsilon_1 + \varepsilon_2 \tag{8.4}$$

From Eqs. (8.1), (8.2), and (8.4)

$$\dot{\varepsilon} = \frac{1}{E}\dot{\sigma} + \frac{1}{\eta}\sigma \tag{8.5}$$

This is the governing equation of the Maxwell model. It is interesting to consider the response that this model predicts under three common-time-dependent modes of deformation.

Creep

If a constant stress σ_0 is applied then Eq. (8.5) becomes

$$\dot{\varepsilon} = \frac{1}{\eta}\sigma_0 \tag{8.6}$$

which indicates a constant rate of increase of strain with time.

From Fig. 8.21 it has seen that for the Maxwell model, the strain at any time t, after the application of a constant stress σ_0, is given by

$$\varepsilon(t) = \frac{\sigma_0}{E} + \frac{\sigma_0}{\eta}t \tag{8.7}$$

Hence, the creep modulus, $E(t)$, is given by

$$E(t) = \frac{\sigma_0}{\varepsilon(t)} = \frac{E\eta}{\eta + Et} \tag{8.8}$$

Relaxation

If the strain is held constant then Eq. (8.5) becomes

$$0 = \frac{1}{E}\dot{\sigma} + \frac{1}{\eta}\sigma \tag{8.9}$$

Solving this differential equation with the initial condition $\sigma = \sigma_0$ at $t = t_0$ then,

$$\sigma(t) = \sigma_0 e^{-\frac{E}{\eta}t} \tag{8.10}$$

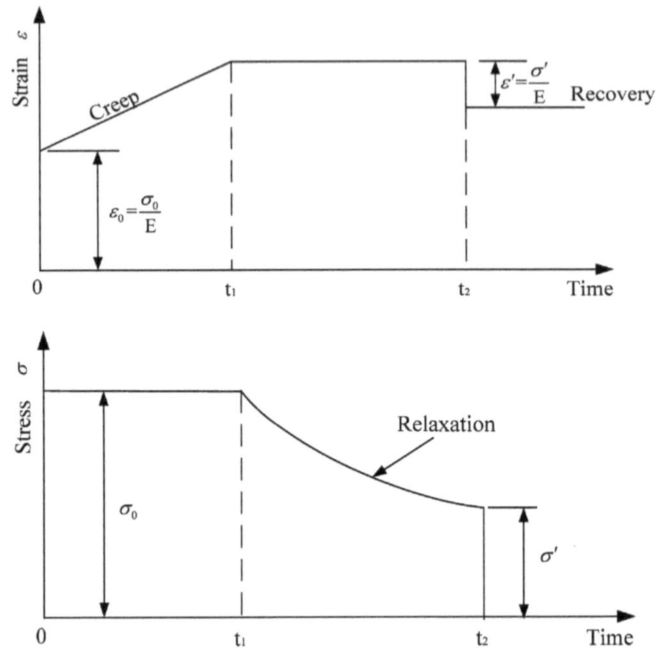

Figure 8.21 Response of Maxwell model.

$$\sigma(t) = \sigma_0 e^{-t/\tau_s} \tag{8.11}$$

where $\tau_s = \frac{\eta}{E}$ is referred to as the relaxation rime.

This indicates that the stress decays exponentially with a time constant of $\frac{\eta}{E}$.

Recovery

When the stress is removed there is an instantaneous recovery of the elastic strain, and then, as shown by Eq. (8.5), the strain rate is zero so that there is no further recovery (see Fig. 8.21).

It can be seen therefore that although the relaxation behavior of this model is acceptable as a first approximation to the actual materials response, it is inadequate in its prediction for creep and recovery behavior.

8.3.1.2 The Kelvin model

In this model the spring and dashpot elements are connected in parallel as shown in Fig. 8.20B.

For equilibrium of forces it can be seen that the applied load is supported jointly by the spring and the dashpot, so

$$\sigma = \sigma_1 + \sigma_2 \tag{8.12}$$

In this case the total strain is equal to the strain in each of the elements,

$$\varepsilon = \varepsilon_1 = \varepsilon_2 \tag{8.13}$$

From Eqs. (8.1), (8.2), and (8.12)

$$\sigma = E\varepsilon_1 + \eta\dot{\varepsilon}_2 \tag{8.14}$$

or using Eq. (8.13)

$$\sigma = E\varepsilon + \eta\dot{\varepsilon} \tag{8.15}$$

This is the governing equation for the Kelvin model, and it is interesting to consider its predictions for the common-time-dependent deformations.

Creep

If a constant stress, σ_0, is applied then Eq. (8.15) becomes

$$\sigma_0 = E\varepsilon + \eta\dot{\varepsilon} \tag{8.16}$$

and this differential equation may be solved for the total strain, ε, to give

$$\varepsilon(t) = \frac{\sigma_0}{E}\left(1 - e^{-\frac{E}{\eta}t}\right) \tag{8.17}$$

This indicates an exponential increase in strain from zero up to the value, σ_0/E, that the spring would have reached if the dashpot had not been present. This is shown in Fig. 8.22. As for the Kelvin model, the creep modulus may be determined as

$$E(t) = \frac{\sigma_0}{\varepsilon(t)} = E\left(1 - e^{-t/\tau_s}\right)^{-1} \tag{8.18}$$

Relaxation

If the strain is held constant then Eq. (8.16) becomes

$$\sigma = E\varepsilon \tag{8.19}$$

That is, the stress is constant and supported by the spring element so that the predicted response is that of an elastic material, no relaxation (see Fig. 8.22).

Recovery

If the stress is removed, then Eq. (8.16) becomes

$$0 = E\varepsilon + \eta\dot{\varepsilon} \tag{8.20}$$

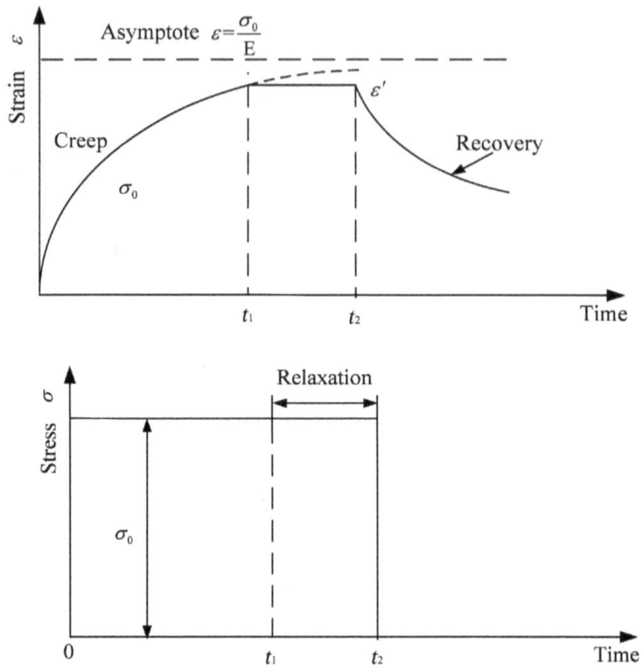

Figure 8.22 Response of Kelvin model.

Solving this differential equation with the initial condition $\varepsilon = \varepsilon'$ at the time of stress removal, then

$$\varepsilon(t) = \varepsilon' e^{-\frac{Et}{\eta}} \tag{8.21}$$

This represents an exponential recovery of strain which is a reversal of the predicted creep.

8.3.1.3 The Burger model

It can be seen that the simple Kelvin model gives an acceptable first approximation to creep and recovery behavior but does not account for relaxation. The Maxwell model can account for relaxation but was poor in relation to creep and recovery. It is clear therefore that some compromise may be achieved by combining the two models. The Burger model is shown in Fig. 8.20C, where a Maxwell and a Kelvin model are connected in a series. In this case, the stress−strain relations are again given by Eqs. (8.1) and (8.2). The geometry of deformation yields. Total strain

$$\varepsilon = \varepsilon_1 + \varepsilon_2 + \varepsilon_k \tag{8.22}$$

where ε_k is the strain response of the Kelvin model. From Eqs. (8.1), (8.2), and (8.22)

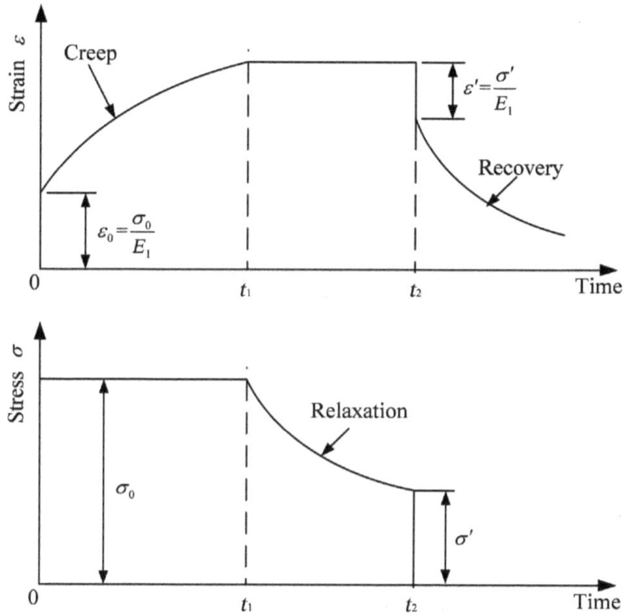

Figure 8.23 Response of Burger model.

$$\varepsilon(t)=\frac{\sigma_0}{E_1}+\frac{\sigma_0 t}{\eta_1}+\frac{\sigma_0}{E_2}\left(1-e^{-\frac{E_2}{\eta_2}t}\right) \tag{8.23}$$

From this the strain rate may be obtained as

$$\dot{\varepsilon}=\frac{\sigma_0}{E_1}+\frac{\sigma_0}{\eta_2}e^{-\frac{E_2}{\eta_2}t} \tag{8.24}$$

The response of this model to creep, relaxation, and recovery situations is the sum of the effects described for the previous two models and is illustrated in Fig. 8.23. It can be seen that although the exponential responses predicted in these models are not a true representation of the complex viscoelastic response of polymeric materials, the overall picture is, for many purposes, an acceptable approximation to the actual behavior. As more and more elements are added to the model, the simulation becomes better but the mathematics become complex.

8.3.2 Simulation of microstructure molding process

With recent advances in numerical simulation capabilities and computing technology, a FEM can address some of these issues by providing deeper insight into the process and performance prediction. For a reliable simulation model, it is necessary to have accurate material representation with temperature-dependent mechanical and thermal properties. The lens molding process is usually performed at a temperature of 150–200°C above the glass transition temperature (T_g) where the glass viscosity generally lies between $10^{7.6}$ and $10^{9.0}$ P. In this temperature range, also referred to as the transition temperature range, the glass can be described as a viscoelastic material exhibiting stress relaxation. Stress relaxation that influences residual stresses in a molded glass lens is therefore an important technological subject.

The present investigation is hence undertaken to model the stress relaxation phenomenon during lens molding. It is imperative to develop a FEM-based simulation model of microgroove forming, in order to gain a fundamental understanding of the process by evaluating various parameters, which are difficult or impossible to measure during experiments. By creating a simple numerical model, we are able to predict the performance of microgroove forming process and at the same time identify the various parameters that would enable us to model the actual process.

8.3.2.1 2D modeling

Simulations of the glass molding process are conducted using a commercial FEM code MSC. Marc, which is powerful in nonlinear solution for forging and molding process of various materials. The program is capable of simulating large deformations of material flow under isothermal or nonisothermal conditions. A general Maxwell model is set up to describe the deformation during the pressing stage, as shown in Fig. 8.24. The time-dependent response is characterized by the deviatoric terms, as shown in Eq. (8.25):

$$\sigma(t) = \int_0^t G(t-\tau)\frac{d\varepsilon}{d\tau}d\tau \tag{8.25}$$

The above integrals are evaluated for current time t based on past time τ. $G(t-\tau)$ is not a constant value, but is represented by a Prony series as in Eq. (8.26):

$$G(t) = G_0 \sum\nolimits_{i=1}^{n} w_i e^{-\frac{t}{t_{ri}}} \tag{8.26}$$

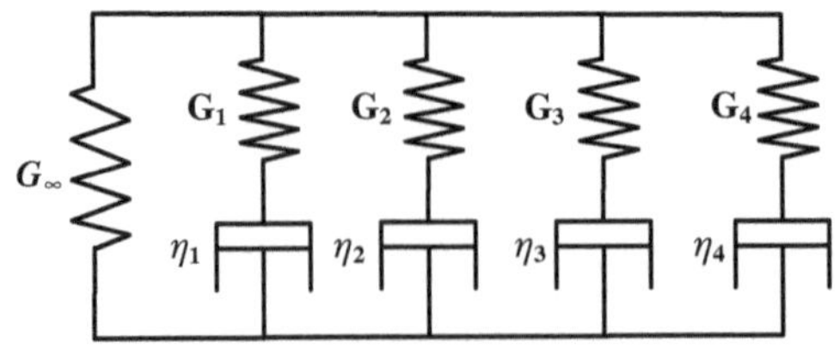

Figure 8.24 General Maxwell model to describe the viscoelasticity of glass in transition region.

where w_i is the relative moduli, t_{ri} is the reduced time, used to describe the shift in time due to temperature. The shift function ($A(T)$) used in this model is the Tool-Narayanaswamy (TN) shift function, as shown in Eq. (8.27):

$$\ln(A(T)) = \frac{H}{R}\left(\frac{1}{T_{\text{ref}}} - \frac{1}{T}\right) \tag{8.27}$$

where T_{ref} is the reference temperature. H is the activation energy, and R is the ideal gas constant.

In an actual microgroove molding process, there are 500 V-grooves on the mold spacing regularly in a 5×5 mm^2 area. For simplification, eight grooves are calculated in this simulation. Fig. 8.25 shows the two-dimensional simulation model of GMP for microgrooves, the upper mold, which is flat and fixed to the top, and the lower mold with microgrooves that move upward to press the softened glass. The glass object is meshed with 48,000, 3-node, triangle, and plane strain solid elements. In order to make sure the FEM simulation convergence in the contact and save the remeshing time at the sharp corner during the simulation, the sharp ridges of the grooves on the top are rounded with a radius of 1 μm on the top, while the sharp angle is held as the actual size of the valley. The molds are both modeled as rigid objects by the two curves. As the glass is pressed at the same temperature with the molds, no heat transfer is considered in this model (Zhou, Yan, Masuda, Owada, & Kuriyagawa, 2009).

Some simulation results are obtained. First, the effects of the molding temperature (T_{mld}) are numerically studied at 560°C, 570°C, 580°C, and 590°C, respectively. The pressing velocity (v_{p}) is set constantly at 0.1 mm/s with a displacement of 15 μm upward. The friction coefficient (μ_{f}) is specified to 0.1. Fig. 8.26 shows the equivalent stress distribution at the displacement of 15 μm at 570°C. The stress distributes bilaterally symmetrically on the whole glass plate, the lowest in the center groove and the highest on the left and right sides. The glass material flows from the center to the two sides and fills the center groove of the mold first; it then fills the outer grooves later. When the displacement reaches 13.74 μm, the V-groove of the mold is entirely replicated on the glass plate. The highest and the lowest stresses distribute more or less at the same area, but the peak-valley value of the stress changes fiercely.

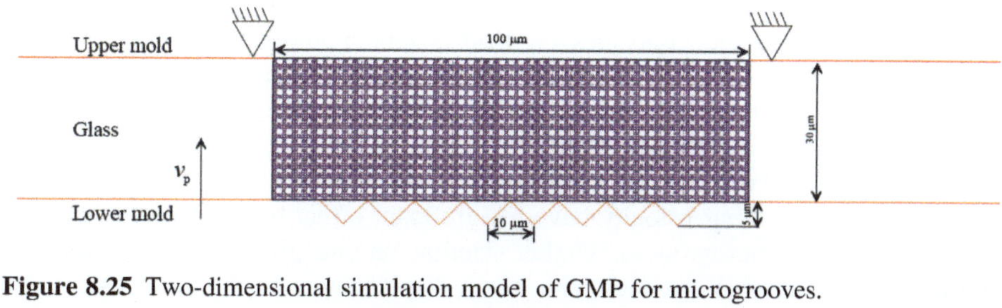

Figure 8.25 Two-dimensional simulation model of GMP for microgrooves.

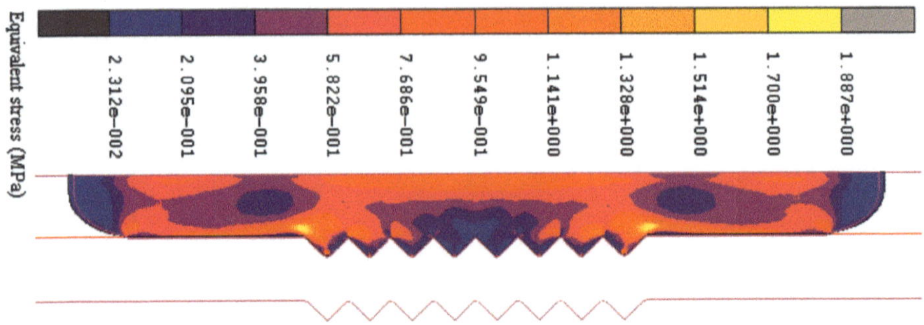

Figure 8.26 Equivalent stress distribution at the molding temperature of 570°C (T_{mld} = 570°C, v_p = 0.1 mm/s, μ_f = 0.1).

Second, Fig. 8.27A shows the maximum and minimum values of the equivalent stress at the temperatures from 560°C to 590°C. The lowest stress is more or less the same at different temperatures, but the highest temperature declines to 1/10 at 590°C, compared with the stress at 560°C. Also, the effects of the pressing velocity are investigated at 0.01, 0.1, and 1 mm/s, respectively. The molding temperature is fixed at 570°C. From Eq. (8.25), we know that the stress will be directly proportional to the strain rate. From the simulation results shown in Fig. 8.27B, we can find that the highest stress rises to 14.32 MPa, which is 7.6 times the highest stress at the pressing velocity of 0.1 mm/s, and 41 times the highest stress at the pressing velocity of 0.01 mm/s.

Third, as the friction coefficient on the glass-mold interface is unavailable, the coefficient is arbitrarily specified to 0.1 in the above simulation. In order to examine the effects of the friction, the GMP process is simulated by varying the friction coefficient from 0.1, 0.3, to 0.5 at the glass-mold interfaces. A true stick-up friction model is adopted. The molding temperature is 570°C. Fig. 8.27C shows the stress change at the friction coefficients of 0.1, 0.3, and 0.5. The stress at a friction coefficient of 0.3 becomes higher than that at 0.1, but does not much change when the coefficient is further increased to 0.5. Another interesting phenomenon is observed. The sharp angled V-groove can be formed at a level of less displacement when the friction becomes higher. A displacement of 13.7 μm is needed at a friction coefficient of 0.1, but only 10.8 μm at 0.3, and 10.1 μm at 0.5. The reason can be attributed to the different flow velocities of glass in the horizontal and longitudinal directions. The glass material is hard to flow in the horizontal direction at a high-friction force, and the increased pressing load results in an easy flow in the longitudinal direction. Consequently, the glass is squeezed to fill up the V-grooves with less displacement of the lower mold at a higher-friction coefficient.

A comprehensive analysis of these effects should be considered to optimize the molding condition. From the simulation results, it is shown that the molding temperature and the pressing velocity have a significant effect on the stress during the GMP process for microgrooves. Higher friction on the glass-mold interface can help to form the sharp V-grooves with a less displacement of the mold, but it will

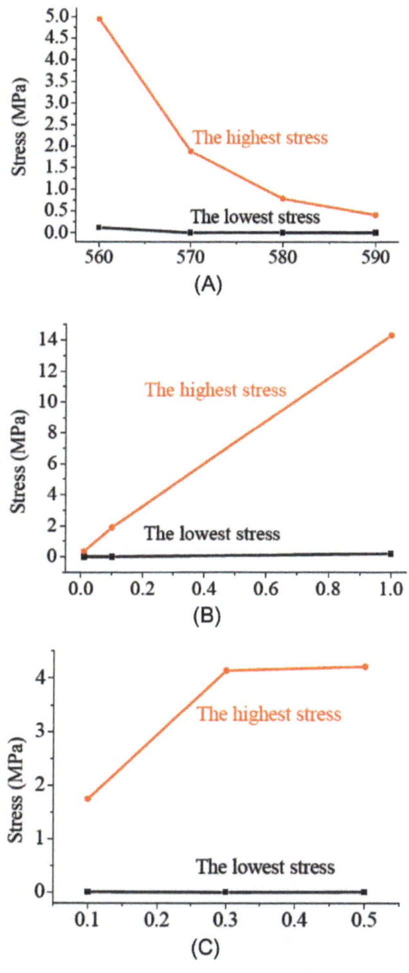

Figure 8.27 Stress change at different (A) molding temperatures, (B) pressing velocities, (C) friction coefficients.

increase the stress. In fact, the friction coefficient is almost decided by the surface roughness of the mold. A lower stress can reduce the deformation of the mold and achieve a higher form accuracy. From this viewpoint, an intermediate molding temperature of 570°C, and a slow pressing velocity of 0.1 mm/s, are selected as the optimal pressing conditions in the glass molding process (GMP) experiments.

8.3.2.2 3D modeling

It is clear from the above that 2D numerical simulation is used to illustrate the details of the GMP process for microstructure, and that the molding condition has been optimized. However, the 2D simulation is not able to reveal the process of the

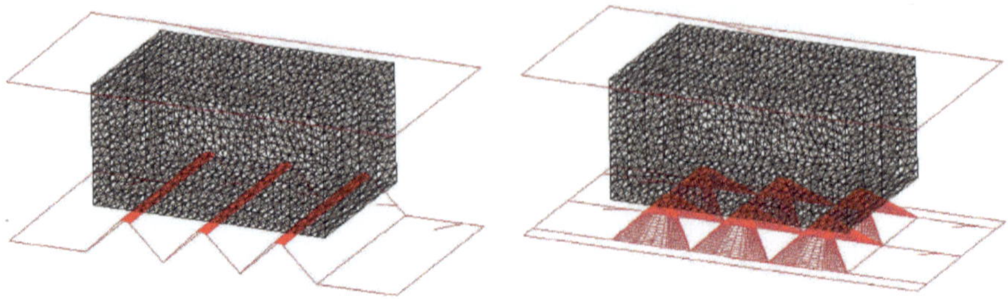

Figure 8.28 Models of the GMP for microstructures: (A) microgrooves, (B) micropyramids.

glass material flow among microstructures forming, such as microgroove and micropyramid, because the cross-section of the V-groove is the same as that of the pyramid at the vertex. Therefore, 3D simulation is necessary in the numerical analyses of micropyramid forming and the General Maxwell model is extended to the 3D simulation. Here is an example of 3D numerical simulations that are carried out to study the forming process of microgroove array and micropyramid array (Zhou, Yan, & Kuriyagawa, 2010).

Simulations of the forming process of microgroove array and micropyramid array are conducted using a commercial FEM code MSC. Marc, which is powerful in nonlinear solutions for the forging and molding process of various materials. The 3D models of the GMP for microgrooves and micropyramids are shown in Fig. 8.28A and B, respectively.

In a real molding process, there are 500 microgrooves in a microgroove array and 500×500 micropyramids in a micropyramid array on the mold. In order to save the computing time of the FEM simulation, only 3 grooves and 3×2 pyramids are modeled in the microgroove forming and the micropyramid forming in the respective simulations. The upper flat mold, which is fixed on the top, and the lower mold with microgrooves, will move upward to press the softened glass. The glass object is meshed with 514,084-node, tetrahedral solid elements. In order to make sure of the FEM simulation convergence in the contact deformation and save the remeshing time at the sharp corner during the simulation, the sharp ridges of the microgrooves and the shape vertexes of the micropyramids are rounded with a radius of 1 μm, while the sharp angle of the valleys is held as the actual size. The molds are both modeled as rigid objects by two surfaces. As the glass is pressed at the same temperature with the molds, no heat transfer is considered in this model.

Some simulation results such as stress and strain distribution are achieved. They can be used to facilitate the comparison between microgroove forming and micropyramid forming; the 3D stress distributions and strain distributions in the glass in the GMP for both microgrooves and micropyramids are shown in the same view angle, so that the differences can be easily distinguished by comparing the corresponding parts with each other.

Fig. 8.29A and B show the stress distributions during the deformation at the displacement of 6 μm during the pressing for microgrooves and micropyramids, respectively. It demonstrates that the stress on the peaks of the microgrooves is a little higher than that on the peaks of the micropyramids, but much lower on the valleys. Therefore, a higher pressing load is needed in the GMP for microgrooves than that in the GMP for micropyramids. Furthermore, the stress concentrates on the area of the microgrooves in the microgroove forming, but the stress in the micropyramid forming scatters to the whole part of the glass.

Fig. 8.30A is the cross-section view of the molded microgrooves, which shows the strain distribution at the displacement of 6 μm during the pressing for microgrooves, and Fig. 8.30B is the cross-section view of the strain distribution for micropyramids. The strain on the peak of the microgroove is larger than that of the micropyramids. Comparing the molded microgrooves with the molded micropyramids, it shows that the horizontal deformation of the glass in the microgroove forming is larger than that in the micropyramid forming, especially on the left and right tips. Moreover, the filling ratio of the GMP for microgroove forming is a little larger than that for micropyramids.

Some reasons for these differences can be generalized due to the simulation results above. First, the glass material can only flow along the two symmetric ramps of a microgroove during the pressing, but the glass material can flow along the four

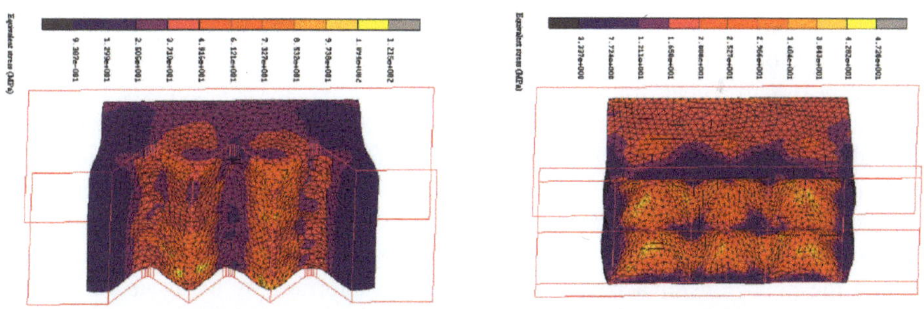

(A) (B)

Figure 8.29 Stress distribution in the glass during pressing: (A) microgrooves, (B) micropyramids.

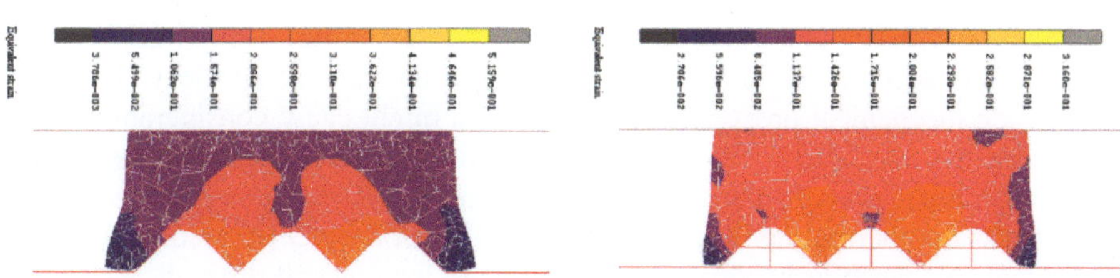

(A) (B)

Figure 8.30 Strain distribution in the glass during pressing: (A) microgrooves, (B) micropyramids.

equal ramps of a micropyramid, just like an awl is much easier to pierce into the wood than an ax at the same pressing load. Second, the micropyramid array has almost twice the space of cavities to accommodate more glass material than the microgroove array during the pressing, so the amount of glass material flowing in the horizontal direction in the micropyramid forming process is smaller than that in the microgroove forming process. The different filling ratios of the GMP for these two kinds of microstructures have been explained by the simulation results.

8.3.3 GMP simulation coupling heat transfer and viscous deformation

In GMP, a glass preform is pressed at a temperature tens of degrees centigrade above its transition temperature to replicate the shape of the mold to the glass surface. At the molding temperature, the glass shows viscosity. Two issues are practically important in the simulation of glass molding process: one is to model heat-transfer phenomena considering the temperature dependence of specific heat and thermal conductivity of glass, and the other is to make clear the viscosity of glass near the softening point.

Since glass is transparent in infrared light, it cannot be directly heated by the infrared lamp, but instead has to be indirectly heated by the molds and the surrounding gas. Therefore, heat transfer in this case involves heat conduction, heat convection, and radiation. Also, heat expansion coefficient, heat conductivity, heat capacity and other parameters of glass are also changing with temperature. From these aspects, the heat transfer in glass molding is a very complicated issue. However, at present, precise measurement of the heat transfer in a glass lens is still difficult, although Field and Viskanta (1990) demonstrated the possibility of an experimental measurement of temperature distributions in soda-lime glass plates.

FEM has shown to be an effective approach to simulate a forming process and visualize the heat transfer in glass. For example, Wilson, Schmid, and Liu (2004) studied the heat transfer across a tool-workpiece interface by using an FEM code, DEFORMTM-2D. Viskanta and Lim (2001) proposed a physical model for internal heat transfer in glass and heat exchange across glass-mold interface in one dimension. Yi and Jain (2005) applied DEFORMTM-2D to the simulation of aspherical glass lens molding.

However, most of these studies analyzed the heating process by separately considering heat transfer and glass deformation. The numerical models they used did not take into account the time dependent and temperature-dependent changes of thermal and mechanical properties of glass, which might cause considerable simulation errors. To date, there is no available literature on the dynamic modeling of a high-temperature glass forming process by comprehensively considering the heat transfer and the thermal deformation of glass. Therefore, precise prediction of process parameters, such as heating time and pressing load, is still difficult.

In this section, a series of thermo-mechanical models coupling heat transfer and viscous deformation are established for the glass molding process, from heating, pressing, annealing to cooling, to enable FEM simulation and visualization of the process, and accurate prediction of optimal process parameters. Firstly, thermal phenomenon during the glass molding process is analyzed by considering heat transfer within glass and at the interface among glass, molds, and the surrounding gas environment. Secondly, the high-temperature viscosity of glass near the softening point is measured using an ultraprecision glass molding machine by uniaxially pressing cylindrical glass preforms between a pair of flat molds. Thirdly, the developed numerical models and the measured viscosity properties are incorporated into coupled thermo-mechanical FEM simulation of the glass pressing process based on a viscous fluid model. It shows that the developed thermal and mechanical models coupling heat transfer and viscous deformation can be used to precisely predict important process parameters of glass molding (Yan, Zhou, Masuda, & Kuriyagawa, 2009).

8.3.3.1 Theoretical models of heat transfer and viscous deformation

Thermal expansion of glass

Thermal expansion is an important issue in lens molding because it greatly influences the form accuracy of the lens. The thermal expansion of glass is strongly temperature dependent. The expansion coefficient is approximately linear at low temperature (below T_g), and becomes nonlinear in the high-temperature range. The thermal expansion coefficient increases significantly from T_g to A_t, then a negative expansion (namely contraction) occurs above A_t. Ohlberg and Woo (1973) proposed a complicated formula to describe the change of glass expansion coefficient, but it is still hard to precisely describe the thermal expansion coefficient in a single function. For simplification, linear approximations are adopted to represent the changes of thermal expansion coefficient in different temperature ranges. To begin with, we consider the one-dimensional thermal expansion problem. The linear thermal expansion coefficient (α) can be given by Eq. (8.25):

$$\alpha = \frac{1}{l} \times \frac{dl}{dT} = \begin{cases} \alpha_0 & -30 \leq T < 70 \\ \alpha_1 & 70 < T < T_g \\ \alpha_1 + \alpha_2(T - T_g) & T_g \leq T < A_t \end{cases} \tag{8.28}$$

where l is the overall length of material in the direction being measured; T is the instantaneous temperature in °C; a_0 is the constant thermal expansion coefficient in the working temperature range of glass; a_1 is the constant thermal expansion coefficient below T_g; and a_2 is the average gradient of the increase of thermal expansion coefficient between T_g and A_t. The glass material used in Section 3.3 is L-BAL42, the thermal characteristics of which are listed in Table 8.3. Other glasses show similar general trends in the volume–temperature curves. According to the data provided by the glass manufacturer, we obtain that $a_0 = 7.2 \times 10^{-6}$ L/°C, $a_1 = 8.8 \times 10^{-6}$ L/°C, and $a_2 = 3.0 \times 10^{-8}$ L/°C2 in Eq. (8.28).

Next, we consider volume expansion in three-dimensional glass molding. The volume thermal expansion coefficient (β) is approximately three times the linear

Table 8.3 Thermal expansion characteristics of glass L-BAL42

Characteristic point in the volume–temperature curve	Temperature (°C)
Strain point (*StP*)	476
Annealing point (*AP*)	494
Transition temperature (T_g)	506
Yielding temperature (A_t)	538
Softening point (*SP*)	607

thermal expansion coefficient, thus it can be defined by Eq. (8.29). The temperature-dependent volume ($V(T)$) can then be calculated using Eq. (8.30).

$$\beta \approx 3\alpha \tag{8.29}$$

$$V(T) = V_0\left(1 + \int_0^T \beta dT\right) = V_0\left(1 + 3\int_0^T \alpha dT\right) = V_0 + \Delta V \tag{8.30}$$

where V_0 is the volume of glass at the reference temperature (0°C), and ΔV is the swelling increment in volume after heating to temperature T. In the FEM simulation of molding process, a module is activated to restore the volume change due to thermal expansion during remeshing, and the coefficient of thermal expansion is used to define the volumetric strain due to temperature changes.

Heat transfer models

On most occasions, glass molding processes are based on the "isothermal molding" method where pressing is performed after the glass preform reaches the same temperature as the molds. However, due to the fact that the heat absorption rate of glass is distinctly different from that of the mold materials, such as tungsten carbide (WC), silicon carbide (SiC), titanium carbide (TiC), nickel–phosphorous (NiP) plated steels, and other alloys, even the glass preform and the molds are heated together by the same infrared lamp, there is a significant delay of temperature rise within the glass preform. Most of the heat transferred to the glass preform is from the lower mold and surrounding nitrogen during the soaking time. However, it is difficult to directly measure the temperature change in the glass preform even if the temperature of the molds can be easily monitored by thermocouples. Therefore, modeling of heat transfer phenomenon in glass molding is very important for working out heat balance.

Initially, we consider the heat transfer within a glass preform. The governing equation for heat conduction within an incompressible glass material is given by Eq. (8.31):

$$\rho C_p \frac{\partial T}{\partial t} = k\nabla^2 T \tag{8.31}$$

where ρ is the density; C_p is the specific heat; k is the thermal conductivity of glass; and t is the heating time. Most of the work on simulation of glass molding and

Table 8.4 Composition ratio of glass L-BAL42

Glass composition	Content (wt%)
SiO_2	40–50 (≈45)
BaO	20–30 (≈25)
B_2O_3	2–10 (≈6)
Al_2O_3	2–10 (≈6)
ZnO	2–10 (≈6)
Others	0–2 (≈2)

other forming processes treated the specific capacity C_p and thermal conductivity k as constants (Choi, Ha, Kim, & Grandhi, 2004; Zhou & Li, 2005). However, in fact these two parameters are temperature-dependent, as experimentally demonstrated by early researchers (Moser & Kruger, 1968; Richet, Bottinga, & Tequi, 1984).

As glass is a compound material, the specific heat of glass is known to vary with its composition and temperature. An empirical equation, namely, Sharp–Ginther equation (Sharp & Ginther, 1951), has been proposed to express the mean specific heat (C_m) of glass, as shown in Eq. (8.32).

$$C_m = \frac{a_3 T + C_0}{0.00146T + 1} \times 4.168 \times 10^3 \tag{8.32}$$

where a_3 is a constant of a glass material; C_0 is the true specific heat at 0°C; and T is the temperature in °C. a_3 and C_0 are both decided by the glass compositions. This equation can be applied to a wide temperature range and has been widely used by later researchers.

The composition percentages of the glass samples offered by the glass manufacturer are shown in Table 8.4. An intermediate value of each composition is used to calculate the constants a_3 and C_0 in Eq. (8.23). The calculated values are $a_3 = 0.000408$ J/°C^2 and $C_0 = 0.144$ J/°C, respectively. It was reported that oxides BaO and ZnO influence the thermal properties of glass (Moore & Sharp, 1958). However, from experimental results in the present study, we cannot find obvious change of heat capacity due to BaO and ZnO and so we ignore their effects.

The true specific heat C_p in Eq. (8.31) then can be easily calculated from the differential equation as follows:

$$C_p = \frac{d(TC_m)}{dT} \tag{8.33}$$

In Eq. (8.31), the thermal conductivity k is a variable of temperature in a complex manner due to the effects of high-temperature radiation. For simplicity, many investigators treated the high-temperature irradiation in glass as an equivalent thermal conductivity problem, and a temperature-dependent thermal conductivity ($k(T)$) was proposed for glass materials by Mann, Field, and Viskanta (1992). Based on

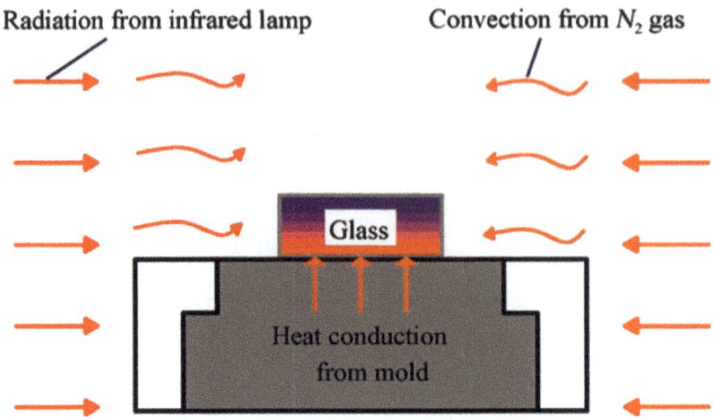

Figure 8.31 Schematic presentation of heat transfer phenomenon during heating process.

the glass property data provided by the manufacturer, a modified formula is given in Eq. (8.34) to represent the thermal conductivity of glass L-BAL42.

$$k(T) = 1.028 + 0.000624T \tag{8.34}$$

Next, we consider the heat transfer from the mold and the nitrogen environment to the glass preform. As shown in Fig. 8.31, conduction, convection and radiation are three basic modes of heat transfer. In glass molding, thermal conduction at the glass-mold interface and thermal convection between the glass preform and the flowing nitrogen gas are primary contributions to the temperature rise of glass preform. The effect of infrared irradiation on glass is insignificant so that usually it can be ignored, although McGraw (1961) proved that the effect of radiation on the temperature distribution within glass can be described by equivalent thermal conductivity.

The thermal boundary conditions of the glass preform during heating can be given by Eq. (8.35):

$$\begin{aligned} -k\frac{\partial T}{\partial n} &= h_{\mathrm{M}}(T - T_{\mathrm{M}}) \\ -k\frac{\partial T}{\partial n} &= h_{\mathrm{N}}(T - T_{\mathrm{N}}) \end{aligned} \tag{8.35}$$

where h_{M} is the interface heat transfer coefficient between the mold and glass; h_{N} is the heat transfer coefficient between the surrounding nitrogen gas and glass; h_{M} and h_{N} are the temperatures of the mold and the nitrogen gas, respectively. In Eq. (8.35), h_{M} is a function of interface pressure, thickness of air gap, amount of sliding, interface temperature, etc., and h_{N} is affected by chamber geometry, gas blowing velocity, flowing direction, and so on. Therefore, strictly speaking, h_{M} and

h_N are both variables. However, precise measurement of the changes of these coefficients is technically impossible under the present conditions. As a result, like in most of the previous studies (Yi & Jain, 2005), constant values of h_M (2800 W/(m2K)) and h_N (20 W/(m2K)) are assigned in the FEM simulation.

High-temperature viscosity of glass

When the temperature is below T_g, glass is deemed as solid, which can be treated as a rigid or rigid-plastic material. However, when temperature is above the softening point *SP*, glass becomes a viscous liquid which is soft and flowable. Between T_g and *SP*, the properties of glass change greatly with temperature and it behaves as a viscoelastic/plastic body. To achieve glass deformation at a certain load, an experiment is performed near the softening point *SP* where glass can be approximately treated as a viscous Newtonian fluid. High-temperature viscosity (η') is conventionally measured by parallel-plate viscometers according to Eq. (8.36), as reported by Gent (1960).

$$\eta' = \frac{2\pi F h^5}{3V\dot{h}(2\pi h^3 + V)} \tag{8.36}$$

where F is the pressing load; h is the instantaneous height of the cylindrical glass sample; $\dot{h}$ is the axial deformation rate; and V is the volume of the sample. Instead of a parallel-plate viscometer, we use an ultraprecision molding machine to measure the viscosity of glass. Two flat molds are used as the rigid parallel-plate pair. Pressing load, displacement and speed are recorded by a load cell and a sensing-control system equipped in the machine.

During pressing, glass is assumed to be incompressible. However, the volume of glass will change due to thermal expansion. To improve the accuracy of viscosity measurement, Meister (1982) attempted to make modifications to Eq. (8.30) by considering the thermal expansion. Unfortunately, he made a mistake during formula deduction, which made the modified formula unreliable. Therefore, we correct the formula deduction process as shown in Eq. (8.37).

$$\begin{aligned}\eta &= \frac{2\pi F(h+\Delta h)^5}{3(V+\Delta V)\dot{h}[2\pi(h+\Delta h)^3 + V + \Delta V]} \\ &= \frac{2\pi F h^5(1+\alpha T)^5}{3V(1+3\alpha T)\dot{h}[2\pi h^3(1+\alpha T)^3 + V(1+3\alpha T)]} \\ &\approx \frac{(1+\alpha T)^5}{(1+3\alpha T)^2}\eta' \approx \frac{1+5\alpha T}{1+6\alpha T}\eta'\end{aligned} \tag{8.37}$$

where η is the modified value of viscosity, and Δh is the swelling increment in height.

After the glass viscosity at certain temperature intervals have been experimentally measured, the viscosity between these temperature intervals can be estimated by curve fitting. The function usually used for fitting glass viscosity is the Vogel–Fulcher–Tamman (VFT) equation as shown in Eq. (8.38) (Fulcher, 1925; Kobayashi, Takahashi, & Hiki, 2006).

$$\log \eta = A + \left[\frac{B}{(T - T_0)}\right] \tag{8.38}$$

where A, B, and T_0 are constants.

Stress–strain relationship in viscous deformation

Usually, bulk glass deformation behavior at a low strain rate above yielding point A_t has been modeled by the Newtonian incompressible law, described in terms of equivalent stress σ and strain rate $\dot{\varepsilon}$ as given by Eq. (8.39).

$$\sigma = 3\eta\dot{\varepsilon} \tag{8.39}$$

where η is the viscosity of glass, which can be obtained by the VFT equation. However, Chang et al. (2007) found that flow stress function did not follow the exact Newton's Law for fluids and revised it by modifying the exponent of strain rate based on their experimental results. Similar phenomena is observed. According to the properties of L-BAL42 glass, a modified flow equation is adopted as Eq. (8.40) in the FEM simulation.

$$\sigma = 3\eta\dot{\varepsilon}^{1.2} + 0.3 \tag{8.40}$$

It should be mentioned that during pressing, heat transfer still continues. In this case, heat transfers to the glass preform from both the upper and lower molds. However, if the glass preform has been soaked enough so that a uniform temperature distribution is achieved within the glass preform during the heating stage, heat transfer during pressing will be insignificant and will not affect the constants in Eq. (8.40).

The frictional force between glass and molds during pressing is modeled as a constant shear friction (Li, Jinn, Wu, & Oh, 2001), which can be defined by Eq. (8.41).

$$f_s = m\tau \tag{8.41}$$

where f_s is the frictional stress, τ is the shear yield stress, and m is the friction factor. In this study, a value of 1.0 is assigned to m, assuming complete sticking between glass and molds without slip. Under this condition, the friction stress is a function of the yield stress of the deforming glass.

8.3.3.2 FEM simulations

Computer simulations of the glass molding process are carried out using a commercially available FEM program DEFORM™-3D. The program is capable of simulating large deformation of material flow under isothermal and nonisothermal conditions. The FEM model is shown in Fig. 8.32. In the model, the two molds are assumed as completely rigid bodies; the glass preform is deformable and becomes a Newtonian fluid at the pressing temperature. Two-dimensional rigid wall boundaries are used for the upper mold, and one-dimensional rigid wall boundaries are set to the lower mold. The glass preform is totally free of constraints except its surface contact with the two molds. In order to save computation time and improve simulation accuracy, a calculation is done for one-fourth of the cylindrical glass preform by taking advantage of the geometrical symmetry. The element remeshing is updated automatically based on an optimized Lagrangian scheme. The numerical models and boundary conditions and the experimentally measured results of glass viscosity are incorporated into the FEM simulation, which enable a coupled thermal-mechanical analysis.

To start the simulation, the mold is heated from the room temperature to the pressing temperature (590°C). In simulation, the temperature distribution within the molds is assumed to be uniform. The temperature of the lower mold measured by the thermal couple is used as the temperature boundary condition of the mold. Fig. 8.33 shows temperature distributions in the glass preform during heating. After

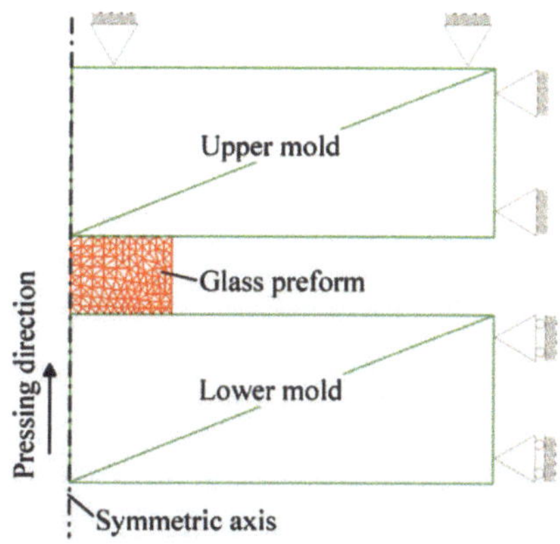

Figure 8.32 FEM simulation model for glass molding tests using flat molds.

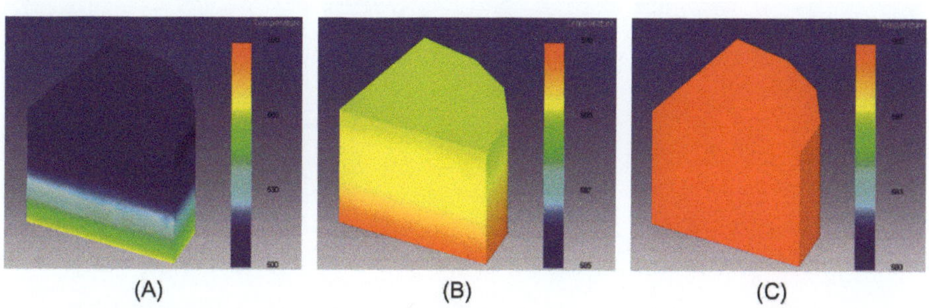

Figure 8.33 Temperature distributions in the glass preform at different heating time: (A) 120 s, (B) 180 s, and (C) 220 s.

heating for 120 s, the temperature of the bottom surface is 66°C higher than that of the top surface (Fig. 8.33A). The temperature difference is decreased to 10°C after heating for 180 s (Fig. 8.33B). Then, after heating for 220 s, the temperature becomes uniform within the whole glass preform and reach the pressing temperature 590°C (Fig. 8.33C).

Another phenomenon caused by nonuniformity of temperature in glass is that the initially cylindrical glass preform, after pressing, will be deformed to be an isosceles trapezoid where the diameter of the bottom surface is bigger than that of the top surface. Fig. 8.34A shows an example of FEM simulated cross-sectional geometry with strain distribution of a glass piece pressed after a heating time of 180 s. The curvature radius of the upper corner is apparently larger than that of the lower corner. This is due to the fact that the temperature of the upper surface is lower than that of the lower surface in the beginning of press, which cause a higher viscosity at the upper part. Although during pressing the heat transfer from the upper mold can finally eliminate the temperature difference, the deformation of the upper part has been delayed compared to the lower part. This effect finally leads to the trapezoidal geometry of the glass piece. Similarly, in lens molding, nonuniformity of temperature will cause lens form error. Fig. 8.34B shows the equivalent stress (von Mises stress) distribution under the same condition as Fig. 8.34A. The high-stress region is not symmetrical to the horizontal centerline but tends to be lower. This kind of stress distribution may cause deflection of the lens and nonuniformity in its optical properties. Fig. 8.34C shows an instantaneous velocity distribution of element nodes in glass at the end of the first press. It can be seen that near the symmetrical center, velocity vectors are vertically directed; while in the outer region, material flow direction tends to be horizontal. It is also noteworthy that the material in the lower corner is completely stationary at this moment, and material flow can be found only in the upper region.

Fig. 8.35A is a simulated cross-sectional geometry with strain distribution of the glass piece pressed after a heating time of 220 s. In this case, the temperature within the glass has become uniform. The curvature radius of the upper corner of the glass piece is completely the same as the lower corner. Fig. 8.35B shows the corresponding equivalent stress distribution. It is seen that the stress concentration region is

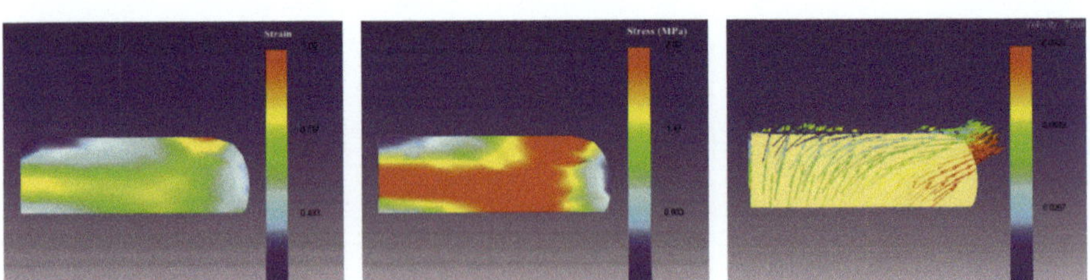

Figure 8.34 FEM simulated results of (A) cross-sectional geometry with strain distribution, (B) equivalent stress distribution and (C) velocity distribution in the glass piece pressed after a heating time of 180 s.

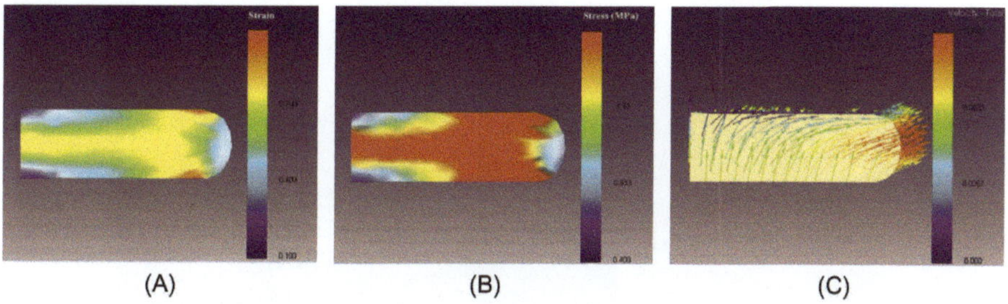

Figure 8.35 FEM simulated results of cross-sectional geometry with strain distribution (A), equivalent stress distribution (B), and velocity distribution (C) in the glass piece pressed after a heating time of 220 s.

basically symmetrical to the horizontal centerline. Fig. 8.35C shows the velocity distribution in the glass at the end of the press. Obviously, velocity vector distribution in the outer region of the glass piece is very uniform, from the upper corner to the lower corner. From this point, we can say that choosing a suitable heating time is not only an important issue for prolonging the service life of molds, but also an essential step for improving form accuracy and optical property of the molded lenses.

High-temperature heat transfer and viscous deformation of glass in a lens molding process have been studied through theoretical analysis and FEM simulations. Thermal expansion and heat transfer phenomena in the glass molding process have been modeled by considering the temperature dependence of thermal properties of glass and interfacial conditions among glass, molds, and environmental gas. The minimum heating time is predicted by FEM simulation. In addition, high-temperature material flow of glass is also simulated by a modified Newtonian fluid model.

8.4 Glass molding process for microstructures

8.4.1 Glass molding machine

In order to conduct glass molding process (GMP) at high temperatures, different molding machine are widely used. There are several commercial machines available, including two main machines that will be discussed below. Both of these machines offer the capability and flexibility required for both scientific research and industrial practice, which enables precise control over the mold position, load, and temperature, while incorporating an extremely flexible design where myriad tests can be accommodated.

8.4.1.1 Glass molding machine PFLF7-60A

The photograph and basic structure and functional features of glass molding machine PFLF7-60A is shown in Figs. 8.36 and 8.37.

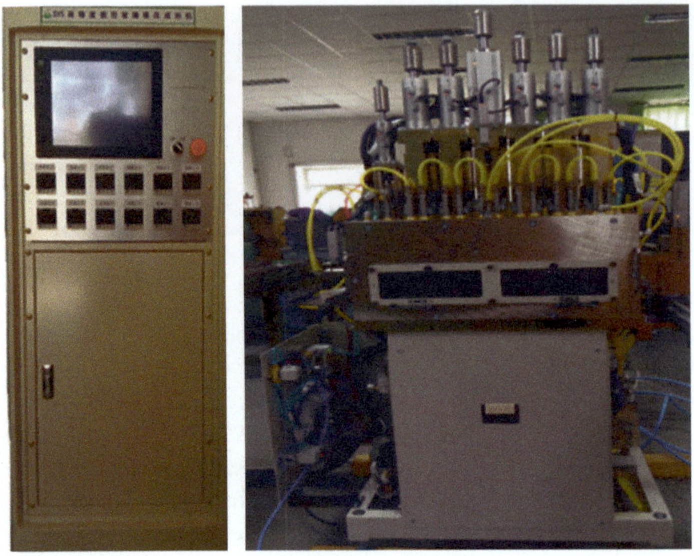

Figure 8.36 Photograph of glass molding machine PFLF7-60A (SYS Co., Ltd., Japan).

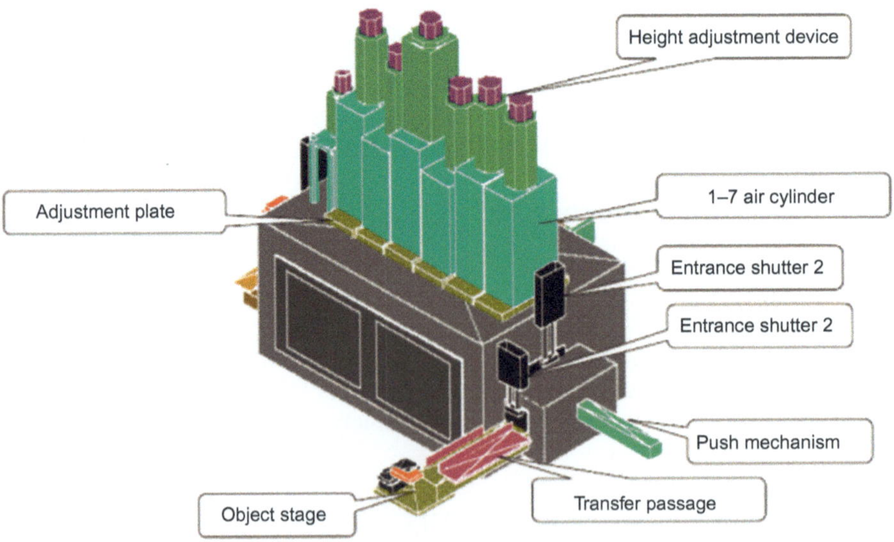

Figure 8.37 The basic structure and functional features of glass molding machine PFLF7-60A.

Figure 8.38 The seven cylinder of glass molding machine PFLF7-60A.

The machine is equipped with a drive system, force adaptive control, precision position control and data collection. Some basic adjustment of the machine will be introduced here.

Adjustment of cylinder (see Fig. 8.38)

Cylinder 1 (heating 1) Cylinder 1 increases the mold temperature with little or without any pressure and it adjusts upward and downward positions by the weight of the cylinder. It is also equipped with two regulators, pressure adjustment and an adjustment for contacting the mold. In this working position, the lens is supposed to be heated to high temperature preliminary and ready for the next position.

Cylinder 2–3 (heating 2–3) Cylinder 2 and 3 are primarily intended to apply slight pressure and to raise the mold temperature at the same time. So a proper pressure control needs to be done. In these two working positions, the lens is supposed to be heated to the molding temperature.

Cylinder 4 (pressing) Cylinder 4 is primarily intended to carry out actual molding. It is necessary to perform optimum pressure control on the molds with the lens inside. Cylinder 4 is also equipped with a two-control system consisting of a electro-pneumatic regulator and manual regulator.

Cylinders 5–6 (cooling 1–2) Cylinders 5 and 6 are primarily intended to apply slight pressure and to lower the mold temperature. In these two working positions, the temperature of lens is supposed to be reduced at a low speed so as to complete the annealing process.

Cylinder 7 (cooling 3) Cylinder 7 is to decrease the mold temperature with little or without any pressure and the primary purpose is lowering the mold temperature. In this working position, the lens is supposed to be cooled to around 200°.

Other adjustments

Cooling water flow adjustment The flow rate can be confirmed at a flow rate meter of cooling water. In addition, the cooling water should flow and keep the operating temperature of 20° during heating.

Figure 8.39 Photograph of glass molding machine GMP211 (Toshiba Machine Co., Ltd., Japan).

Nitrogen flow adjustment There is a flow meter that checks the nitrogen pressure and adjusts the nitrogen flow rate with the appropriate value by specific methods. Moreover, nitrogen flows should never stop through the driving time of the heating.

Air adjustment Each cylinder operating air are managed by one of the original regulator. In addition, it is better to adjust to 0.5 MPa in normal operation.

8.4.1.2 Glass molding machine GMP211

The photograph of glass molding machine GMP211 is shown in Fig. 8.39.

Some typical molded products by glass molding machine from Toshiba Machine Co., Ltd. are shown in Table 8.5.

8.4.2 Molding quality control

8.4.2.1 Temperature control

Glass is a strongly temperature-dependent material. At room temperature, glass is a highly hard and brittle material; at a high temperature, however, it becomes a visco-elastic body or a viscous liquid. Thermal expansion of glass is also significant. For example, the volume-temperature relationship of a commonly used glass L-BAL42 (Ohara Corp., Kanagawa, Japan) is plotted in Fig. 8.40. Softening point (*SP*) is defined as the temperature at which the glass deforms under its own weight and behaves as liquid. Yielding point (A_t), also called "deformation point," is a temperature at which glass reaches its maximum expansion and a relatively low plasticity and starts shrinking. When such a melt is gradually cooled, the volume decreases abruptly down to a transition temperature (T_g) below which volume shrinkage occurs at a slower rate. Pressing of a glass lens is always performed above T_g. Annealing point (*AP*) is the upper end of the annealing range for the pressed glass lens, at which the internal stress is reduced to a practically acceptable value over a short period.

Table 8.5 Typical molded products from Toshiba Machine Co., Ltd.

<table>
<tr>
<td>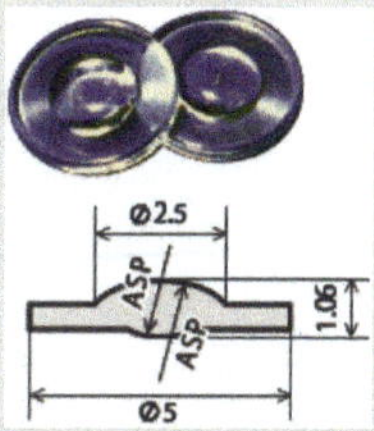

Molding machine: GMP311V
Mold: Tungsten carbide alloy
Glass material: L-BAL42
Molding temperature: 570°C</td>
<td>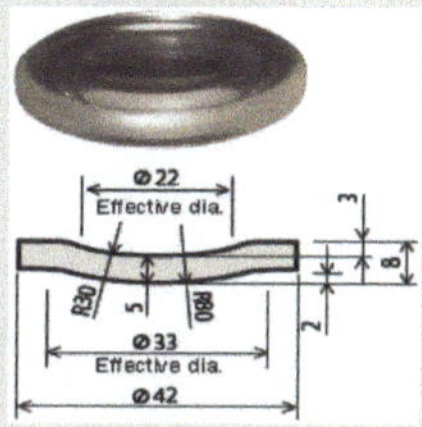

Molding machine: GMP311V
Mold: Ni alloy
Glass material: K-PSFn214
Molding temperature: 470°C</td>
<td>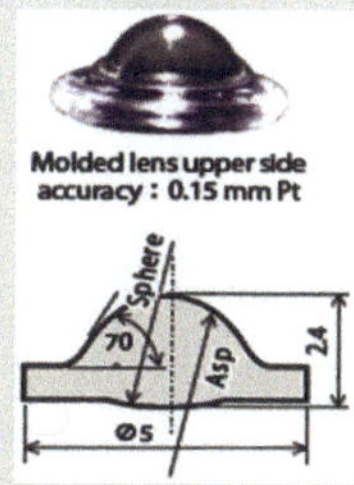

Molding machine: GMP311V
Mold: Tungsten carbide alloy
Glass material: L-BAL42
Molding temperature: 550°C</td>
</tr>
<tr>
<td>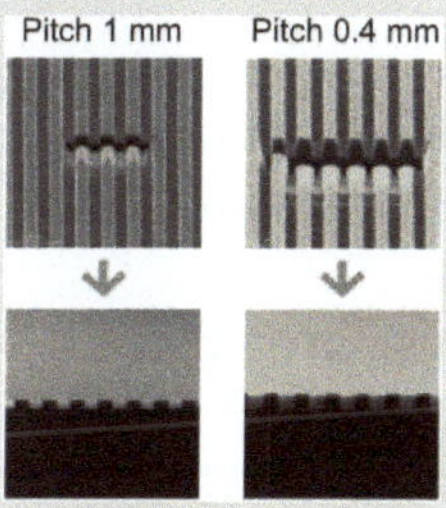

Molding machine: GMP207
Mold: Amorphous carbon
Glass material: Quartz glass
Molding temperature: 1400°C</td>
<td>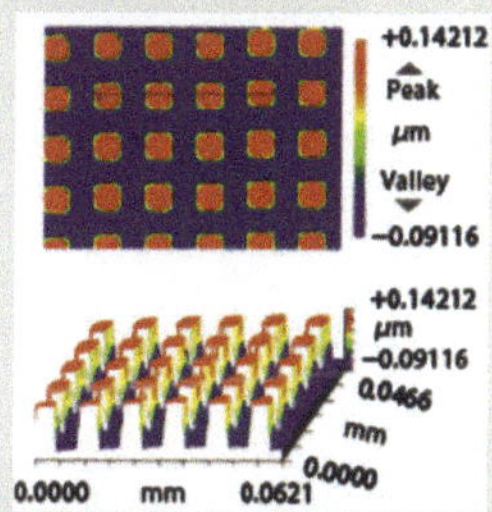

Molding machine: GMP207
Mold: Amorphous carbon
Glass material: Quartz glass
Molding temperature: 1350°C</td>
<td>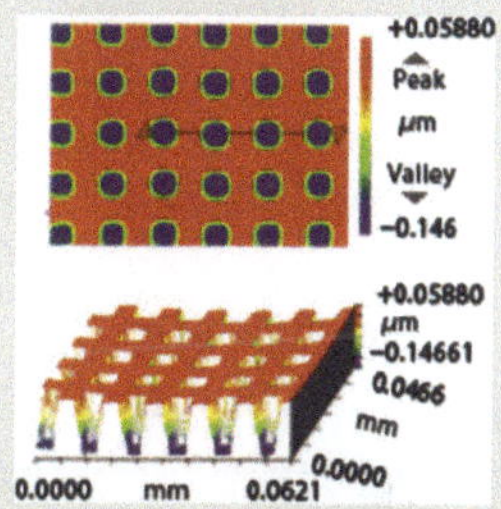

Molding machine: GMP207
Mold: Amorphous carbon
Glass material: Quartz glass
Molding temperature: 1350°C</td>
</tr>
</table>

Strain point (*StP*) represents the lower end of the annealing temperature range. It is also the upper limit of service temperature of a glass component. During annealing, glass is slowly cooled down from *AP* to somewhat below *StP*.

Determination of a suitable temperature for pressing is an essential issue for glass molding. If pressing is performed above A_t and held on to keep the shape of the lens during cooling, the volume expansion around A_t will lead to a sharp increase in pressing load, and in turn, adhesion of glass to molds. On the contrary, if pressing is done below A_t, a high pressing load will be required since glass is not sufficiently softened at this temperature range. In this case, significant residual stresses will occur in the glass lens, and the high pressing load may also shorten the service life of the molds. This problem is a critical one especially when molding micro Fresnel lenses and DOEs where the molds have extremely fine grooves on

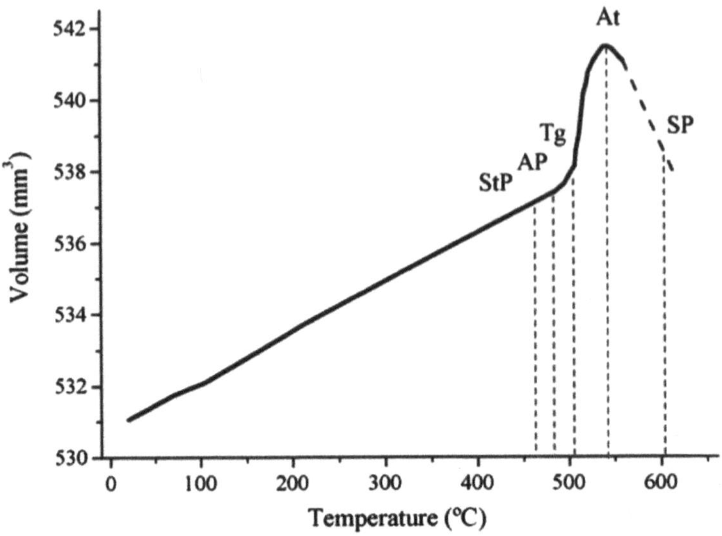

Figure 8.40 Plot of volume change against temperature for a typical optical glass L-BAL42, showing strongly temperature-dependent thermal expansion characteristics. *SP*, softening point; A_t, yielding point; T_g, transition point; *AP*, annealing point; *StP*, strain point.

Table 8.6 **Advantages and disadvantages of vacuum and nitrogen environment**

Molding environment	Advantages	Disadvantages
Vacuum environment	Better effect on prevent oxidation	High cost
Nitrogen environment	Easy operating	Air exhaust system

the surface. Therefore a suitable temperature for pressing is truly necessary in glass molding process for microstructures.

8.4.2.2 Mold oxidation prevention

In a lens molding process, a glass blank or gob is initially heated to a temperature above its transition and subsequently pressed between the two mold halves into a lens shape. During the whole molding process, especially the heating and pressing process, the glass and mold must be heated in oxygen free environment to prevent oxidation that may damage the surface quality of glass and reduce the service life of mold.

In present glass molding machines, they generally offer two approaches to isolate from oxygen. One is a vacuum environment and the other is a nitrogen environment. For instance, glass molding machine GMP211 by Toshiba Machine Co., Ltd. provides a vacuum environment to fabricate the lens, while glass molding machine PFLF7-60A by SYS Co., Ltd. offers a nitrogen environment. There are advantages and disadvantages in both methods to avoid oxidation, as shown in Table 8.6.

8.4.2.3 Methods to increase mold life

Improve pressing method

Some steps can be added into the molding process to increase the mold life in condition of no bad effect on form accuracy. Based on the conventional method, we propose a two-step pressing method. The first pressing step is done at a high temperature near the softening point (*SP*), so that glass behaves as a viscous liquid and most of the material deformation can be achieved at a low load. After the first step, the pressing load is reduced and the temperature is lowered to be between T_g and A_t. Then the second pressing is done at this temperature to close the molds and obtain the final shape of the lens. This two-step pressing method can achieve both high lens accuracy and long mold life.

Coating

One crucial requirement of a mold for glass lenses is the demolding ability. That is to say, the finished lens must be easily removed from the molds. In high-temperature glass molding, the surface of the mold tends to oxidize and the oxidized material diffuses into the glass, causing adhesion of the glass lens to the mold. Therefore, a release agent is usually coated onto the surface of the mold to prevent oxidation. It is generally considered that the lifetime of a mold is almost completely determined by that of the release agent coating. Wearing and flaking of this coating reduces the lifetime of the mold. Moreover, deterioration of the oxidation resistance of the release agent due to in-diffusion of elements from the mold leads to an increased level of wettability by the glass which again reduces the mold lifetime. Thus, the lifetime of the mold is influenced both by interactions between the release agent and the mold and the release agent and the glass. Consequently, the choice of an appropriate release agent for the mold substrate is critical.

Mold equipment diagnosis techniques

As for the measures to achieve the long life of mold, there is an improvement in software in addition to the aforementioned improvements in hardware. The improvement in software mentioned here means application of an equipment diagnosis technique to find out the deterioration of equipment quantitatively, predict mold life up to the allowable criterion, and replace the mold at the proper timing just before the criterion is reached.

8.4.3 Molding defects

8.4.3.1 Incomplete filling

Conventionally, the accuracy is low in microgroove forming due to the incomplete filling of the material in the cavity of microgroove mold. It makes incomplete filling become one of most important molding defects in the glass molding process. When the stress applied to the material sample is constant, the dashpot representing Newtonian fluid features plays a dominant role during deformation. The energy loss caused by Newtonian fluids reduces the filling rate after a period of deformation because the material flow rate decreased by time.

We should look into the reasons for the forming error of the molded microstructures. First, the GMP is carried out at a high temperature of 570°C. After pressing, the formed microstructure arrays are cooled to room temperature. The formed microstructures will contract and reduce the filling ratio. Therefore, a low molding temperature is preferred in the microstructure forming for higher filling ratio of glass.

The filling ratios of the glass at different friction coefficients are studied here by using a commercial FEM code ABAQUS. We can see from Fig. 8.41 that the filling ratios of the glass tend to become lower with the declining of a friction coefficient. Therefore, it is appropriate to avoid the incomplete filling by increasing the friction coefficient between glass and molds.

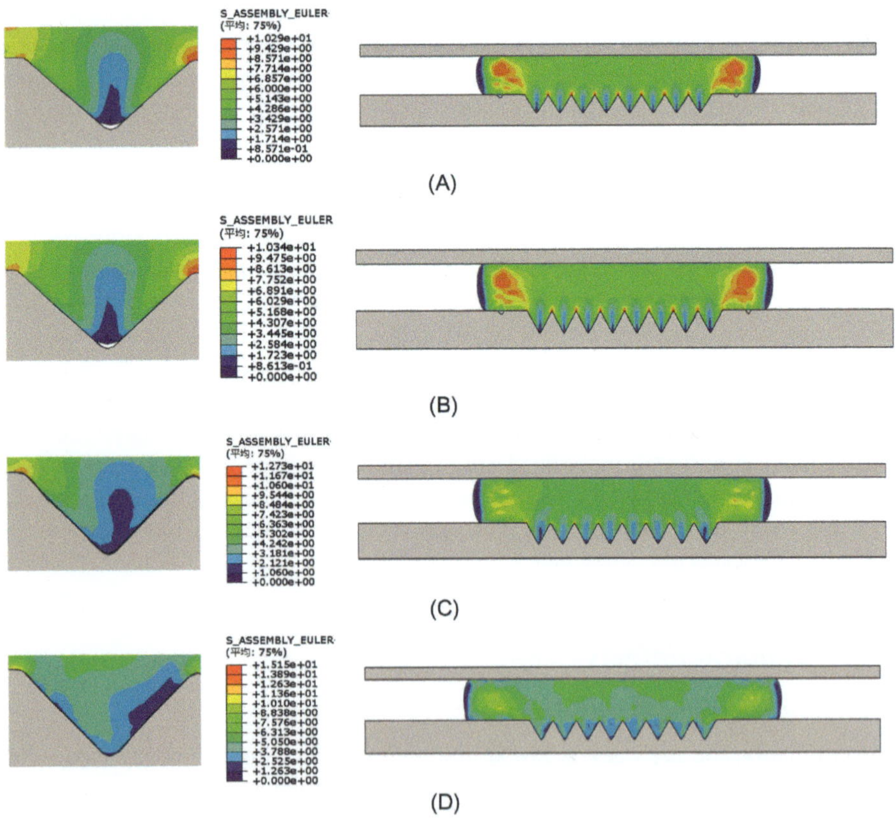

Figure 8.41 Filling ratios of the glass at different friction coefficients: (A) 0.5; (B) 0.3; (C) 0.1; (D) 0.

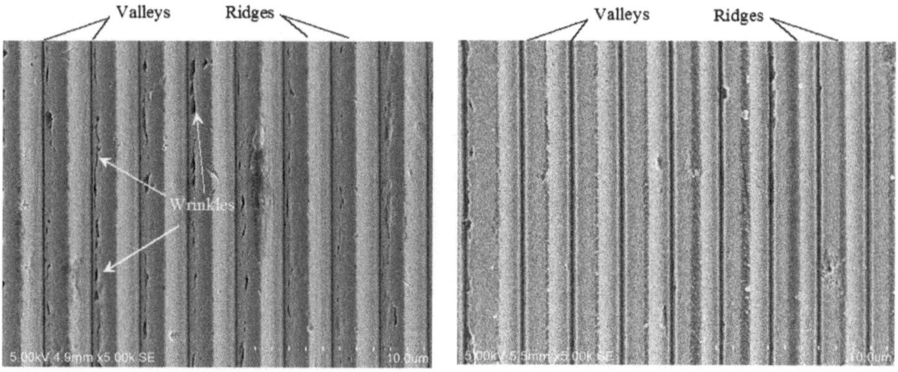

(A) (B)

Figure 8.42 Formed microgrooves on the resin surface without (A) and with (B) ultrasonic vibration.

8.4.3.2 Surface defects

Wrinkles

Microgrooves were fabricated by a hot pressing process. Fig. 8.42A is the SEM photograph of the microgrooves pressed without ultrasonic vibration, and Fig. 8.42B is the SEM photograph of the microgrooves fabricated by ultrasonic vibration assisted pressing. We can find that many irregular wrinkles appear on the slope surface of the microgrooves, especially near the valleys of the microgrooves, which is consistent with stress concentration area appears in FEM simulation, when they are pressed without ultrasonic vibration. When ultrasonic vibration is applied, wrinkles almost disappear, and a smooth surface is achieved.

Dents and pores

There are some dents and pores on the surface of aspherical lenses, as shown in Fig. 8.43. The possible reasons could be various, including improper molding temperature and molding pressure. Also, it could be caused by the defect on molds. Fig. 8.44 shows the SEM photographs of microgrooves with side burrs at different magnifications. The shape errors or other surface defects of the microstructures on the mold fabricated by microcutting can be easily replicated to the microstructures on the glass surface in the GMP process. Therefore, the shape accuracy and surface defects of the microstructures on the mold should be precisely generated above all.

Glass adhesion

Fig. 8.45 shows the micrographs of microgrooves on the mold before GMP and after GMP. It is clear from the figure that some glass material has adhered to the molds. Process conditions during glass molding such as high temperature, long contact time and high mechanical load often lead to strong adhesion between the inorganic glass piece and the molds. This sticking behavior that could result in the surface quality deterioration of the molds and potentially destroy the mold has not

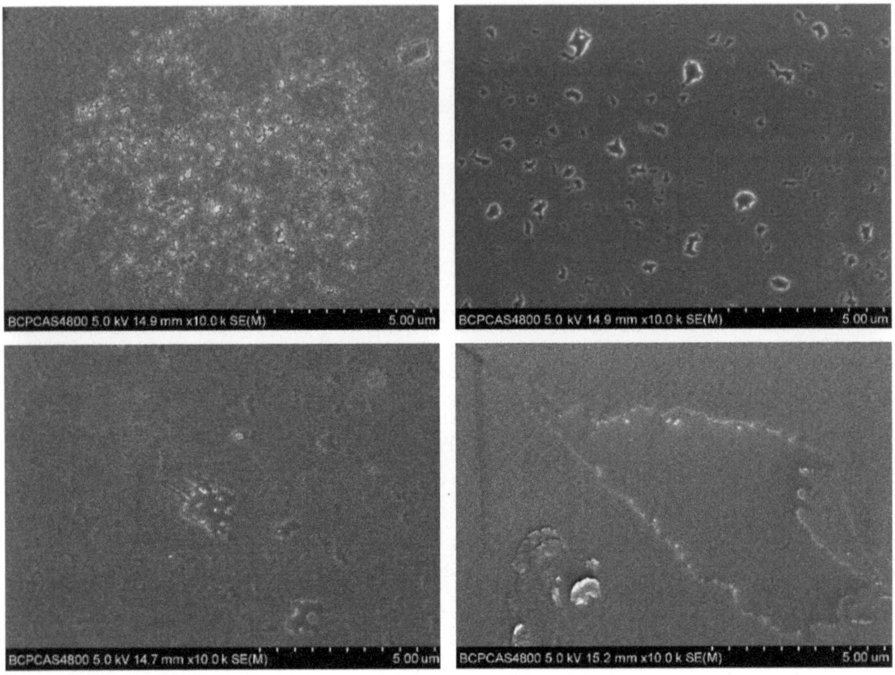

Figure 8.43 Photograph of surface defects on aspherical lenses.

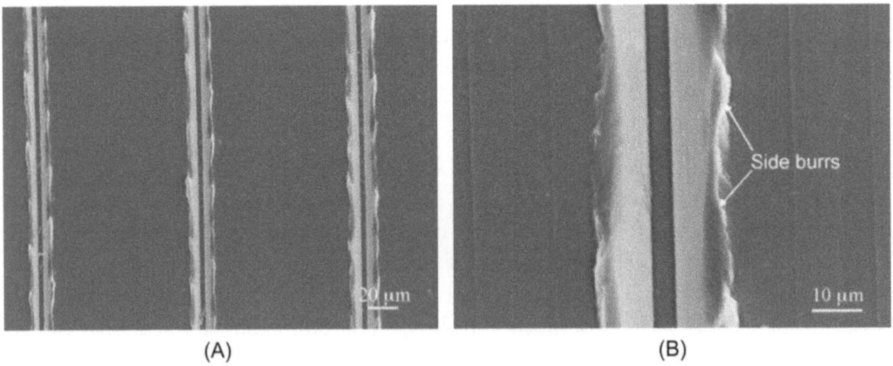

Figure 8.44 SEM photographs at different magnifications showing microgrooves with side burrs (Yan, Oowada et al., 2009).

been properly understood. The majority of the discrepancies can be attributed to the lack of proper understanding of the glass-mold interaction (Fischbach et al., 2010).

Experiments were performed to search for how glass-mold interfaces impact sticking in glass mold systems. It shows that the sticking of glass to the mold can be partially controlled by three methods: decreasing surface roughness value,

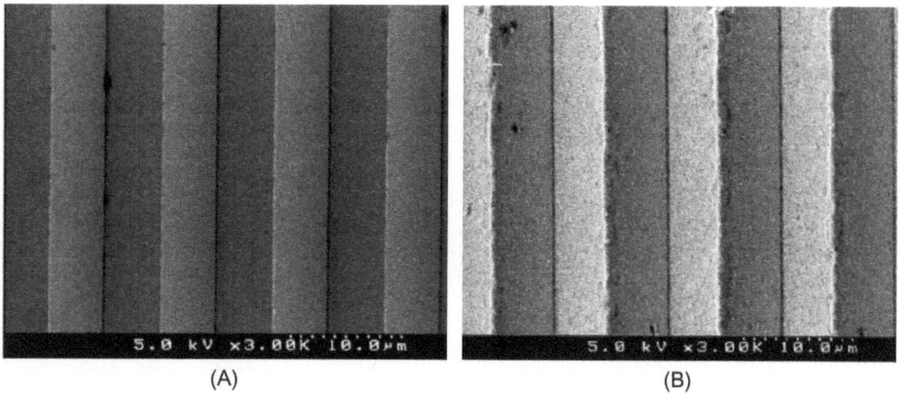

Figure 8.45 SEM photographs of microgrooves on the mold: before (A) and after (B) GMP (Zhou et al., 2011).

lubricating the surface, and pre-oxidizing the surface. In the case of precision glass molding, lubrication and increased surface roughness cannot be used to reduce the sticking force because of the very nature of high surface quality required for optical performance. Some attempts are made to create a new testing method for hot forming processes where the characterization of the sticking behavior of mold materials and coatings is possible and the corrosion and wear behavior of mold material can be quantified. The sticking characteristics of the molds and coatings could be described by the "lower" and the "upper" sticking temperatures. Although the results of their experiment appear to be valid, the use of molten glass is a significant deviation from the process conditions associated with precision glass molding, where the glass never reaches its melting temperature but remains a viscoelastic solid throughout the process.

8.5 Summary

This chapter mainly introduces the glass molding process (GMP) for microstructures, which include microgrooves, micropatterns, microprisms and microlenses. It describes the basic working principle of microstructures in many fields as well as the fundamentals of glass molding techniques. This chapter mainly concentrates on modeling and simulation of microstructure molding process and offers specific methods to conduct simulation calculations of coupling heat transfer and viscous deformation. Also, some typical problems, such as molding quality control and molding defects prevention in the glass molding process, are discussed in detail.

References

Cardarelli, F. (2008). *Materials handbook: A concise desktop reference*. London: Springer Science & Business Media.

Chang, S. H., Lee, Y. M., Jung, T. S., Kang, J. J., Hong, S. K., Shin, G. H., & Heo, Y. M. (2007). Simulation of an aspheric glass lens forming behavior in progressive GMP process. *AIP Conference Proceedings, 908*, 1055–1060.

Choi, J., Ha, D., Kim, J., & Grandhi, R. V. (2004). Inverse design of glass forming process simulation using an optimization technique and distributed computing. *Journal of Materials Processing Technology, 148*, 342–352.

Crawford, R., & Crawford, R. J. (1998). *Plastics engineering*. Amsterdam and Boston, MA: Butterworth-Heinemann.

Field, R. E., & Viskanta, R. (1990). Measurement and prediction of the dynamic temperature distributions in soda-lime glass plates. *Journal of the American Ceramic Society, 73*(7), 2047–2053.

Fischbach, K. D., Georgiadis, K., Wang, F., Dambon, O., Klocke, F., Chen, Y., & Allen, Y. Y. (2010). Investigation of the effects of process parameters on the glass-to-mold sticking force during precision glass molding. *Surface and Coatings Technology, 205*(2), 312–319.

Fulcher, G. S. (1925). Analysis of recent measurements of the viscosity of glasses. *Journal of the American Ceramic Society, 8*, 339–355.

Gent, A. N. (1960). Theory of the parallel plate viscometer. *British Journal of Applied Physics, 11*(2), 85.

Gobet, F., Caron, J., Bessieres, I., Bonnet, T., Aléonard, M. M., Antoine, M., & Hannachi, F. (2016). Experimental and Monte Carlo absolute characterization of a medical electron beam using a magnetic spectrometer. *Radiation Measurements, 86*, 16–23.

Groover, M. P. (2007). *Fundamentals of modern manufacturing: Materials processes, and systems*. New York, NY: John Wiley & Sons.

Gruger, H., Wolter, A., Schuster, T., Schenk, H., & Lakner, H. K. (2003). Performance and applications of a spectrometer with micromachined scanning grating. In Integrated optoelectronics devices (pp. 284–291). International Society for Optics and Photonics.

Kimura, A., et al. (2012). A sub-nanometric three-axis surface encoder with short-period planar gratings for stage motion measurement. *Precision Engineering, 36*(4), 576–585.

Kobayashi, H., Takahashi, H., & Hiki, Y. (2006). Temperature dependence of the viscosity through the glass transition in metaphosphate glasses and polystyrene. *Materials Science and Engineering, 442*, 263–267.

Kolomiets, B. T. (1964a). Vitreous semiconductors (I). *Physica Status Solidi (b), 7*(2), 359–372.

Kolomiets, B. T. (1964b). Vitreous semiconductors (II). *Physica Status Solidi (b), 7*(3), 713–731.

Li, G., Jinn, J. T., Wu, W. T., & Oh, S. I. (2001). Recent development and applications of three dimensional finite element modeling in bulk forming processes. *Journal of Materials Processing Technology, 113*, 40–45.

Madanipour, K., & Tavassoly, M. T. (2010). Moiré fringes as two-dimensional autocorrelation of transmission function of linear gratings and its application for modulation transfer function measurement. *Optics and Lasers in Engineering, 48*(1), 43–47.

Mann, D., Field, R. E., & Viskanta, R. (1992). Determination of specific heat and true thermal conductivity of glass from dynamic temperature data. *Heat and Mass Transfer, 27*, 225–231.

McGraw, D. A. (1961). Transfer of heat in glass during forming. *Journal of the American Ceramic Society, 44*, 353–363.

Meister, R. (1982). Parallel plate viscometer and its characteristic. *Bulletin of the Research Institute of Mineral Dressing and Metallurgy* (Vol. 38, pp. 1–10). Sendai, Japan: Tohoku University.

Moore, J., & Sharp, D. E. (1958). Note on calculation of effect of temperature and composition on specific heat of glass. *Journal of the American Ceramic Society*, *41*, 461–463.
Moser, J. B., & Kruger, O. L. (1968). Thermal conductivity and heat capacity of the monophosphide and monosulfide of plutonium. *Journal of the American Ceramic Society*, *51*, 369–373.
Ohlberg, S. M., & Woo, T. C. (1973). Thermal stress analysis of glass with temperature dependent coefficient of expansion. *Rheologica Acta*, *12*, 345–348.
Rekhson, S. M. (1986). Viscoelasticity of glass. *Glass Science and Technology*, *3*, 1–117.
Richet, P., Bottinga, Y., & Tequi, C. (1984). Heat capacity of sodium silicate liquids. *Journal of the American Ceramic Society*, *67*, C6–C8.
Sharp, D. E., & Ginther, L. B. (1951). Effect of composition and temperature on the specific heat of glass. *Journal of the American Ceramic Society*, *34*, 260–271.
Slóely, A. (2001). Materials handbook: A concise desktop reference. *Chemical Engineering Progress*, *Vol. 5*, 639–640.
Smith, W. F., & Hashemi, J. (2011). *Foundations of materials science and engineering*. Singapore and London: McGraw-Hill.
Sutton, N., et al. (1997). A novel instrument for studying the flow behaviour of erythrocytes through microchannels simulating human blood capillaries. *Microvascular Research*, *53* (3), 272–281.
Takahashi, M., Murakoshi, Y., Maeda, R., & Hasegawa, K. (2007). Large area micro hot embossing of Pyrex glass with GC mold machined by dicing. *Microsystem Technologies*, *13*(3–4), 379–384.
Takahashi, M., Sugimoto, K., & Maeda, R. (2005). Nanoimprint of glass materials with glassy carbon molds fabricated by focused-ion-beam etching. *Japanese Journal of Applied Physics*, *44*(7S), 5600.
Tseng, S. F., Chen, M. F., Hsiao, W. T., Huang, C. Y., Yang, C. H., & Chen, Y. S. (2014). Laser micromilling of convex microfluidic channels onto glassy carbon for glass molding dies. *Optics and Lasers in Engineering*, *57*, 58–63.
Tsukada, K., Sekizuka, E., Oshio, C., & Minamitani, H. (2001). Direct measurement of erythrocyte deformability in diabetes mellitus with a transparent microchannel capillary model and high-speed video camera system. *Microvascular Research*, *61*(3), 231–239.
Viskanta, R., & Lim, J. M. (2001). Theoretical investigation of heat transfer in glass forming. *Journal of the Merican Ceramic Society*, *84*(10), 2296–2302.
Wilson, W. R. D., Schmid, S. R., & Liu, J. (2004). Advanced simulations for hot forging: Heat transfer model for use with the finite element method. *Journal of Materials Processing Technology*, *155*, 1912–1917.
Yan, J., Oowada, T., Zhou, T., & Kuriyagawa, T. (2009). Precision machining of microstructures on electroless-plated NiP surface for molding glass components. *Journal of Materials Processing Technology*, *209*(10), 4802–4808.
Yan, J., Zhou, T., Masuda, J., & Kuriyagawa, T. (2009). Modeling high-temperature glass molding process by coupling heat transfer and viscous deformation analysis. *Precision Engineering*, *33*(2), 150–159.
Yi, A. Y., & Jain, A. (2005). Compression molding of aspherical glass lenses—A combined experimental and numerical analysis. *Journal of the American Ceramic Society*, *88*(3), 579–586.
Youn, S. W., Takahashi, M., Goto, H., & Maeda, R. (2006). Microstructuring of glassy carbon mold for glass embossing—Comparison of focused ion beam, nano/femtosecond-pulsed laser and mechanical machining. *Microelectronic Engineering*, *83*(11), 2482–2492.

Zhou, H., & Li, D. (2005). Mold cooling simulation of the pressing process in TV panel production. *Simulation Modelling Practice and Theory*, *13*, 273–285.

Zhou, T., Yan, J., Masuda, J., Owada, T., & Kuriyagawa, T. (2009). Numerical and experimental investigation on the transferability of ultraprecision glass molding press for microgrooves. In The 3rd international conference of Asian society for precision engineering and nanotechnology (pp. 1A1–10A).

Zhou, T., Yan, J., Masuda, J., Oowada, T., & Kuriyagawa, T. (2011). Investigation on shape transferability in ultraprecision glass molding press for microgrooves. *Precision Engineering*, *35*(2), 214–220.

Zhou, T. F., Liang, Z. Q., Wang, X. B., & Kuriyagawa, T. (2013). Experiment on glass microgroove molding by using polycrystalline nickel phosphorus mold. *Advanced Materials Research*, *797*, 483–488.

Zhou, T. F., Yan, J. W., & Kuriyagawa, T. (2010). Comparing microgroove array forming with micro-pyramid array forming in the glass molding press. *Key Engineering Materials.*, *447*, 361–365.

Index

Note: Page numbers followed by "*f*" and "*t*" refer to figures and tables, respectively.

FSC
www.fsc.org
MIX
Papier | Fördert gute Waldnutzung
FSC® C083411

Kolibri 360 GmbH
Ferdinand-Jühlke-Straße 7
99095 Erfurt, Deutschland
produktsicherheit@kolibri360.de